Adult Stem Cells

Adult Stem Cells

Edited by

Kursad Turksen

Hormones, Growth, and Development Program,
Ottawa Health Research Institute, Ottawa Hospital,
Ottawa, Ontario, Canada

HUMANA PRESS ✳ TOTOWA, NEW JERSEY

For additional copies, pricing for bulk purchases, and/or information about other Humana titles, contact Humana at the above address or at any of the following numbers: Tel.: 973-256-1699; Fax: 973-256-8341; E-mail: humana@humanapr.com

Cover Illustration: From Fig. 4 in Chapter 14, "Stem Cell Biology of the Inner Ear and Potential Therapeutic Applications," by Thomas R. Van De Water, Ken Kojima, Ichiro Tateya, Juichi Ito, Brigitte Malgrange, Philippe P. Lefebvre, Hinrich Staecker, and Mark F. Mehler

Production Editor: Wendy S. Kopf.
Cover design by Patricia F. Cleary.

Printed in the United States of America. 10 9 8 7 6 5 4 3 2 1

e-ISBN 1-59259-732-7

Library of Congress Cataloging in Publication Data

Adult stem cells / edited by Kursad Turksen.
    p. ; cm.
  Includes bibliographical references and index.
  ISBN 1-58829-152-9 (alk. paper)
    1. Cell differentiation. 2. Stem cells.   I. Turksen, Kursad.
    [DNLM: 1. Stem Cells--physiology.   QH 581.2 A243 2004]
  QH607.A28 2004
  571.8'35--dc22

2003014358

Preface

Studies on stem cells have been attracting intense scientific and public attention, not only because of controversies surrounding the use of embryonic stem cells but also because of very provocative data that have been emerging on adult stem cells. Much of the public attention and debate has been focused on the possibility that adult stem cells may be used as a substitute for human embryonic stem cells or as a justification for stopping work on them. This has somewhat diminished attention on very heated scientific debates that take us to the very heart of how the concept of stem cells is perceived. To this author, the latter debates have not been unlike certain philosophical debates of the last century.

Since the seminal studies of Till and McCulloch in the 1960s, the popular paradigm on adult stem cells has been that lineage-restricted stem cells are derived from pluripotent stem cells very early during development. To many, and consistent with much data, the restriction to particular lineages was considered absolute. In other words, there was a sense of determinism in the stem quality of particular stem cells: once they were allocated, they were programmed to specific roles in a given tissue. Furthermore, some adult tissues were considered devoid of detectable stem cell presence or activity. During the last decade, new challenges to our previous notions about stem cells have arisen, one example being the demonstration of stem cells in adult neuronal tissue where they had been said not to exist. Our certainty about stem cell biology has been challenged even further by recent reports that previously designated tissue-restricted adult stem cells might not only be multipotent but also pluripotent. In essence, the debate has become similar to the that between Cartesian and Existentialist philosophers many decades ago. Are stem cells fated to be particular stem cells determined to particular lineage(s) or do they have they the capacity to actualize diverse potentials in diverse environments? In other words, do stem cells exercise "free will"? In a sense, we are debating in a cellular context whether "essence precedes existence" or "existence precedes essence" of stem cells.

In *Adult Stem Cells*, the authors have made an effort, if not to enter the philosophical debate, at least to contribute to current understanding of the potential of several adult stem cell types and their regulation. The debate is certainly still heated and ongoing, and we are confronting new challenges to our understanding of stem cell biology on a weekly basis. Nevertheless, it is hoped that this volume will challenge all of us interested in stem cells to dream about, and to discriminate between, the "essence" and the "existence" of stem cells.

I would like to express my appreciation to all contributors for their unique contributions to this volume. I would also like to thank Elyse O'Grady for supporting this project from its inception during a brief conversation that we had at an ASCB meeting. I also acknowledge the Humana Press staff for doing such an excellent job in publishing this volume.

I would like to acknowledge Dr. Jane E. Aubin for her continuing support and encouragement and to Dr. Aubin and N. Urfe for stimulating discussions. Finally, a special thank you is due to Ms. Tammy Troy for her unquenchable enthusiasm and support for our research and for this project.

Kursad Turksen

Contents

Contributors

ELIZABETH ANDERSON • *Clinical Research Department, Christie Hospital NHS Trust, Manchester, UK*

TAKAYUKI ASAHARA • *Cardiovascular Research Program, St. Elizabeth's Medical Center, Boston, MA; Kobe Institute of Biomedical Research and Innovation/RIKEN Center of Development Biology, Kobe, Japan; and Department of Physiology, Tokai University School of Medicine, Tokai, Japan*

MICHAEL B. CHANCELLOR • *Departments of Urology and Obstetrics and Gynecology, McGowan Institute for Regenerative Medicine, University of Pittsburgh School of Medicine, Pittsburgh, PA*

NATASHA CHERMAN • *Craniofacial and Skeletal Diseases Branch, National Institute of Dental and Craniofacial Research, National Institutes of Health, Bethesda, MD*

CHRISTOPHER CHERMANSKY • *Department of Urology, University of Pittsburgh School of Medicine, Pittsburgh, PA*

ROBERT B. CLARKE • *Clinical Research Department, Christie Hospital NHS Trust, Manchester, UK*

WILLIAM B. COLEMAN • *Department of Pathology and Laboratory Medicine, University of North Carolina School of Medicine, Chapel Hill, NC*

ANNE T. COLLINS • *Prostate Research Group, Department of Surgery, School of Surgical Sciences, University of Newcastle upon Tyne, Newcastle upon Tyne, UK*

DIRK G. DE ROOIJ • *Department of Endocrinology, Faculty of Biology, Utrecht, The Netherlands*

JOE W. GRISHAM • *Department of Pathology and Laboratory Medicine, University of North Carolina School of Medicine, Chapel Hill, NC*

STAN GRONTHOS • *Division of Hematology, Institute of Medical and Veterinary Science, Adelaide, South Australia, Australia*

MASATOSHI HARUTA • *Translational Research Center, Kyoto University Hospital, Kyoto, Japan*

JOHNNY HUARD • *Departments of Molecular Genetics and Biochemistry and Orthopedic Surgery, McGowan Institute for Regenerative Medicine, University of Pittsburgh School of Medicine, Pittsburgh, PA*

JEFFREY M. ISNER (DECEASED) • *Cardiovascular Research Program, St. Elizabeth's Medical Center, Boston, MA*

JUICHI ITO • *Department of Otolaryngology, Head and Neck Surgery, Graduate School of Medicine, Kyoto University, Kyoto, Japan*

ALI KHADEMHOSSEINI • *Biological Engineering Division, Massachusetts Institute of Technology, Cambridge, MA*

KEN KOJIMA • *Departments of Integrative Brain Science and Otolaryngology, Head and Neck Surgery, Graduate School of Medicine, Kyoto University, Kyoto, Japan*

MITSUKO KOSAKA • *Japan Science and Technology Corporation/PRESTO, Okayama Technology Center, Okayama, Japan*

PHILIPPE P. LEFEBVRE • *Center for Cellular and Molecular Neurobiology and Department of Otolaryngology and Audiophonology, University of Liege, Liege, Belgium*

EDWARD M. LEVINE • *Department of Ophthalmology and Visual Science, University of Utah School of Medicine, Salt Lake City, UT*

BRIGITTE MALGRANGE • *Center for Cellular and Molecular Neurobiology and Department of Otolaryngology and Audiophonology, University of Liege, Liege, Belgium*

NADIA N. MALOUF • *Department of Pathology and Laboratory Medicine, University of North Carolina School of Medicine, Chapel Hill, NC*

MARK F. MEHLER • *Departments of Neurology, Neuroscience and Psychiatry and Behavioral Science, Albert Einstein College of Medicine, Bronx, NY*

DAVID E. NEAL • *Prostate Research Group, Department of Surgery, School of Surgical Sciences, University of Newcastle upon Tyne, Newcastle upon Tyne, UK*

ANNA POLESSKAYA • *Oncogenes, Differentiation, and Signal Transduction, Centre National de la Recherche Scientifique, Villejuif, France*

JOSEF PRILLER • *Department of Neurology, Charité, Humboldt-University, Berlin, Germany*

PAMELA GEHRON ROBEY • *Craniofacial and Skeletal Diseases Branch, National Institute of Dental and Craniofacial Research, National Institutes of Health, Bethesda, MD*

HANS-REIMER RODEWALD • *Department of Immunology, University of Ulm, Ulm, Germany*

MICHAEL RUDNICKI • *Laboratory of Stem Cell Research, Ottawa Health Research Institute, Ontario, Canada*

SONGTAO SHI • *Craniofacial and Skeletal Diseases Branch, National Institute of Dental and Craniofacial Research, National Institutes of Health, Bethesda, MD*

DARRYL SHIBATA • *Norris Cancer Center, School of Medicine, University of Southern California, Los Angeles, CA*
WILLIAM B. SLAYTON • *Department of Pediatrics, College of Medicine, University of Florida, Gainesville, FL*
GERALD J. SPANGRUDE • *Division of Hematology, Departments of Oncological Sciences, Pathology, and Medicine, University of Utah School of Medicine, Salt Lake City, Utah*
HINRICH STAECKER • *Division of Otolaryngology, School of Medicine, University of Maryland, Baltimore, MD*
GUANGWEI SUN • *Japan Science and Technology Corporation/PRESTO, Okayama Technology Center, Okayama, Japan*
MASAYO TAKAHASHI • *Translational Research Center, Kyoto University Hospital, Kyoto, Japan*
ICHIRO TATEYA • *Department of Otolaryngology, Head and Neck Surgery, Graduate School of Medicine, Kyoto University, Kyoto, Japan*
KURSAD TURKSEN • *Hormones, Growth, and Development Program, Ottawa Health Research Institute, Ottawa Hospital, Ottawa, Ontario, Canada*
THOMAS R. VAN DE WATER • *University of Miami Ear Institute, Department of Otolaryngology, University of Miami School of Medicine, Miami, FL*
MONICA L. VETTER • *Department of Neurobiology and Anatomy, University of Utah School of Medicine, Salt Lake City, UT*
PETER W. ZANDSTRA • *Biotechnology Process Engineering Center, Massachusetts Institute of Technology, Cambridge, MA, and Institute of Biomaterials and Biomedical Engineering, University of Toronto, Toronto, Canada*

Color Plates

Color plates 1–10 appear in an insert following p. 82.

Adult Stem Cell Plasticity

William B. Slayton and Gerald J. Spangrude

1. INTRODUCTION

Although modern medicine has provided the ability to cure infections and malignancy, the ability to repair damaged organs is less advanced. Solid organ transplantation has been performed successfully, but is fraught with problems such as rejection, infection, and secondary malignancy from immunosuppression. Organ shortages create ethical issues with respect to the equitable distribution of donated tissues. Regenerative medicine, the field devoted to rebuilding damaged organs from stem cells, may provide alternatives to solid organ transplantation. However, the field of regenerative medicine is in its infancy. The potential sources of the tissues to regenerate organs include cloned cells, embryonic or fetal stem cells, or adult stem cells. Although each of these sources of stem cells has potential biological advantages and disadvantages, ethical and legal concerns have been raised by cloning *(1–4)* and the use of embryonic and fetal stem cells *(5)*.

Adult stem cells might provide medical solutions that avoid the ethical and legal problems of cloning and fetal stem cell approaches. Until recently, stem cells from adult tissues were believed restricted in their capacity to produce tissues other than the tissue from which they arose. A number of studies have challenged this view. Specifically, these studies have suggested that adult stem cells from various organs are *plastic*, meaning that they can differentiate not only into their original source tissue, but also into cells of unrelated tissue.

Bone marrow transplant has been used to treat nonhematopoietic disorders such as osteogenesis imperfecta *(6)* and metachromatic leukodystrophy. However, in the case of osteogenesis imperfecta, transplanted mesenchymal stem cells are believed to be the source of reparative osteocytes. In metachromatic leukodystrophy, the mechanism of improvement is unknown, but is thought to be related to a bystander effect of cells, with normal aryl sulfatase circulating past diseased neurons *(7,8)*. The role of the hematopoietic stem cell as a replacement for diseased osteocytes or neurons

From: *Adult Stem Cells*
Edited by: K. Turksen © Humana Press Inc., Totowa, NJ

has not been advocated as a possible mechanism for transplant-induced improvement in these disorders.

Adult stem cell plasticity might allow, for instance, use of bone marrow stem cells to replace damaged myocardial cells following ischemic damage, pancreatic islet cells to cure insulin-dependent diabetes, or cells from the substantia nigra to cure Parkinson's disease. However, as exciting as the prospect is for adult stem cells to solve some of our most daunting medical challenges, newer studies have challenged the interpretation of some of the pioneer studies that generated this excitement. This chapter is an overview of the current controversies in adult stem cell biology.

2. EVOLUTIONARY PERSPECTIVE

Limb and organ regeneration is common among organisms. Alvarado wrote an excellent review of the evolutionary aspects of regeneration *(9)*. Stem cell activity and regeneration can be studied at the most basic level in simple organisms such as the planarian *(10)* (Fig. 1). In the planarian, the molecular mechanisms underlying asexual modes of reproduction are indistinguishable from mechanisms of regeneration following injury. In the hydra, similar molecular messages that stimulate asexual reproduction are triggered by injury. Primitive organisms capable of regenerating damaged body parts include hydra, planarian, mollusks, insects, crustaceans, and echinoderms (starfish). Chordates that can regenerate include amphibians such as frogs and salamanders.

Almost every phylum has species that are able to regenerate lost body parts *(9)*. Regeneration in these organisms requires the ability of cells in the injured tissue to "dedifferentiate," which requires the ability of these organisms to regulate pluripotentiality. Sites of injury in chordates that regenerate form an area of dedifferentiated cells called the "regeneration blastema," and the cells within this structure recapitulate molecular developmental processes that occur during embryogenesis *(9)*.

In summary, the ability to regenerate is a common trait shared by many species. The reason some classes of animals have lost the ability to regenerate is unclear. Alvarado hypothesized that the ability to regenerate confers neither a positive nor negative evolutionary bias, allowing this trait to disappear in many classes of animals, including mammals *(9)*. However, reversal of cell fates and stem cell plasticity may be vestiges of these evolutionarily ancient processes.

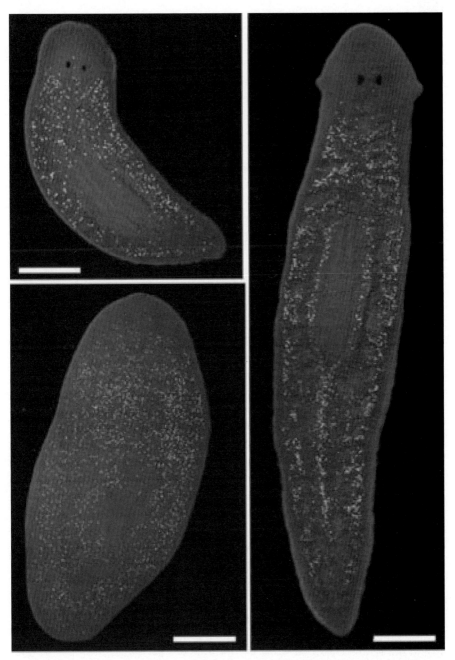

Fig. 1. Bromodeoxyuridine labeling of regenerative stem cells in planarians *Phagocata* sp. (*upper left*); *Girardia dorotocephala* (*lower left*); and *Schmidtea mediterranea* (*right*). Scale bars: A, 150 microns; B, 300; C, 450. (From ref. *10*. Photo courtesy Dr. Alejandro Alvarado. Used with permission of Academic Press.) (*See* color plate 1 in the insert following p. 82.)

3. MODELS OF PLASTICITY

One way that stem cells might achieve plasticity prior to generating heterologous cell types is by reversion to a state similar to the embryonic stem cell *(11)*. The fact that the differentiated state is reversible was first inferred from heterokaryon experiments *(12)*. In these studies, cells from mature muscle tissue were fused with cells of various mature phenotypes. Muscle-specific genes were induced from nuclei of hepatocytes, keratinocytes, and fibroblasts *(13,14)*. The process of cell culture prior to transplant, as performed in many of the studies involving brain or muscle, may allow for a "dedifferentiation" process.

The concept of transdetermination that developed through experiments performed in *Drosophila melanogaster* fits this model of plasticity. Imaginal disks are areas of tissue within the fly larva that eventually develop into adult cuticular structures such as antennae, legs, and wings. During metamorphosis, these cells synthesize pigment and secrete cuticle for specific fly structures. When transplanted prior to metamorphosis, these disks still make the part they would have made if not transplanted. However, imaginal disks can be broken apart and transplanted into the abdominal cavity of adult flies, where regenerative growth can occur. When subsequently transplanted into the body cavity of a host larva, these cells will enter metamorphosis synchronously with the host larva and produce the appendage that is appropriate to the location of migration *(11)*. The primary mechanism by which neural stem cells (NSCs) acquire the ability to produce hematopoietic cells seems to require a period in culture when such dedifferentiation may take place.

Another model of plasticity contends that stem cells are common to all tissues, but are limited in their ability to differentiate based on aspects of the microenvironment *(15)*. Supportive of this model is the fact that stem cells from various tissues express numerous common genes. Common expression of subsets of genes found in cDNA (complementary deoxyribonucleic acid) libraries generated from hematopoietic stem cells and neurospheres has been reported *(16)*. However, it is unclear why hematopoietic stem cells lose their ability to self-renew when cultured, whereas NSCs grow well in culture and maintain stem cell function. If all stem cells were equal, this would not be the case.

In summary, the mechanisms by which stem cells from one tissue can produce mature cells of another tissue have not been clearly established and may vary depending on the particular conditions of the experimental system.

4. PROVING PLASTICITY

A number of recent reviews have outlined the current controversies in stem cell plasticity *(12,17–19)*. Problems with initial studies in this field are numerous. First, plasticity has primarily been inferred from the behavior of undefined mixtures of cells. It is therefore unclear which cells in these mixtures produce the cells that give rise to the original and new phenotypes and whether separate cell lineages arise from the same cell. Second, cell populations have been transplanted following time in tissue culture, and it is unclear whether the stem cells as originally isolated had the ability to produce heterologous tissue or whether epigenetic modification occurred because of the culture period. Third, most studies have not demonstrated the ability of transdifferentiating stem cells to self-renew. Finally, most studies have not demonstrated functionality of the progeny of transdifferentiated stem cells.

These criteria are the measure by which all further studies that claim to demonstrate stem cell plasticity should be evaluated. Hematopoietic stem cell biologists have developed a number of approaches to identify and characterize the behavior of putative stem cells, and by using modifications of these approaches, many of the controversial questions in this field will be resolved. In summary, the science that has been performed to date suggesting stem cell plasticity has not clearly established that adult stem cells are plastic.

Technical limitations to each approach used to measure the presence of donor-derived cells in recipient tissue following transplantation can also lead to misleading results. For instance, two recent studies have demonstrated that embryonic stem cells will fuse with hematopoietic cells or with NSCs when cultured together, and that these chimeric cells will display the phenotype of both original cell types *(20,21)*. The possibility of such fusion events would call into question plastic behavior observed following a culture period, especially when different cell types were mixed. When using the Y chromosome or β-galactosidase (β-Gal) staining to detect donor-derived cells, controls that measure background staining within specific tissue types are crucial *(18)*. Autofluorescence can be mistaken for the fluorescence of green fluorescent protein. When evaluating studies that argue that stem cells are plastic, a careful review of the model system, discovery whether prior culturing of the cells was performed, and a review of experimental controls are essential.

5. BLOOD TO LIVER

Several studies have now demonstrated that bone marrow-derived cells can produce hepatocytes. Petersen et al. demonstrated the contribution of hematopoietic stem cells to oval cells, which are believed to be the resident stem cell of the liver *(22)*. Donor contribution to host liver was shown in three ways:

1. Male donor marrow was infused into female recipients.
2. Cells obtained from dipeptidyl peptidase IV (DPP-IV)–positive male rats were transplanted into DPP-I–negative female rats.
3. Whole liver transplants were performed using Lewis rats expressing the L21-6 antigen as recipients and Brown-Norway rats not expressing this antigen as the allogeneic bone marrow transplant donors *(23)*.

In bone marrow transplant models, unmanipulated bone marrow cells were transplanted into lethally irradiated hosts. Following engraftment, the animals were given 2-acetylaminofluorene, which blocks hepatocyte proliferation, and liver damage was induced using carbon tetrachloride. Two weeks following injury, donor contribution to the organ was determined. Between 0.1 and 0.15% of the hepatocytes were donor derived, and approx 0.1% of oval cells were donor derived (Fig. 2).

Theise et al. *(24)* showed that this phenomenon occurred in animals that had not incurred massive hepatic injury. A cohort of female mice received whole bone marrow cells from male mice following lethal irradiation. Mice were sacrificed at various times following transplant. In addition to looking at the effect of unseparated bone marrow, the investigators also looked at animals that had received sorted $CD34^{pos}Lin^{neg}$ cells, a subset of bone marrow that includes hematopoietic stem cells. All animals had up to 2% donor-derived hepatocytes at 2 mo or longer posttransplant. Fluorescent *in situ* hybridization (FISH) analysis for the Y chromosome of the donor cells was utilized along with another probe to mRNA (messenger RNA) for albumin to ensure that the signal came from hepatocytes. This study demonstrated that the phenomenon of hepatic cell replacement was not completely dependent on tissue damage.

An additional study by Theise et al. *(25)* showed that this phenomenon is present following sex-mismatched human bone marrow transplant. Archived fixed tissue was obtained following two bone marrow transplants involving female recipients of bone marrow from male donors. In addition, tissue from patients who had undergone liver transplant for which the recipients were male and the donors were female were studied. Y chromosome–positive hepatocytes and cholangiocytes ranged from 4 to 43% and 4 to 38%, respec-

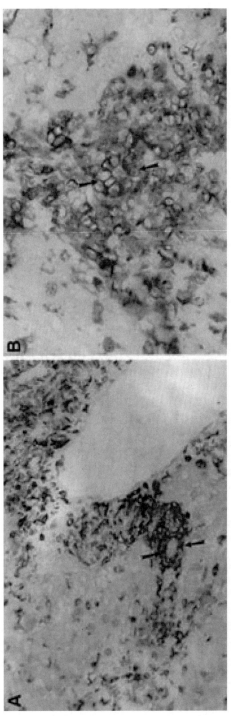

Fig. 2. Hepatic reconstitution of bone marrow derived cells. Brown Norway L21-negative liver was transplanted into a Lewis animal (L21 positive). At 75 d posttransplantation, liver damage was induced with carbon tetrachloride. The liver was harvested 15 d later and stained for L21. (Photomicrograph courtesy Bryon Petersen.) (*See* color plate 2 in the insert following p. 82.)

tively, with the highest levels seen in a patient with severe hepatic damage secondary to recurrent hepatitis C disease. This important study suggested that the phenomenon seen by Petersen and colleagues and Theise and coworkers in rodents also occurs in humans.

Lagasse et al. *(26)* published a study suggesting that hepatocytes derived from transplanted bone marrow cells are functional. In this study, a targeted mutation of fumarylacetoactetate hydrolase (FAH$^{-/-}$) in the mouse, an animal model of type I tyrosinemia, was utilized. These mice have an inborn error of metabolism that causes liver failure and necrosis. Provision of 2-(2-nitro-4-trifluoro-methylbenzyol)-1-3-cyclohexanedione (NTBC) protects the animals from liver failure and renal tubular damage. In one set of experiments, 1 million bone marrow cells were transplanted from male ROSA26/129SvJ mice, wild-type for FAH and transgenic for the lacZ gene, the *Escherichia coli* enzyme for β-Gal. The NTBC treatment was discontinued 3 wk after transplantation to give transplanted cells a selective advantage. After several months, 20 to 50% of the liver mass of FAH$^{-/-}$ mice transplanted in this manner contained lacZ staining cells.

In the second set of experiments *(26)*, small numbers of donor cells that were highly enriched for hematopoietic stem cells based on expression of stem cell–specific antigens were injected along with 200,000 FAH$^{-/-}$ cells to promote survival from radiation conditioning. Animals that received 50 or more stem cells engrafted and had hepatic reconstitution, suggesting that the cells that repair the liver in this model are contained within the hematopoietic stem cell pool. In this study, they used the Y chromosome and β-Gal expression to establish donor identity. Bone marrow cells lacking expression of stem cell–specific antigens did not give rise to hematolymphoid engraftment or to hepatocytes, which demonstrates that the hematopoietic and hepatic activity is localized in the stem cell pool. Recently, however, two subsequent reports demonstrated that the rescue of hepatocyte function in this model system is a result of cell fusion rather than plasticity *(26a,26b)*. This observation raises the intriguing possibility that primitive stem cells may be more fusogenic than other types of cells.

Krause et al. *(27)* performed a study to address the issue of the clonal origin of transdifferentiated tissue. Stem cells were first enriched from bone marrow based on negative selection for maturation antigens. These cells were then stained with a membrane-bound dye (PKH26) and transplanted into irradiated recipient mice. Two days later, PKH26-fluorescent cells were isolated from the bone marrow of transplanted animals, and single cells that had homed to the bone marrow were transplanted into secondary recipients. Of 30 recipient animals, 5 demonstrated engraftment of lymphoid and my-

cells mobilized from the bone marrow by cytokine exposure. However, neither of these studies utilized adequately purified hematopoietic stem cells to rule out the possibility that nonhematopoietic cells that reside in the bone marrow and are mobilized by cytokine treatment are responsible for cardiac myocyte repair. Jackson et al. established that the bone marrow side population cells were capable of ischemic damage repair in capillaries, larger blood vessels, and the myocardial cells themselves *(35)*.

In summary, the potential of muscle satellite cells to provide a source of hematopoietic stem cells in disorders such as aplastic anemia or following depletion of stem cells by chemotherapy and the prospect of the ability of hematopoietic stem cells to repair skeletal and cardiac muscular damage from disorders such as Duchenne's muscular dystrophy are exciting potential clinical outcomes from this work. However, the hematopoietic activity in muscle is likely of bone marrow origin. So far, no clonal studies have yet clearly established that the same cell that reconstitutes the hematopoietic system also is involved in cardiac repair.

7. BRAIN TO BLOOD, BLOOD TO BRAIN

Adult neurons were at one time thought to have a limited ability to be replaced. However, recent studies have suggested that the adult central nervous system has a considerable capacity to repair itself following injury. In 1999, Doetsch et al. identified NSCs in the subventricular zone, which produces neuroblasts that migrate to the olfactory bulb *(36)*. This region has been described as *brain marrow* because, similar to bone marrow, it is a region of cell proliferation and neurogenesis *(37)*. Murine NSCs give rise to all germ layers when injected into chick blastocysts, suggesting that these stem cells have a wide repertoire of possible fates *(38)*. In addition, oligodendrocyte precursor cells, responding to external signals in culture, have been shown to revert to cells with the phenotype of neutral stem cells *(39)*.

Unlike hematopoietic stem cells, which do not replicate in culture and lose their ability to self-renew, NSCs can be grown in culture, where they produce structures called *neurospheres (40)*. Neurospheres contain cells that can produce all of the different cell types that constitute normal brain. In studying the behavior of neurospheres, the capacity of the cells from these structures to produce blood was observed.

In one such study by Bjornson et al. *(41)*, cells for transplant were derived from tissue containing NSCs isolated from fetal brain and cultured and, in separate experiments, from clonally derived NSC cell lines. Neurospheres from prospectively isolated fetal NSCs were grown in epidermal growth factor and basic fibroblast growth media. Donor animals and NSC cell lines were derived from ROSA26 animals, that are transgenic for lacZ. Contribution of the donor NSCs to the recipient blood was determined using H-2k^b, which is expressed by hematopoietic cells of donor origin (ROSA26), but not of recipient (BALB/c) origin. Donor-derived engraftment was observed in 100% of bone marrow recipients, 100% of embryonic NSC recipients, 70% of adult NSC recipients, and 63% of the clonal adult NSC recipients. Between 35 and 65% of CD45-positive hematopoietic cells were donor derived, regardless of whether initial tissue came from adult brain or embryonic brain. Donor-derived hematopoietic cells were present in the blood 5 to 12 mo posttransplant. Repopulation of the immune system after neural cell transplant took an average of 3 wk longer than after bone marrow transplant.

The authors (40,41) proposed that the conversion of neural cells to hematopoietic lineages does not occur immediately, or that NSCs proliferate more slowly than hematopoietic stem cells. However, these studies involved the culture of heterogeneous tissue following transplant, and it is unclear which cell from this mixture produced blood. A subsequent attempt to replicate this result was unable to show production of blood from NSCs (neurospheres), and the authors suggested that epigenetic changes or mutations need to occur for these cells to have hematopoietic potential that exhibits plasticity (42).

Other studies have focused on the ability of bone marrow-derived cells to contribute to the brain. In one such study, Eglitis and Mezey (43) showed data to suggest that unfractionated bone marrow contributes to micro- and macroglia following bone marrow transplant in mice. Donor contribution was determined either by retrovirally tagged cells or by the Y chromosome in sex-mismatched transplants. Recipient animals in this study had defective steel-factor receptors and, as a result, defective hematopoietic stem cells. This provides the transplanted stem cells with a competitive advantage for engraftment. In this study, donor-derived glial cells were present as early as 7 d following transplant, and these cells increased in number during the posttransplant period. Previously, it was unknown whether these glial cells were derived from neural progenitors or whether they had a hematologic origin, so it is unclear whether this activity should be considered plasticity. Furthermore, it is unclear whether these cells were functional, and whether the mutant background of the recipient animals had any effect on the engraftment of these cells within the brain.

In a study that similarly used mutant recipient animals, Mezey et al. (44) studied the ability of bone marrow-derived cells to produce neurons in PU.1 null mice following bone transplant. PU.1 is a transcription factor expressed

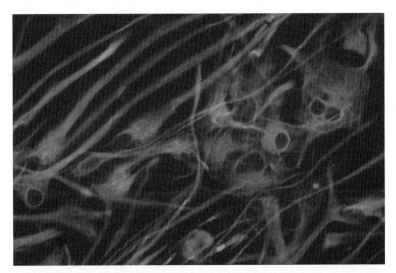

Fig. 3. Human multipotent astrocytic stem cells. (Figure courtesy Dennis Steindler.)

exclusively in blood. Knockout mice fail to produce macrophages, neutrophils, mast cells, osteoclasts, and B and T cells at birth. These animals require bone marrow transplant in the first 48 h of life to survive and develop. FISH analysis for Y chromosome, with concomitant immunohistochemistry to identify cells containing NeuN, a nuclear protein found exclusively in neurons, were used to determine donor-derived contribution to neurons. At 1–4 mo of age, animals were sacrificed and examined for donor-derived tissue. Overall, between 2 and 4% of cells in the brain were Y chromosome-positive, with less than 1% of the neurons donor derived. Areas with the highest frequency of donor-derived cells were within the choroid plexus, ependyma, and subarachnoid space, suggesting that the site of entry of these cells from the bloodstream is the cerebrospinal fluid.

Questions of brain–blood plasticity would benefit from clonal studies that can clearly demonstrate hematopoietic and neuronal potential from a single, uncultured cell. The recent ability to isolate and purify NSCs prospectively will allow such clonal studies to be performed *(45,46)* (Fig. 3). The ability of these cells to produce hematopoietic cell fate without prior culture should be determined because it is important to know whether this plasticity is inherent to NSCs or somehow conferred on these cells by a period in culture. Furthermore, it is important to determine the karyotype of the donor-derived cells that result from neurospheres produced in culture because the hematopoietic activity may be the result of a fusion event within the cultured

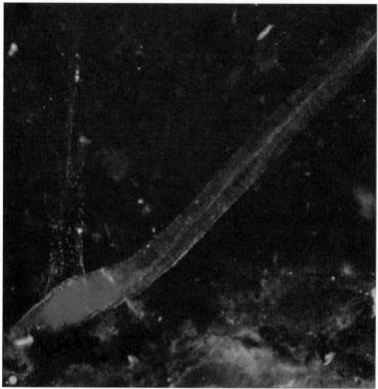

Fig. 4. Functional blood vessels derived from highly purified, bone marrow-derived stem cells. (Courtesy Edward Scott.) (*See* color plate 3 in the insert following p. 82.)

tissues. A further question concerns the origin of adult NSCs and whether there is trafficking between the brain marrow and bone marrow *(47)*.

8. REESTABLISHING A BLOOD SUPPLY

Critical to the regenerative process is the ability to provide nutrients to healing tissue, and this is dependent on the development of new blood vessels. The origin of vascular endothelial cells has long been thought to be the bone marrow, possibly from a precursor that was a predecessor of the hematopoietic stem cell, a cell termed the *hemangioblast (48,49)*. Grant et al. *(50)* demonstrated both endothelial and blood reconstitution following single-cell transplant of hematopoietic stem cells (Lin^{neg} Sca-1^{pos} c-Kit^{pos}) from mice expressing green fluorescent protein into lethally irradiated recipients. One month after the mice were hematopoietically reconstituted, a retinal injury was induced with a laser. In every mouse that had hematopoietic reconstitution, the regenerating blood vessels expressed GFP (Fig. 4) *(50)*.

9. CONCLUSION

Adult stem cell biology is at the forefront of the emerging field of regenerative medicine, offering a source of cells to generate tissues that lack some of the ethical and political impediments inherent in embryonic, fetal, and cloned cells. However, much additional work is necessary to confirm that the cells producing the unexpected plastic behavior are the same cells we call stem cells, and that they fulfill the definition of these cells in every way. The mechanisms by which blood stem cells can regenerate liver and brain need further definition. It is important to know that cells that have adopted a new fate will maintain that fate throughout the life of the organism and not revert to their tissue of origin. It is also important to determine the risks of transformation, particularly involving strategies that employ in vitro manipulation of stem cells.

It is possible that focusing primarily on adult stem cells and ignoring the clinical potential of therapeutic cloning or embryonic stem cells may lead to missed opportunities to capitalize on the biological differences between these stem cell sources. Each approach may be advantageous in certain clinical situations. There will likely be a flurry of scientific activity to determine the true potency of various types of adult stem cells.

ACKNOWLEDGMENT

This work was supported by the National Institutes of Health (K08 HL03962 and R01 DK57899).

REFERENCES

1. Smaglik, P. (2001). Fears of cults and kooks push Congress towards cloning ban. Nature 410, 617.
2. Morality, prejudice and cloning. Nature 2002, 415, 349.
3. Check, E. (2002). Cloning agenda "skewed" by media frenzy. Nature 415, 722.
4. Beyond the cloning debate. Nature 2002, 416, 109.
5. Antoniou, M. (2001). Embryonic stem cell research. The case against. Nat Med 7, 397–399.
6. Horwitz, E. M., Prockop, D. J., Fitzpatrick, L. A., et al. (1999). Transplantability and therapeutic effects of bone marrow-derived mesenchymal cells in children with osteogenesis imperfecta. Nat Med 5, 309–313.
7. Kaufman, C. L., and Ildstad, S. T. (1999). Leukodystrophy and bone marrow transplantation: role of mixed hematopoietic chimerism. Neurochem Res 24, 537–549.

8. Krivit, W., Peters, C., and Shapiro, E. G. (1999). Bone marrow transplantation as effective treatment of central nervous system disease in globoid cell leukodystrophy, metachromatic leukodystrophy, adrenoleukodystrophy, mannosidosis, fucosidosis, aspartylglucosaminuria, Hurler, Maroteaux-Lamy, and Sly syndromes, and Gaucher disease type III. Curr Opin Neurol 12, 167–176.

9. Alvarado, A. S. (2000). Regeneration in the metazoans: why does it happen? Bioessays 22, 578–590.

10. Newmark, P. A., and Alvarado, A. S. (2000). Bromodeoxyuridine specifically labels the regenerative stem cells of planarians. Dev Biol 220, 142–153.

11. Wei, G., Schubiger, G., Harder, F., and Muller, A. M. (2000). Stem cell plasticity in mammals and transdetermination in *Drosophila*: common themes? Stem Cells 18, 409–414.

12. Blau, H. M., Brazelton, T. R., and Weimann, J. M. (2001). The evolving concept of a stem cell: entity or function? Cell 105, 829–841.

13. Blau, H. M., Chiu, C. P., and Webster, C. (1983) Cytoplasmic activation of human nuclear genes in stable heterocaryons. Cell 32, 1171–1180.

14. Blau, H. M., Pavlath, G. K., Hardeman, E. C., et al. (1985). Plasticity of the differentiated state. Science 230, 758–766.

15. Spradling, A., Drummond-Barbosa, D., and Kai, T. (2001). Stem cells find their niche. Nature 414, 98–104.

16. Terskikh, A. V., Easterday, M. C., Li, L., et al. (2001). From hematopoiesis to neuropoiesis: evidence of overlapping genetic programs. Proc Natl Acad Sci U S A 98, 7934–7939.

17. Anderson, D. J., Gage, F. H., and Weissman, I. L. (2001). Can stem cells cross lineage boundaries? Nat Med 7, 393–395.

18. Wulf, G. G., Jackson, K. A., and Goodell, M. A. (2001). Somatic stem cell plasticity: current evidence and emerging concepts. Exp Hematol 29, 1361–1370.

19. Lemischka, I. (2001). Stem cell dogmas in the genomics era. Rev Clin Exp Hematol 5, 15–25.

20. Terada, N., Hamazaki, T., Oka, M., et al. (2002). Bone marrow cells adopt the phenotype of other cells by spontaneous cell fusion. Nature 416, 542–545.

21. Ying, Q. L., Nichols, J., Evans, E. P., and Smith, A. G. (2002). Changing potency by spontaneous fusion. Nature 416, 545–548.

22. Petersen, B. E., Bowen, W. C., Patrene, K. D., et al. (1999). Bone marrow as a potential source of hepatic oval cells. Science 284, 1168–1170.

23. Petersen, B. E. (2001). Hepatic "stem" cells: coming full circle. Blood Cells Mol Dis 27, 590–600.

24. Theise, N. D., Badve, S., Saxena, R., et al. (2000). Derivation of hepatocytes from bone marrow cells in mice after radiation-induced myeloablation. Hepatology 31, 235–240.

25. Theise, N. D., Nimmakayalu, M., Gardner, R., et al. (2000). Liver from bone marrow in humans. Hepatology 32, 11–16.

26. Lagasse, E., Connors, H., Al-Dhalimy, M., et al. (2000). Purified hematopoietic stem cells can differentiate into hepatocytes in vivo. Nat Med 6, 1229–1234.

26a. Wang, X., Willenbring, H., Akkari, Y., et al. (2003) Cell fusion is the principal source of bone marrow derived hepatocytes. Nature 422, 897–901.

26b. Vassilopoulos, G., Wang, P. R., and Russell, D. W. (2003) Transplanted bone marrow regenerates liver by cell fusion. Nature 422, 901–904.

27. Krause, D. S., Theise, N. D., Collector, M. I., et al. (2001). Multi-organ, multi-lineage engraftment by a single bone marrow-derived stem cell. Cell 105, 369–377.

28. Jackson, K. A., Mi, T., and Goodell, M. A. (1999). Hematopoietic potential of stem cells isolated from murine skeletal muscle. Proc Natl Acad Sci U S A 96, 14,482–14,486.

29. Kawada, H., and Ogawa, M. (2001). Bone marrow origin of hematopoietic progenitors and stem cells in murine muscle. Blood 98, 2008–2013.

30. Ferrari, G., Cusella-De Angelis, G., Coletta, M., et al. (1998). Muscle regeneration by bone marrow-derived myogenic progenitors. Science 279, 1528–1530.

31. Gussoni, E., Soneoka, Y., Strickland, C. D., et al. (1999). Dystrophin expression in the mdx mouse restored by stem cell transplantation. Nature 401, 390–394.

32. Goodell, M. A., Rosenzweig, M., Kim, H., et al. (1997). Dye efflux studies suggest that hematopoietic stem cells expressing low or undetectable levels of CD34 antigen exist in multiple species. Nat Med 3, 1337–1345.

33. Goodell, M. A., Brose, K., Paradis, G., Conner, A. S., Mulligan, R. C. (1996). Isolation and functional properties of murine hematopoietic stem cells that are replicating in vivo. J Exp Med 183, 1797–1806.

34. Orlic, D., Kajstura, J., Chimenti, S., et al. (2001). Bone marrow cells regenerate infarcted myocardium. Nature 410, 701–705.

35. Jackson, K. A., Majka, S. M., Wang, H., et al. (2001). Regeneration of ischemic cardiac muscle and vascular endothelium by adult stem cells. J Clin Invest 107, 1395–1402.

36. Doetsch, F., Caille, I., Lim, D. A., Garcia-Verdugo, J. M., and Alvarez-Buylla, A. (1999). Subventricular zone astrocytes are neural stem cells in the adult mammalian brain. Cell 97, 703–716.

37. Scheffler, B., Horn, M., Blumcke, I., et al. (1999). Marrow-mindedness: a perspective on neuropoiesis. Trends Neurosci 22, 348–357.

38. Clarke, D. L., Johansson, C. B., Wilbertz, J., et al. (2000). Generalized potential of adult neural stem cells. Science 288, 1660–1663.

39. Kondo, T., and Raff, M. (2000). Oligodendrocyte precursor cells reprogrammed to become multipotential CNS stem cells. Science 289, 1754–1757.

40. Reynolds, B. A., and Weiss, S. (1992). Generation of neurons and astrocytes from isolated cells of the adult mammalian central nervous system. Science 255, 1707–1710.

41. Bjornson, C. R., Rietze, R. L., Reynolds, B. A., Magli, M. C., and Vescovi, A. L. (1999). Turning brain into blood: a hematopoietic fate adopted by adult neural stem cells in vivo. Science 283, 534–537.

42. Morshead, C. M., Benveniste, P., Iscove, N. N., and van der Kooy, D. (2002). Hematopoietic competence is a rare property of neural stem cells that may depend on genetic and epigenetic alterations. Nat Med 8, 268–273.

43. Eglitis, M. A., and Mezey, E. (1997). Hematopoietic cells differentiate into both microglia and macroglia in the brains of adult mice. Proc Natl Acad Sci USA 94, 4080–4085.

44. Mezey, E., Chandross, K. J., Harta, G., Maki, R. A., and McKercher, S. R. (2000). Turning blood into brain: cells bearing neuronal antigens generated in vivo from bone marrow. Science 290, 1779–1782.

45. Uchida, N., Buck, D. W., He, D., et al. (2000). Direct isolation of human central nervous system stem cells. Proc Natl Acad Sci U S A 97, 14,720–14,725.

46. Morrison, S. J., White, P. M., Zock, C., and Anderson, D. J. (1999). Prospective identification, isolation by flow cytometry, and in vivo self-renewal of multipotent mammalian neural crest stem cells. Cell 96, 737–749.

47. Steindler, D. A., and Pincus, D. W. (2002). Stem cells and neuropoiesis in the adult human brain. Lancet 359, 1047–1054.

48. Shi, Q., Rafii, S., Wu, M. H., et al. (1998). Evidence for circulating bone marrow-derived endothelial cells. Blood 92, 362–367.

49. Choi, K., Kennedy, M., Kazarov, A., Papadimitriou, J. C., and Keller, G. (1998). A common precursor for hematopoietic and endothelial cells. Development 125, 725–732.

50. Grant, M., May, W. S., Caballero, S., et al. (2002). Adult hematopoietic stem cells provide functional hemagioblast activity during retinal neovascularization. Nat Med, 8, 607–612.

2

Spermatogonial Stem Cells

Dirk G. de Rooij

1. THE SEMINIFEROUS EPITHELIUM

Spermatogenesis takes place in the seminiferous tubules (reviewed in ref. *1*) (Fig. 1A). These tubules are lined by the peritubular myoid cells and contain both germ cells at all steps of their development into spermatozoa and a somatic cell type, the Sertoli cells (Fig. 1B). The Sertoli cells are situated on the basal membrane of the tubules, but have their ramifications throughout the epithelium up to the tubule lumen. These cells support the seminiferous epithelium by the secretion of growth factors *(2)* and other factors like lactate and pyruvate, which some types of germ cells need for their metabolism *(3)*. In between the tubules, there is interstitial tissue containing Leydig cells, which produce testosterone.

The process of spermatogenesis starts with mitotic divisions carried out by a cell type called *spermatogonia*. The last mitotic division renders the next cell type, the spermatocytes that go through S phase, then pass through the lengthy prophase of the first meiotic division, and subsequently carry out the two meiotic divisions to give rise to haploid spermatids. Initially, spermatids have a round shape, but then elongate to become spermatozoa, which leave the seminiferous tubules through the tubule lumen.

The mitotic divisions are carried out by spermatogonia on the basal membrane of the seminiferous tubules. Various subsequent types of spermatogonia can be distinguished and are discussed below. Because there is a considerable difference between primate and nonprimate mammals, nonprimates are discussed first, and primates are described separately.

2. SPERMATOGONIAL STEM CELLS IN RODENTS AND OTHER NONPRIMATE MAMMALS

As in all renewing tissues, stem cells are at the basis of the spermatogenic process. Spermatogonial stem cells are single cells located on the basal membrane of the seminiferous tubules and are called A-single (A_s) spermatogo-

From: *Adult Stem Cells*
Edited by: K. Turksen © Humana Press Inc., Totowa, NJ

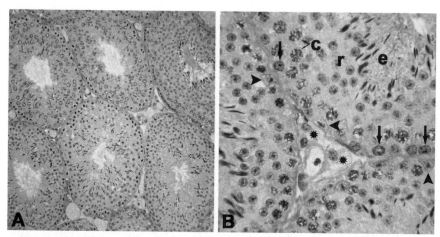

Fig. 1. (A) Section of a mouse testis showing cross sections through seminiferous tubules in the wall where spermatogenesis takes place. In between the seminiferous tubules, there is the interstitial tissue. **(B)** Part of a seminiferous tubule and interstitial tissue showing the cell types in the testis: arrowheads, Sertoli cells; asterisks, Leydig cells in the interstitial tissue; arrows, spermatogonia; c, spermatocytes; r, round spermatids; e, elongated spermatids.

nia *(4–7)*. On division, spermatogonial stem cells give rise either to two new single cells or to a pair of daughter cells (A_{pr} spermatogonia) that do not complete cytokinesis and stay connected by an intercellular bridge *(8,9)*. In all further divisions starting with the pair, cytokinesis will also be incomplete, leading to the formation of increasingly large syncytia of germ cells. Hence, all differentiating progeny of a spermatogonial stem cell will stay connected by intercellular bridges, which is a unique characteristic compared to other renewing tissues. As A_{pr} spermatogonia are morphologically similar to A_s spermatogonia, the intercellular bridge is the first visible expression of the entrance into the differentiation pathway.

It is not known yet whether the divisions of spermatogonial stem cells are symmetrical *(7)*. If the stem cells produce either two new stem cells or a differentiating pair, the divisions can be called *symmetrical*. However, preceding such a division, there might be one in which one of the daughter cells remains in the stem line and the other may already be predestined to produce a pair at its next division.

3. DIFFERENTIATING PROGENY OF STEM CELLS

As described above, the first cells in the spermatogenic lineage destined to develop into spermatozoa are the A_{pr} spermatogonia. These A_{pr} sper-

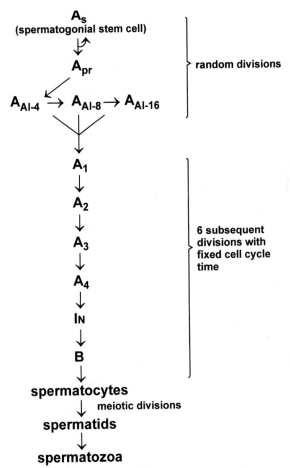

Fig. 2. Scheme of the subsequent types of cells in the spermatogenic lineage in the mouse and rat and the supposed way spermatogonial multiplication and stem cell renewal takes place.

matogonia divide further to form chains of 4, 8, and occasionally up to 32 so-called A-aligned (A_{al}) spermatogonia (Fig. 2). The A_{al} spermatogonia can go through a differentiation step and become so-called A_1 spermatogonia. This differentiation step involves slight morphological changes *(10)* and brings about a change in cell cycle characteristics of the spermatogonia *(11–14)*. Although A_s, A_{pr}, and A_{al} spermatogonia cycle more or less at random, the A_1 spermatogonia and following generations of spermatogonia proliferate with a very strict cell cycle time (e.g., about 30 h in the mouse) *(15)*. In most nonprimate mammals, there are six divisions following the formation of A_1 spermatogonia; the last gives rise to spermatocytes. In rodents, there are about 10 spermatogonial divisions between the spermatogonial stem cells and the formation of spermatocytes *(7)*.

4. CYCLICITY OF THE SPERMATOGENIC PROCESS

Spermatogenesis is a cyclic process. During a species-specific period of time, each area of the seminiferous epithelium goes through a strictly regulated sequence of events and at the end returns to the starting situation. This is called the *cycle of the seminiferous epithelium* (a review is given in ref. *1*). This cycle also affects the spermatogonial stem cells because they go through a period with relatively high proliferative activity, during which they divide two or three times, and then these cells are quiescent for some time *(16)*. During the whole cycle of the seminiferous epithelium, the numbers of stem cells and A_{pr} spermatogonia remain about the same *(17,18)*. Some of the A_s spermatogonia divide into A_{pr}; the accompanying stem cell loss is balanced by self-renewing divisions of other A_s spermatogonia. As A_{pr} spermatogonia divide, they become A_{al} spermatogonia; consequently, during the period of active proliferation, more and more A_{al} spermatogonia are formed. Once every epithelial cycle, the A_{al} spermatogonia differentiate into A_1 spermatogonia, which start the series of (mostly six) divisions to become spermatocytes.

5. CHARACTERISTICS OF SPERMATOGONIAL STEM CELLS

5.1. Morphological and Cell Cycle Characteristics of Spermatogonial Stem Cells

Spermatogonial stem cells lie amid the other types of spermatogonia, newly formed spermatocytes, and Sertoli cells on the basal membrane of the tubules. These stem cells can be recognized morphologically using a special technique. Pieces of seminiferous tubules are prepared in their entirety to produce whole mounts of these structures *(19)*. These whole mounts of seminiferous tubules enable study of the topography of the spermatogonia lying on the basal membrane and to distinguish singles, pairs, and chains of these cells (Fig. 3). Hence, differential cell counts of stem cells and A_{pr} and A_{al} spermatogonia can be carried out. Furthermore, a method was developed to perform autoradiography on these whole mounts *(20)*. Using ^{3}H-thymidine and the labeled mitoses technique, it was found that spermatogonial stem cells have a relatively long cell cycle time of at least 56 h in the rat *(11)* and 90 h in the Chinese hamster *(14)*. These cell cycle times resemble those of A_{pr} and A_{al} spermatogonia, but are longer than in subsequent types of spermatogonia *(12,13)*.

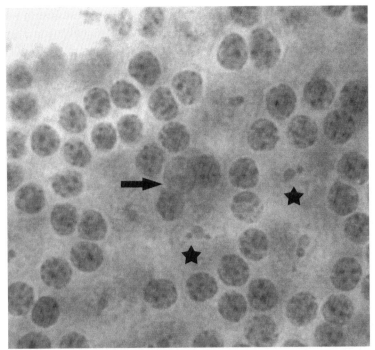

Fig. 3. View on the basal membrane of a mouse seminiferous tubule. Cells present in this particular area are Sertoli cells (indicated by asteriks), preleptotene spermatocytes, and an A_s spermatogonium (indicated by the arrow), the spermatogonial stem cell.

5.2. Purification of Spermatogonial Stem Cells

In one adult mouse testis, there are about 35,000 stem cells, which is only 0.03% of all germ cells *(18).*Various techniques have been developed to purify the total population of A spermatogonia, achieving a purity varying between 85 and 98% *(21–23)*. Unfortunately, in the mouse only about 3% of the A spermatogonia were calculated to be stem cells *(18)*, and it will not likely be much different in other animals. Hence, although a 100-fold enrichment of stem cells can be achieved by purifying A spermatogonia, the purity is still very low. To increase the purity, a method has been developed to isolate spermatogonia from rats deficient in vitamin A *(24)*. In these rats and mice, spermatogenesis is arrested at the differentiation step of A_{al} into A_1 spermatogonia, and the testes of these animals only contain A_s, A_{pr}, and A_{al} spermatogonia *(25)*. A cell population containing roughly 10% stem cells can be obtained using animals deficient in vitamin A *(18,24)*.

Certain biochemical markers have been used to enrich spermatogonial stem cells *(26,27)*. Using anti-β(1)- and anti-α(6)-integrin and negatively selecting for the c-kit receptor, which is not present on spermatogonial stem cells *(28)*, a 40-fold enrichment of spermatogonial stem cells from testicular germ cells could be accomplished *(26)*.

Taken together, the state of the art in the purification of spermatogonial stem cells has not yet reached any further than purity of about 10% at most. More specific membrane markers for these cells will have to be found for further progress in this field.

5.3. Functional Test for Spermatogonial Stem Cells: Spermatogonial Stem Cell Transplantation

The presence of spermatogonial stem cells or their functionality can be checked by the spermatogonial transplantation technique developed by Brinster and coworkers *(29,30)*. In this technique, germ cells of one mouse are transplanted into the testes of a recipient mouse, the endogenous spermatogenesis of which is depleted by treatment with the alkylating agent busulfan. Also, mice carrying the *Wv/Wv* mutation can be used since their testes do not contain germ cells. The donor stem cells repopulate the seminiferous epithelium of the recipient mice.

Interestingly, rat spermatogonial stem cells also are able to repopulate the mouse testes and produce normal rat spermatogenesis in the mouse *(31,32)*. However, stem cells from other species transplanted into mouse testes either produce defective spermatogenesis (for hamster, *see* ref. *33*) or initiate repopulation by spermatogonia only (for rabbit and dog, *see* ref. *34*; for the bull, *see* ref. *35*). Spermatogonial cells from rabbit, dog, and bull form pairs and chains of A spermatogonia, indicating that these stem cells do produce differentiating A_{pr} spermatogonia, but these cells fail to develop further into A_1.

6. REGULATION OF STEM CELL RENEWAL AND DIFFERENTIATION

Like all other renewing tissues, the seminiferous epithelium is able to react to cell loss and especially stem cell loss by mechanisms that initiate enhanced stem cell renewal to replace the lost cells. The potential of recovery of spermatogonial stem cells has been studied extensively at the cellular level by studying the reaction of the seminiferous epithelium to irradiation. In addition, data have become available on the genes involved in the regulation of spermatogonial stem cell renewal and differentiation.

6.1. Regulation at the Cellular Level

The seminiferous epithelium has several levels of response to cell loss or insufficient cell production, such as after irradiation or treatment with cyto-

Table 1
Ways the Seminiferous Epithelium Can Cope
With Various Degrees of Cell Loss

Degree of cell loss	Epithelial reaction
Minor shortage in spermatogonial	Less density-related apoptosis of spermatogonia
Greater than 50% loss of spermatogonia	Enhanced proliferation of A_s, A_{pr}, and A_{al} spermatogonia
Very severe (stem) cell loss	Stem cells only self-renew during at least the first six divisions after cell loss

static agents (Table 1). At the first level, small local shortages in cell numbers are abolished in a way made possible by the fact that the stem cell compartment generally produces too many differentiating cells. It has become clear that stem cell density varies considerably in different areas of seminiferous tubules; consequently, the number of differentiating cells produced also varies considerably *(36,37)*. This varying density of the differentiating cells evens out by density-dependent apoptosis of spermatogonia. In high-density areas, many spermatogonia undergo apoptosis; only few or none do so in low-density areas *(38)*. The result is an even distribution of spermatocytes over the seminiferous epithelium. At the same time, however, this density regulation mechanism serves as a mechanism to deal with relatively minor dips in local cell production and occasional cell loss. Second, although in the normal seminiferous epithelium stem cells and A_{pr} and A_{al} spermatogonia only proliferate during a restricted part of the epithelial cycle, after cell loss the inhibition of the proliferative activity at the end of the proliferation period does not take place *(39,40)*. Prolonged proliferation can then help enhance production of differentiating cells as well as stem cells.

In case of severe cell loss, there is a third level of response. It was found that, after a high dose of irradiation, surviving spermatogonial stem cells only self-renew during at least their first six divisions, leading to a rapid recovery of stem cell numbers in those areas where one or more stem cells did survive *(41)*. Nothing is known yet about the triggers involved in preventing stem cell differentiation or enhancing self-renewal in such a situation.

6.2. Stem Cell Niches in the Seminiferous Epithelium

In several renewing tissues, stem cells were found to occupy specific areas. For example, in the intestine, stem cells reside near the bottom of the crypts *(42)*, and stem cells in the bone marrow are supposed to occupy specific niches *(43)*. Until very recently, in the seminiferous epithelium no such niches were found for spermatogonial stem cells. Now, it has become clear that most spermatogonial stem cells are present in those areas of seminiferous tubules that border interstitial tissue *(10)* (Fig. 4). Apparently, the interstitial tissue affects stem cell behavior. One could speculate that this is caused by the high testosterone levels present in these areas. In this respect, it is interesting that high testosterone levels have been found to prevent the differenti-ation of A_{al} spermatogonia into A_1 spermatogonia *(44–46)*. Possibly, testosterone also has a role in regulating stem cell behavior. However, it has to be kept in mind that germ cells do not possess androgen receptors, so tes-tosterone can only indirectly affect spermatogonia via peritubular myoid cells or (more likely) Sertoli cells, which both express this receptor.

6.3. Genes Involved in the Regulation of Stem Cell Behavior

Although from the above it is clear that the ratio between self-renewal and differentiation of spermatogonial stem cells is under the control of regulatory mechanisms, the genes involved in such mechanisms are largely unknown. However, recent data indicate that glial cell line derived neurotrophic factor (GDNF) is involved (Fig. 4). GDNF normally is secreted by Sertoli cells *(47)*; a subset of spermatogonia, Ret and GDNF family receptor α-1 (GFR-α1), expresses the receptors for this growth factor *(48)*. On ectopic expression of GDNF in spermatogonia, large clusters of single type A spermatogonia are formed, and normal spermatogenesis is suppressed. Moreover, in mice overexpressing GDNF in spermatogonia, seminomatous germ cell tumors are formed at about 1 yr of age *(49)*. GDNF-deficient mice die during the first postnatal day *(50)*, whereas the heterozygotes survive. In these mice, spermatogenesis deteriorates with age as spermatogonia are depleted *(48)*. It was concluded that GDNF has a role in the regulation of self-renewal and differentiation of spermatogonial stem cells. Too high levels of GDNF prevent differentiation and cause accumulation of stem cells, and low levels favor differentiation over self-renewal and cause stem cell depletion.

Furthermore, very recent data indicate that, in the classical spontaneous mouse mutant *luxoid*, adult males exhibit a progressive loss of spermatogonial stem cells *(51)*. Apparently, the as-yet-unknown gene involved in this mutation also has a role in the regulation of spermatogonial stem cell renewal and differentiation (Fig. 5).

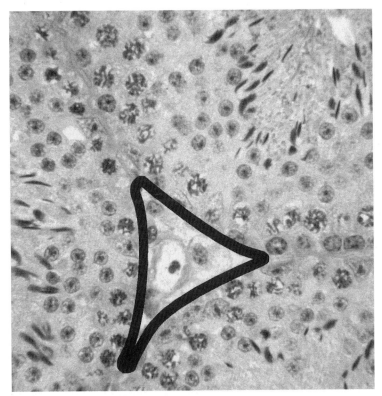

Fig. 4. Testis section as in Fig. 1B. A dark line is drawn over those parts of the tubule membrance on which most stem cells are expected to be present.

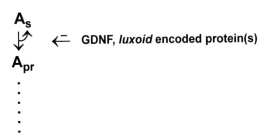

Fig. 5. Molecular regulation of spermatogonial stem cell renewal and differentiation.

7. SPERMATOGONIAL STEM CELLS IN THE HUMAN AND OTHER PRIMATES

7.1. Spermatogonial Cell Types in Primates

Spermatogonial multiplication and stem cell renewal in primates has been described in much less detail than in rodents. As in rodents, in primates there are A and B spermatogonia, but the composition of the type A spermatogonial population is less clear. The A spermatogonia have been subdivided into A_{pale} and A_{dark} spermatogonia according to their staining with hematoxylin *(52)*. The A_{pale} spermatogonia divide once every cycle of the seminiferous epithelium. In the human, this means that they divide only once every 16 d *(53,54)*. A_{dark} spermatogonia in the normal situation are quiescent cells and are supposed to be reserve (stem) cells.

7.2. Nature of the A_{pale} and A_{dark} Spermatogonia

An important question is whether A_{pale} and A_{dark} are spermatogonial stem cells. To answer this question, an obvious approach is to look at the topographical arrangement of these spermatogonia on the basal membrane. In the normal seminiferous epithelium, the density of A_{pale} and A_{dark} spermatogonia is relatively high. Therefore, the clones are too close together to decide whether A_{pale} and A_{dark} spermatogonia consist of clones of 1 or 2^n cells comparable to spermatogonia in nonprimate mammals. However, such a clonal arrangement could be observed for both A_{pale} and A_{dark} spermatogonia during repopulation after irradiation, when spermatogonial density is much lower *(55)*. In that situation, singles, pairs, and chains of A_{pale} and A_{dark} spermatogonia can be observed in tubule whole mounts. As it seems unlikely that the spermatogonial compartment would be principally different between primates and nonprimate mammals, it has been hypothesized that only the single cells among the A_{pale} and A_{dark} spermatogonia have stem cell properties *(55,56)*.

7.3. Difference Between A_{pale} and A_{dark} Spermatogonia

Another question is what the principal difference, except for the difference in proliferative activity, between A_{pale} and A_{dark} spermatogonia is. It has generally been assumed that A_{dark} spermatogonia are reserve (stem) cells. The first real evidence for such a reserve function came from a study in irradiated monkeys. In rhesus monkeys, shortly after irradiation, it was found that the number of A_{pale} spermatogonia decreased to a minimum at about d 9 after irradiation; there was no change in the number of A_{dark} spermatogonia *(57)*. This pattern can be explained by the different proliferative activity of these cells. Irradiation kills the cells when they divide; consequently, the proliferating A_{pale} die, and the quiescent A_{dark} survive. However, after longer intervals than 9 d, the A_{dark} spermatogonia also decreased in number concomitant with a transient rise in the number of A_{pale}. Finally, both cell types dropped to very low numbers. The transient nature of the increase in A_{pale} after more than 9 d after irradiation can be explained by assuming that, after the decline in A_{pale} numbers, the A_{dark} spermatogonia are activated. The activation first causes the A_{dark} to acquire the A_{pale} appearance and accompanying proliferative activity. Then, having become A_{pale}, they try to divide, but because of the lethal radiation damage acquired during the time they were A_{dark}, proper division fails, and the cells enter apoptosis.

7.4. Conclusion

The A_{pale} and A_{dark} spermatogonia in primates are topographically arranged in singles, pairs, and chains of spermatogonia. Unless spermatogonial renewal and multiplication in primates is totally different from that in nonprimate mammals, only the singles among them are stem cells and are comparable to the A_s in nonprimates (Fig. 6). Only the A_{pale} spermatogonia are able to proliferate and do so once every epithelial cycle. From this, it can be deduced that the formation of a chain of eight from a differentiating stem cell will take three cycles (i.e., 48 d in the human). The A_{dark} spermatogonia are quiescent and only are activated after cell loss. When these cells are activated, they become A_{pale} spermatogonia first and then start to proliferate. During repopulation by surviving stem cells, new A_{dark} spermatogonia are set aside again *(55)*. In comparing primate to nonprimate spermatogenesis, it is clear that there is much less proliferative activity of the stem cells in the primate seminiferous epithelium, even by the active stem cells, the A_{pale}. A quiescent stem cell compartment, the A_{dark}, is missing in nonprimate mammals. The low proliferative activity of spermatogonial stem cells in primates seems advantageous because this lowers the chance of errors in deoxyribonucleic acid (DNA) duplication during S phase. The less stem cells divide, the better *(58)*.

8. DIFFERENTIATION CAPACITY OF SPERMATOGONIAL STEM CELLS AND THEIR PRECURSORS

Spermatogonial stem cells originate from primordial germ cells (PGCs), which derive from epiblast cells (embryonal ectoderm) *(59)*. During fetal development, the PGCs proliferate and migrate to the genital ridges, where they become enclosed in the seminiferous cords formed by Sertoli cell precursors. Once in the seminiferous cords, the cells are called *gonocytes*, which

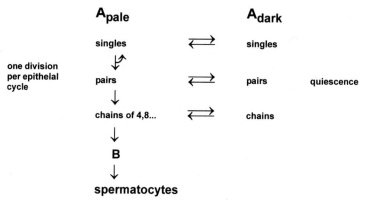

Fig. 6. Proposed scheme of spermatogonial multiplication and stem cell renewal in primates.

are morphologically different from the PGCs *(60–62)*. Gonocytes proliferate for a while and then become quiescent. In mice and rats, gonocytes start spermatogenesis shortly after birth and give rise to spermatogonial stem cells as well as the first A_1 spermatogonia *(see* review in ref. *63)*.

PGCs are single cells that, in culture, can form colonies of cells that morphologically resemble undifferentiated embryonic stem cells (ES cells; *see* ref. *64)*. On feeder layers, these cells can be maintained for a long period of time and can give rise to embryoid bodies and to various cell types in monolayer culture. ES cells that are primordial germ cell derived can contribute to chimeras when injected into host blastocysts *(64)*. Hence, PGCs are stem cells that still have the capacity to differentiate in various directions.

The next cell type in the spermatogenic cell lineage, the gonocytes, can only be cultured in the presence of Sertoli cells *(65)*; PGCs can be cocultured with other types of somatic cells. Furthermore, it has been shown that, in gonocytes, cytokinesis is not completed, leading to the formation of intercellular bridges between daughter cells *(66)*. As described above, in the adult testes intercellular bridge formation is a sign of differentiation; consequently, many gonocytes may already be destined to differentiate along the spermatogenic lineage. From these data, it can be concluded it is likely that in the spermatogenic lineage the multipotentiality of the stem cells is lost at the transition from PGCs to gonocytes.

9. FUTURE DEVELOPMENTS IN SPERMATOGONIAL STEM CELL RESEARCH

Although until recently the emphasis of spermatogonial research was more on the regulation of the A_s, A_{pr}, and A_{al} spermatogonia as a group, now specific data on the molecular regulation of the spermatogonial stem cells themselves are rapidly emerging. GDNF and its receptors and the gene involved in the *luxoid* mutation seem directly involved in the regulation of stem cell behavior. Furthermore, testosterone clearly has an indirect role and may even be responsible for the intriguing fact that spermatogonial stem cells are preferably present in those areas of seminiferous tubules that border interstitial tissue. These findings and the possibility to perform functional tests of stem cells by way of the spermatogonial stem cell transplantation technique open this field for new approaches to establish the regulatory mechanisms that govern spermatogonial stem cell renewal and differentiation.

ACKNOWLEDGMENTS

I am grateful to Laura B. Creemers, Majken Nielsen, and Ans M. M. van Pelt for critical reading of the manuscript and M. Pesci for preparing the figures.

REFERENCES

1. Russell, L. D., Ettlin, R. A., Hikim, A. P. S., and Clegg, E. D. (1990). Histological and Histopathological Evaluation of the Testis. Clearwater, FL: Cache River Press.
2. Jegou, B. (1993). The Sertoli–germ cell communication network in mammals. Int Rev Cytol 147, 25–96.
3. Jutte, N. H., Jansen, R., Grootegoed, J. A., Rommerts, F. F., Clausen, O. P., and van der Molen, H. J. (1982). Regulation of survival of rat pachytene spermatocytes by lactate supply from Sertoli cells. J Reprod Fertil 65, 431–438.
4. Huckins, C. (1971). The spermatogonial stem cell population in adult rats. I. Their morphology, proliferation and maturation. Anat Rec 169, 533–557.
5. Oakberg, E. F. (1971). Spermatogonial stem-cell renewal in the mouse. Anat Rec 169, 515–531.
6. de Rooij, D. G. (1973). Spermatogonial stem cell renewal in the mouse. I. Normal situation. Cell Tissue Kinet 6, 281–287.
7. de Rooij, D. G., and Russell, L. D. (2000). All you wanted to know about spermatogonia but were afraid to ask. J Androl 21, 776–798.
8. Fawcett, D. W., Ito, S., and Slautterback, D. L. (1959). The occurrence of intercellular bridges in groups of cells exhibiting synchronous differentiation. J Biophys Biochem Cytol 5, 453–460.
9. Weber, J. E., and Russell, L. D. (1987). A study of intercellular bridges during spermatogenesis in the rat. Am J Anat 180, 1–24.
10. Chiarini-Garcia, H., Hornick, J. R., Griswold, M. D., and Russell, L. D. (2001). Distribution of type A spermatogonia in the mouse is not random. Biol Reprod 65, 1170–1178.
11. Huckins, C. (1971). The spermatogonial stem cell population in adult rats. II.

A radioautographic analysis of their cell cycle properties. Cell Tissue Kinet 4, 313–334.

12. Huckins, C. (1971). Cell cycle properties of differentiating spermatogonia in adult Sprague-Dawley rats. Cell Tissue Kinet 4, 139–154.

13. Lok, D., and de Rooij, D. G. (1983). Spermatogonial multiplication in the Chinese hamster. I. Cell cycle properties and synchronization of differentiating spermatogonia. Cell Tissue Kinet 16, 7–18.

14. Lok, D., Jansen, M. T., and de Rooij, D. G. (1983). Spermatogonial multiplication in the Chinese hamster. II. Cell cycle properties of undifferentiated spermatogonia. Cell Tissue Kinet 16, 19–29.

15. Monesi, V. (1962). Autoradiographic study of DNA synthesis and the cell cycle in spermatogonia and spermatocytes of mouse testis using tritiated thymidine. J Cell Biol 14, 1–18.

16. Lok, D., and de Rooij, D. G. (1983). Spermatogonial multiplication in the Chinese hamster. III. Labelling indices of undifferentiated spermatogonia throughout the cycle of the seminiferous epithelium. Cell Tissue Kinet 16, 31–40.

17. Lok, D., Weenk, D., and de Rooij, D. G. (1982). Morphology, proliferation, and differentiation of undifferentiated spermatogonia in the Chinese hamster and the ram. Anat Rec 203, 83–99.

18. Tegelenbosch, R. A., and de Rooij, D. G. (1993). A quantitative study of spermatogonial multiplication and stem cell renewal in the C3H/101 F1 hybrid mouse. Mutat Res 290, 193–200.

19. Clermont, Y., and Bustos-Obregon, E. (1968). Re-examination of spermatogonial renewal in the rat by means of seminiferous tubules mounted "in toto." Am J Anat 122, 237–247.

20. Huckins, C., and Kopriwa, B. M. (1969). A technique for the radioautography of germ cells in whole mounts of seminiferous tubules. J Histochem Cytochem 17, 848–851.

21. Bellve, A. R., Cavicchia, J. C., Millette, C. F., et al. (1977). Spermatogenic cells of the prepuberal mouse. Isolation and morphological characterization. J Cell Biol 74, 68–85.

22. Morena, A. R., Boitani, C., Pesce, M., De Felici, M., and Stefanini, M. (1996). Isolation of highly purified type A spermatogonia from prepubertal rat testis. J Androl 17, 708–717.

23. Dirami, G., Ravindranath, N., Pursel, V., and Dym, M. (1999). Effects of stem cell factor and granulocyte macrophage-colony stimulating factor on survival of porcine type A spermatogonia cultured in KSOM. Biol Reprod 61, 225–230.

24. van Pelt, A. M. M., Morena, A. R., van Dissel-Emiliani, F. M. F., et al. (1996). Isolation of the synchronized A spermatogonia from adult vitamin A–deficient rat testes. Biol Reprod 55, 439–444.

25. van Pelt, A. M. M., and de Rooij, D. G. (1990). The origin of the synchronization of the seminiferous epithelium in vitamin A–deficient rats after vitamin A replacement. Biol Reprod 42, 677–682.

26. Shinohara, T., Avarbock, M. R., and Brinster, R. L. (1999). $\beta(1)$- and $\alpha(6)$-

Integrin are surface markers on mouse spermatogonial stem cells. Proc Natl Acad Sci U S A 96, 5504–5509.

27. Shinohara, T., and Brinster, R. L. (2000). Enrichment and transplantation of spermatogonial stem cells. Int J Androl 23(Suppl. 2), 89–91.

28. Schrans-Stassen, B. H. G. J., van de Kant, H. J. G., de Rooij, D. G., and van Pelt, A. M. M. (1999). Differential expression of c-kit in mouse undifferentiated and differentiating type A spermatogonia. Endocrinology 140, 5894–5900.

29. Brinster, R. L., and Avarbock, M. R. (1994). Germline transmission of donor haplotype following spermatogonial transplantation. Proc Natl Acad Sci U S A 91, 11,303–11,307.

30. Brinster, R. L., and Zimmermann, J. W. (1994). Spermatogenesis following male germ-cell transplantation. Proc Natl Acad Sci U S A 91, 11,298–11,302.

31. Clouthier, D. E., Avarbock, M. R., Maika, S. D., Hammer, R. E., and Brinster, R. L. (1996). Rat spermatogenesis in mouse testis. Nature 381, 418–421.

32. Russell, L. D., and Brinster, R. L. (1996). Ultrastructural observations of spermatogenesis following transplantation of rat testis cells into mouse seminiferous tubules. J Androl 17, 615–627.

33. Ogawa, T., Dobrinski, I., Avarbock, M. R., and Brinster, R. L. (1999). Xenogeneic spermatogenesis following transplantation of hamster germ cells to mouse testes. Biol Reprod 60, 515–521.

34. Dobrinski, I., Avarbock, M. R., and Brinster, R. L. (1999). Transplantation of germ cells from rabbits and dogs into mouse testes. Biol Reprod 61, 1331–1339.

35. Izadyar, F., Creemers, L. B., den Ouden, K., and de Rooij, D. G. (2001). Culture and transplantation of bovine spermatogonial stem cells. In: Robaire, B., Chemes, H. E., and Morales, C. R., eds., Andrology in the 21st Century. Englewood, NJ: Medimond, pp. 149–155.

36. de Rooij, D. G., and Lok, D. (1987). Regulation of the density of spermatogonia in the seminiferous epithelium of the Chinese hamster: II. Differentiating spermatogonia. Anat Rec 217, 131–136.

37. de Rooij, D. G., and Janssen, J. M. (1987). Regulation of the density of spermatogonia in the seminiferous epithelium of the Chinese hamster: I. Undifferentiated spermatogonia. Anat Rec 217, 124–130.

38. de Rooij, D. G., and Grootegoed, J. A. (1998). Spermatogonial stem cells. Curr Opin Cell Biol 10, 694–701.

39. van Keulen, C. J., and de Rooij, D. G. (1974). The recovery from various gradations of cell loss in the mouse seminiferous epithelium and its implications for the spermatogonial stem cell renewal theory. Cell Tissue Kinet 7, 549–558.

40. de Rooij, D. G., Lok, D., and Weenk, D. (1985). Feedback regulation of the proliferation of the undifferentiated spermatogonia in the Chinese hamster by the differentiating spermatogonia. Cell Tissue Kinet 18, 71–81.

41. van Beek, M. E. A. B., Meistrich, M. L., and de Rooij, D. G. (1990). Probability of self-renewing divisions of spermatogonial stem cells in colonies, formed after fission neutron irradiation. Cell Tissue Kinet 23, 1–16.

42. Potten, C. S. (1998). Stem cells in gastrointestinal epithelium: numbers, characteristics and death. Philos Trans R Soc Lond B Biol Sci 353, 821–830.

43. Schofield, R. (1983). The stem cell system. Biomed Pharmacother 37, 375–380.

44. Shuttlesworth, G. A., de Rooij, D. G., Huhtaniemi, I., Reissmann, T., Russell, L. D., Shetty, G., et al. (2000). Enhancement of A spermatogonial proliferation and differentiation in irradiated rats by GnRH antagonist administration. Endocrinology 141, 37–49.

45. Tohda, A., Matsumiya, K., Tadokoro, Y., et al. (2001). Testosterone suppresses spermatogenesis in juvenile spermatogonial depletion (jsd) mice. Biol Reprod 65, 532–537.

46. Shetty, G., Wilson, G., Huhtaniemi, I., Boettger-Tong, H., and Meistrich, M. L. (2001). Testosterone inhibits spermatogonial differentiation in juvenile spermatogonial depletion mice. Endocrinology 142, 2789–2795.

47. Trupp, M., Ryden, M., Jornvall, H., et al. (1995). Peripheral expression and biological activities of GDNF, a new neurotrophic factor for avian and mammalian peripheral neurons. J Cell Biol 130, 137–148.

48. Meng, X., Lindahl, M., Hyvonen, M. E., et al. (2000). Regulation of cell fate decision of undifferentiated spermatogonia by GDNF. Science 287, 1489–1493.

49. Meng, X. J., de Rooij, D. G., Westerdahl, K., Saarma, M., and Sariola, H. (2001). Promotion of seminomatous tumors by targeted overexpression of glial cell line-derived neurotrophic factor in mouse testis. Cancer Res 61, 3267–3271.

50. Pichel, J. G., Shen, L., Sheng, H. Z., et al. (1996). Defects in enteric innervation and kidney development in mice lacking GDNF. Nature 382, 73–76.

51. Braun, R. E., Nadler, J. J., Buaas, F. W., Morris, J. L., and Connolly, C. M. (2001). Genetic analysis of the male germ line. In: Abstract Book XVIth Testis Workshop Regulatory Mechanisms of Testicular Cell Differentiation. Newport Beach, CA: Serono Symposia USA, p. 28.

52. Clermont, Y. (1966). Spermatogenesis in man. A study of the spermatogonial population. Fertil Steril 17, 705–721.

53. Heller, C. G., and Clermont, Y. (1963). Spermatogenesis in man: an estimate of its duration. Science 140, 184–185.

54. Clermont, Y. (1966). Renewal of spermatogonia in man. Am J Anat 118, 509–524.

55. van Alphen, M. M. A., van de Kant, H. J. G., and de Rooij, D. G. (1988). Repopulation of the seminiferous epithelium of the rhesus monkey after X irradiation. Radiat Res 113, 487–500.

56. de Rooij, D. G. (1983). Proliferation and differentiation of undifferentiated spermatogonia in the mammalian testis. In: Potten, C. S., ed., Stem Cells. Their Identification and Characterization. Edinburgh, Scotland: Churchill Livingstone, pp. 89–117.

57. van Alphen, M. M. A., van de Kant, H. J. G., and de Rooij, D. G. (1988). Depletion of the spermatogonia from the seminiferous epithelium of the rhesus monkey after X irradiation. Radiat Res 113, 473–486.

58. Lajtha, L. G. (1979). Stem cell concepts. Differentiation 14, 23–34.

59. Lawson, K. A., and Pederson, R. A. (1992). Clonal analysis of cell fate during gastrulation and early neurulation in the mouse. In: Ciba Foundation Symposium 165. Post Implantation Development in the Mouse. New York: Wiley, pp. 3–26.
60. Clermont, Y., and Perey, B. (1957). Quantitative study of the cell population of the seminiferous tubules of immature rats. Am J Anat 100, 241–268.
61. Sapsford, C. S. (1962). Changes in the cells of the sex cords and the seminiferous tubules during development of the testis of the rat and the mouse. Aust J Zool 10, 178–192.
62. Huckins, C., and Clermont, Y. (1968). Evolution of gonocytes in the rat testis during late embryonic and early post-natal life. Arch Anat Histol Embryol 51, 341–354.
63. de Rooij, D. G. (1998). Stem cells in the testis. Int J Exp Pathol 79, 67–80.
64. Resnick, J. L., Bixler, L. S., Cheng, L., and Donovan, P. J. (1992). Longterm proliferation of mouse primordial germ cells in culture. Nature 359, 550–551.
65. van Dissel-Emiliani, F. M., de Boer-Brouwer, M., Spek, E. R., van der Donk, J. A., and de Rooij, D. G. (1993). Survival and proliferation of rat gonocytes in vitro. Cell Tissue Res 273, 141–147.
66. Zamboni, L., and Merchant, H. (1973). The fine morphology of mouse primordial germ cells in extragonadal locations. Am J Anat 137, 299–335.

3

Stem Cells in Skeletal Muscle

Anna Polesskaya and Michael Rudnicki

1. INTRODUCTION

Skeletal muscle is responsible for movement in animals. Pathologies of muscle tissue have serious consequences for the patient, such as decreased mobility, paralysis, and, in extreme cases, death. Often, those conditions are refractory to conventional medical treatments. This explains why much effort in fundamental research is directed toward understanding the complex physiology of skeletal muscle, particularly the means of maintenance and repair of this tissue.

Muscle fibers are multinucleated, with the nuclei located just under the plasma membrane. Most of the skeletal muscle cell is occupied by striated, threadlike myofibrils, and within each myofibril, there are dense Z lines. A *sarcomere* (or muscle functional unit) extends from Z line to Z line. Each sarcomere has thick and thin filaments. The thick filaments are made of myosin and occupy the center of each sarcomere; thin filaments are made of actin and anchor to the Z line. Muscles contract by shortening each sarcomere. The sliding filament model of muscle contraction has thin filaments on each side of the sarcomere that slide past each other until they meet in the middle. Myosin filaments have club-shaped heads that project toward the actin filaments and swivel toward the center of the sarcomere in thousands of cycles dependent on adenosine triphosphate (ATP) during each muscle contraction *(1,2)*.

With each skeletal muscle contraction, considerable mechanical stress occurs, leading to damage and wearing out of muscle fibers. How does the skeletal muscle tissue repair itself?

Muscle fibers are mitotically quiescent. Postnatal growth, repair, and maintenance of muscle fibers are carried out by a distinct myogenic lineage of muscle stem cells, or *satellite cells*. Satellite cells, first described in 1961 *(3)*, are the primary means of forming the muscle mass in adults. These quiescent cells adhere to muscle fibers during embryogenesis and are cov-

From: *Adult Stem Cells*
Edited by: K. Turksen © Humana Press Inc., Totowa, NJ

ered by the continuous basal lamina in adult skeletal muscle (4,5). Immediately following injury, trauma, or exercise, satellite cells are activated and undergo multiple rounds of division prior to fusing with existing or new myofibers (6; reviewed in ref. 7). In rodents, the population of satellite cells decreases with age, which is consistent with the role of these cells as muscle stem cells (8). At birth, satellite cells account for about 32% of muscle nuclei, followed by a drop to less than 5% in the adult. In human degenerating muscle diseases, such as Duchenne muscle dystrophy, the number of satellite cells is severely reduced, most likely because of increased demand for regeneration of muscle fibers (9). Therefore, satellite cells play an essential role in postnatal muscle growth, muscle regeneration, and muscle hypertrophy (for review, see ref. 10).

The developmental origin of satellite cells and the ways and means of their self-renewal in adults are two rapidly progressing domains of muscle stem cell research that are constantly yielding new information. In addition to satellite cells, another precursor cell population called adult stem cells has been identified in skeletal muscle and many other tissues. Adult stem cells have been found within many organs or tissue types of adult mammals, such as bone marrow (11), brain (12,13), and skin (14). Skeletal muscle also contains a pool of primitive, nondetermined, even pluripotent stem cells that are quite distinct from satellite cells (15,16; reviewed in ref. 17). In this case, two alternative hypotheses could be considered. Either adult stem cells and satellite cells represent two independent means by which cells contribute to postmitotic growth and maintenance of muscle or adult stem cells are the precursors of satellite cells.

2. ACTIVATION AND DIFFERENTIATION OF SATELLITE CELLS

Activation of satellite cells is regulated by the MyoD family of basic helix–loop–helix (bHLH) transcription factors, also called MRFs (myogenic regulatory factors) (10). This family includes MyoD and Myf5, which are expressed in determined myoblasts, as well as myogenin and MRF4, the factors that regulate terminal myogenic differentiation (reviewed in ref. 18). The expression pattern of MRFs during the activation, proliferation, and differentiation of satellite cells is analogous to the program regulating the embryonic development of skeletal muscle (Fig. 1). Resting satellite cells do not express any of the MRFs, but rapidly following activation, they start expressing either MyoD or Myf5 and follow by coexpression of MyoD and Myf5 soon after (19). Proliferation of activated satellite cells (myogenic precursor cells, MPCs) is followed by expression of myogenin and MRF4 during terminal differentiation (20).

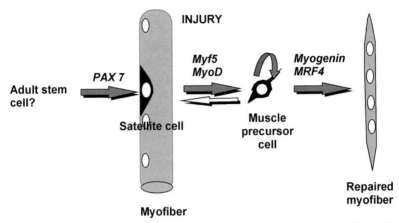

Fig. 1. Activation, proliferation, and differentiation of muscle satellite cells.

Satellite cells appear to be able both to self-renew and to proliferate and contribute to a certain tissue type. These two functions are regulated by distinct mechanisms, as illustrated by the analysis of satellite cells from the MyoD–/– knockout mice *(21)*. Skeletal muscle from these mice has a severely reduced capacity for regeneration, even though there are more satellite cells. These differences become even more striking when the MyoD–/– mice are interbred with mdx mice, the animal model for Duchenne muscular dystrophy. In the double-mutant mice, the number of satellite cells exceeds the wild-type 13-fold, yet there is pronounced muscle atrophy and absence of regeneration *(22)*. These results suggest that, in vivo, upregulation of MyoD is required for satellite cells to enter the MPC proliferative phase that precedes terminal differentiation. In the absence of MyoD, satellite cells seem to demonstrate an increased ability to self-renew and a decreased differentiation potential.

3. DEVELOPMENTAL ORIGIN OF SATELLITE CELLS

In vertebrates, all skeletal muscle cells arise from the *somites*, the temporary mesodermal structures situated next to the neural tube and notochord in embryogenesis. The development of somites is largely regulated by signals from these adjacent structures, which leads to myogenic specification of muscle precursor cells of the somites. Once committed, the myoblasts migrate from the somites and begin the formation of body and limb skeletal muscle. Muscle satellite cells are also supposed to arise, fully or partially, from the somites during embryogenesis. Not all of the determined myogenic cells differentiate immediately and contribute to formation of body and limb muscle. A number of various signals can delay myogenesis, allowing proliferation of a population of muscle progenitor cells that will become satellite cells (reviewed in ref. *3*). These cells are recruited to satellite cell compartments throughout embryogenesis and form a considerable part of skeletal muscle at birth.

Many of the signals that control cell fate in the somite have been identified. Sonic hedgehog is apparently responsible for the early activation of the myogenic bHLH genes MyoD and Myf5 *(24,25)*. Sonic hedgehog also has a role in expanding the population of muscle precursor cells that populate the limb and can generally be described as a survival-promoting factor for myogenic cells *(26)*. Another family of signal transduction proteins that regulates myogenesis, the Wnt family, consists of multiple proteins that activate different signal pathways within myoblast precursors (for review, *see* ref. *27*). The Wnt signaling leads to activation of Myf5 and MyoD synthesis and commitment of the myoblast precursors within somites.

A scenario for exclusively somitic muscle development emerges from the classical studies described above. However, De Angelis and colleagues suggested that satellite cells have nonsomitic origins *(28)*. These authors demonstrated that clonal myogenic MPCs are readily isolated from explants of embryonic aorta. In vitro, these clones exhibit morphology similar to that of satellite cells derived from adult skeletal muscle and coexpress a number of myogenic and endothelial markers. MPCs can also be isolated from limbs of late-stage *c-Met–/–* and *Pax3–/–* mutant embryos. Those mice are severely deficient for determination and migration of myogenic precursors, resulting in complete absence of muscle cells in embryo limbs. However, the vascular system of those embryos is formed normally, suggesting that the MPCs derived from these tissues have endothelial origins. In vivo, aorta-derived MPCs can participate in skeletal muscle regeneration and give rise to many myogenic cells when introduced into skeletal muscle of immunodeficient mice. Importantly, explants of the dorsal aorta in vitro do not give rise to muscle, suggesting that myogenic specification of MPCs requires signals from the surrounding skeletal muscle tissue. The authors hypothesized that, in development, endothelial progenitors from growing blood vessels in skeletal muscle might be recruited to the muscle precursor pool and later contribute to satellite cell population. The fact that muscle satellite cells express such endothelial markers as V-cadherin, vascular endothelial growth factor receptor 2 (VEGFR-2), αM-integrin, β-integrin, P-selectin, smooth α-actin, and PECAM supports this hypothesis. Thus, myogenic specification of pluripotent stem cells is likely to depend on environment-specific, secreted signal transduction proteins.

Embryonic myogenesis is also regulated by transcription factors of the homeobox family, Pax3 and Pax7, two factors of the protein family regulating multiple developmental processes. Early expression of Pax3, but not Myf5 or MyoD, is supposed to be a mechanism that delays terminal differentiation of a population of myogenic precursors, allowing them to expand and form a sufficient pool of myoblasts to form the body muscle *(29)*. In a majority of myoblast precursors, expression of Pax3 precedes that of myogenic bHLHs and has been hypothesized to regulate positively the migration of myoblasts to the limbs and their further terminal differentiation *(30,31)*.

Pax7 is expressed in quiescent and activated satellite cells and is absolutely necessary for the specification of the muscle satellite cell lineage. The skeletal muscle of *Pax7–/–* mice does not contain satellite cells *(32)*, even though the structure of muscle at birth is comparable to that of the wild-type mice. However, in postnatal development, *Pax7* knockout mice grow at a much slower rate than their wild-type siblings and die between 2 and 3 wk of age. The skeletal muscle of *Pax7–/–* mice is small, with 30% reduction in fiber diameter, compared to wild-type mice. Taken together, these data on the roles of Pax proteins in myogenesis suggests an important function for Pax3 in prenatal development, specification, and location of muscle precursor cells, as well as for Pax7 in determining and maintaining the population of satellite cells in adult skeletal muscle.

4. STEM CELLS IN ADULT SKELETAL MUSCLE AND OTHER TISSUES

Until recently, satellite cells were presumed to be the only candidates for the role of stem cells of skeletal muscle. However, multiple groups have demonstrated that stem cells isolated by various techniques either from bone marrow *(15,33)* or from brain *(34,35)* can contribute to myotube formation in vitro or to skeletal muscle repair in vivo. The isolation of a highly enriched population of stem cells from various tissues was based on methods that made use of the ability of stem cells to exclude dyes actively, like Rhodamin 123 and Hoechst 33342 *(11,36)*. The population of cells thus isolated (side population, SP) was characterized with astonishing plasticity.

In addition to the above-mentioned examples of myogenic potential of SP cells from bone marrow and brain, it was also shown that skeletal muscle SP cells can contribute to both muscle and hematopoietic compartments in mice *(15,16,33)*, and that unfractionated bone marrow cells can reconstitute the hepatic cell lineage *(37)* and muscle *(33)*. Strikingly, adult stem cells from brain were reported to repopulate the hematopoietic system of lethally irradiated recipients *(38)*. However, this last result was not reproducible with primary neural stem cells and awaits further investigation *(39)*.

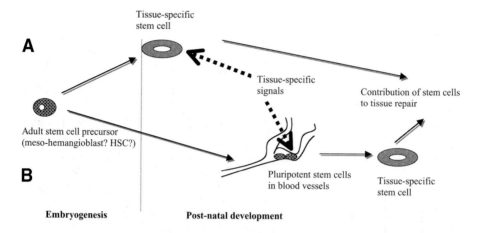

Fig. 2. Hypotheses concerning the origin of adult stem cells. **(A)** Stem cells develop from common primitive precursor (possibly hematopoietic stem cell [HSC]), are trapped in various tissues during embryogenesis or early postnatal stages, and are committed to those tissues by local signals. This theory suggests a certain level of determination of adult stem cells and does not explain their plasticity. **(B)** Stem cells can be recruited to different organs and tissues by specific signaling molecules. In the absence of such signals, the yet-unidentified precursors of stem cells form part of the blood vessel system.

These unexpected findings seriously challenged the classical views of the tissue-restricted fate of stem cells. Therefore, it is very important to raise the issue of possible heterogeneity within the enriched populations of SP cells from various tissues. Indeed, heterogeneity has been reported in the hematopoietic stem cell population *(40)*, and it is presently not known whether fate changes of somatic stem cells reflect the ability of a presently undefined subfraction of a given stem cell population.

The possible solution to the problem of heterogeneity of SP cells, or stem cells from various tissues, is the determination of surface-specific markers characteristic for each stem cell type. This approach, combined with functional assays, will gradually allow the isolation of stem cells with great precision and homogeneity. Given that the plasticity of SP from different tissue types will be confirmed for purified stem cells, how can this plasticity be explained?

One explanation is the existence of one type of primitive stem cell that penetrates all tissue types, perhaps during development, but preserves its pluripotential characteristics *(41)* (Fig. 2A). A variation of this hypothesis is to presume the existence of a number of such stem cell subtypes, each with a potential to contribute to some, but not all, lineages (neurohematopoietic, neuromuscular stem cells).

Another, quite different, explanation is to hypothesize that some cells in an adult organism can be recruited to different tissue types by certain chemokines and signal transduction proteins, but in the absence of such signals, they form part of another tissue (Fig. 2B). Obvious and logical candidates for this role will be cells that form the blood vessels *(42)*. Indeed, as a pluripotent SP cell fraction has been described for multiple tissue types, common features, such as the presence of a network of capillaries, might be explored for these tissue types. Developing blood vessels will be the origin of this common progenitor, an adult "mesohemangioblast."

This hypothesis is partially supported by the study of De Angelis et al. *(28)*, which demonstrated the participation of aorta-derived myogenic progenitors in muscle regeneration. Interestingly, another group of researchers showed that bone marrow-derived hematopoietic stem cells can form smooth muscle cells that contribute to arterial remodeling in vivo *(43)*. This result presents another example of developmental plasticity of adult stem cells and especially the extraordinary plasticity observed between hematopoietic and muscle tissue.

Satellite cells of skeletal muscle coexpress myogenic and endothelial markers, further suggesting that at least some adult stem cells might be derived from vascular lineage. However, the exact identity of these hypothetical blood vessel-based pluripotent precursor cells remains to be established.

5. ADULT SATELLITE CELL SELF-RENEWAL

The mechanisms regulating the self-renewal of satellite cells in adult skeletal muscle are still poorly understood. To proliferate, if satellite cells have to activate the expression of MRFs, it is unclear how they return to the quiescent state after the rounds of proliferation and avoid terminal differentiation. One possible explanation is a possibility of an asymmetric cell division, when a satellite cell would produce one self-renewing pluripotent "daughter" and one committed myogenic precursor. As the activated satellite cells first express either MyoD or Myf5 *(19)* and the MyoD–/– myogenic cells clearly represent an intermediate stage between a quiescent satellite cell and a myogenic precursor *(44)*, it is possible to suggest that expression of Myf5 alone may define a developmental stage during which satellite cells undergo self-renewal.

Another possibility is the dedifferentiation of determined myogenic precursors, which might involve expression of transcription factors, such as msx-1. A study demonstrated that overexpression of msx-1 in differentiated myogenic cell line C2C12 causes dedifferentiation of multinuclear postmitotic myotubes into mononuclear, proliferation myoblasts *(45,46)*. This dedifferentiation was accompanied by decreased expression of myogenic factors MyoD, myogenin, and MRF4. The process of dedifferentiation, however, did not stop at the myoblast stage because the resulting cells were capable of redifferentiation, under suitable culture conditions, into cells that expressed chondrogenic, adipogenic, osteogenic, or myogenic markers.

The possibility of such dedifferentiation as the mechanism of satellite cell self-renewal and subsequent return to the quiescent state looks like an interesting hypothesis. However, there is currently no published evidence that, in vivo, adult satellite cells might be self-renewing by activation, expression of myogenic bHLH proteins, proliferation, and subsequent dedifferentiation. The available data points toward the model of asymmetric division of satellite cells with immediately different profile of gene expression as a more probable scenario. Both these possibilities have to be tested experimentally.

The discovery of the importance of Pax7 for the existence of a satellite cell population in adult skeletal muscle suggested yet another possibility for the ontogeny of satellite cells. The muscle-derived stem cells (currently isolated as SP cells) might represent progenitors for adult satellite cells. There is experimental evidence that muscle-derived stem cells transplanted in mice find their way to the satellite cell compartment in skeletal muscle *(15)*. Another observation to be taken into account is the increased number of hematopoietic precursors in the skeletal muscle Pax7–/– mice *(32)*. The function of Pax7 might be to promote the determination of satellite cells by restricting alternative developmental pathways in pluripotent muscle stem cell precursors.

On the other hand, primitive muscle stem cells might contribute to the repair and maintenance of skeletal muscle directly, without forming satellite cells. In this scenario, regenerating muscle fibers would secrete signal proteins, such as Sonic hedgehog, recruiting muscle stem cells to the site of injury and activating the myogenic determination program in these cells, as described in refs. *47* and *48*. These two mechanisms of skeletal muscle maintenance might also coexist, with muscle stem cells not only contributing to the satellite cell population in healthy muscle, but also having the capability of direct contribution to muscle repair in cases of emergency.

6. MUSCULAR DYSTROPHY AND CELL REPLACEMENT THERAPIES

Primary muscular dystrophies are characterized by a progressive irreversible wasting of skeletal muscle in humans. The most frequent and severe form of this pathology, the Duchenne's muscular dystrophy, is caused by mutations or deletions in dystrophin. This protein normally connects the actin cytoskeleton of myofibers to the extracellular matrix, supporting the myofibers and protecting them from damage during contraction. When dystrophin ceases to function, skeletal muscle deteriorates rapidly, requiring constant regeneration of the fibers by satellite cells. Once the proliferative potential of the satellite cells is exhausted, there is no further regeneration, and the skeletal muscle gradually is replaced by connective tissue *(49)*.

Transplantation of MPCs is one of the most promising therapies for muscular dystrophies. This approach will ideally lead to replenishing of the satellite cell population and restoration of skeletal muscle regenerating abilities. However, cultured myoblasts had proved to be poor tools for transplantation therapies, mainly because of their inability to migrate to appropriate sites of injury and because the majority of transplanted cells have succumbed to immune response of the host *(50)*. Efforts have been directed toward stem cell transplantation therapy for obvious reasons: it is not necessary to understand the process in detail to apply the therapy; it has to be applied only once; and there is only a brief period of attendant toxicities during the period of actual transplantation of progenitor cells. The existing data show that muscle progenitor cells isolated by different techniques, based either on differential adhesion properties of cells *(51)* or on expression on cell surface markers such as CD34 and stem cell antigen 1 (Sca-1) *(52)*, can efficiently participate in muscle regeneration following either intramuscular or intraarterial injection. Whether these transplanted cells truly reconstitute the satellite cell population in the mouse model of muscular dystrophy remains to be established.

An extremely important issue in considering the transplantation of myogenic precursors for treatment of muscular dystrophies is the growth conditions of isolated precursors prior to their introduction into the host. Indeed, the growth factor combination is all important for preserving or changing the qualities of cells in culture, and all the particular conditions for propagating MPCs for transplantation purposes are really not known. Therefore, the molecular mechanisms that direct the myogenic determination or precursor cells in skeletal muscle should be investigated for the ultimate purpose of manipulating and consciously changing the properties of those cells in regenerative medicine.

There are a few approaches to this question: First, the signal transduction proteins implicated in promoting survival, myogenic determination, correct localization, and terminal differentiation in developing skeletal muscle should be addressed. Signal proteins of the Wnt family, as well as Sonic hedgehog and bone morphogenetic proteins (BMPs) 2 and 4, are emerging as appropriate candidates for determining the myogenic destiny of stem cells in skeletal muscle if it is taken for granted that muscle specification in somites during development and myogenic determination of stem cells in adult skeletal muscle are regulated by the same pathways. We know that some of these factors govern the destiny of hematopoietic stem cells *(53)*, and it should be noted that the Wnt–β-catenin pathway seems to be activated permanently in established myogenic cell line C2C12 and is required to preserve its myogenic identity *(54)*. Therefore, all these candidates should be taken into consideration in the search for optimal culture conditions for MPCs.

The second approach is to study the adult skeletal muscle, notably the signaling molecules implicated in regeneration. Presumably, this kind of signaling should recruit precursor cells to the sites of injury and promote their myogenic commitment. Careful analysis of these signaling proteins might yield some unexpected, novel regulation proteins, as well as some already-known signal transduction factors.

7. CONCLUSIONS

It is hardly necessary to point out the importance of studies of origin, functions, and plasticity of adult stem cells. It is a comparatively new, extremely dynamic, and rapidly developing field of research; of course, the interest it holds is primarily connected to possible therapeutic application of stem cells. Indeed, the promise of a successful replacement therapy for such diseases as Duchenne muscular dystrophy is enough of a goal to justify the effort directed toward research of muscle stem cells. The ways and means of manipulation of those cell populations need to be learned, and that means thorough knowledge of signaling proteins that direct proliferation and tissue-specific commitment of adult stem cells. The contribution of stem cells to maintenance of tissues has to be evaluated to understand the physiological relevance of those cells for normal functioning of the organism, as it was done for the hematopoietic system. Are those cells active and functioning within every tissue, and can they restore and repair it after an injury? Some of existing data point to a positive answer to this question.

Adult stem cells can also contribute to serious pathologies, at least in the case of arterial remodeling following angioplasty or in atherosclerosis *(43)*.

Cardiovascular diseases are a serious health problem in the Western world; certainly, a way to control the pathological remodeling of blood vessels by hematopoietic stem cells that form smooth muscle cells will be a major medical achievement. This task requires understanding of mechanisms that "turn blood into muscle," of the extraordinary plasticity between those different tissue types.

Therefore, further research will be directed to analysis of the physiological role of adult stem cells in vivo and creation of experimental systems that allow the reconstruction of distinct developmental events in vitro for more detailed analysis. Understanding the molecular mechanisms that regulate the contribution of pluripotent progenitors to various tissues and organs is a key to successful manipulation, controlled changing of properties, and wide clinical application of adult stem cells.

REFERENCES

1. Huxley, A. F. (2000). Cross-bridge action: present views, prospects, and unknowns. J Biomech 33, 1189–1195.
2. Huxley, A. F. (2000). Mechanics and models of the myosin motor. Philos Trans R Soc Lond B Biol Sci 355, 433–440.
3. Mauro, A. (1961). Satellite cell of skeletal muscle fibers. J Biophys Biochem Cytol 9, 493–495.
4. Bischoff, R. (1990). Interaction between satellite cells and skeletal muscle fibers. Development 109, 943–952.
5. Bischoff, R. and Heintz, C. (1994). Enhancement of skeletal muscle regeneration. Dev Dyn 201, 41–54.
6. Schultz, E. (1989). Satellite cell behavior during skeletal muscle growth and regeneration. Med Sci Sports Exerc 21(5 Suppl.), S181–S186.
7. Grounds, M. D. (1998). Age-associated changes in the response of skeletal muscle cells to exercise and regeneration. Ann N Y Acad Sci 854, 78–91.
8. Gibson, M. C. and Schultz, E. (1983). Age-related differences in absolute numbers of skeletal muscle satellite cells. Muscle Nerve 6, 574–580.
9. Emery, A. E. (1998). The muscular dystrophies. BMJ 317, 991–995.
10. Seale, P. and Rudnicki, M. A. (2000). A new look at the origin, function, and "stem-cell" status of muscle satellite cells. Dev Biol 218, 115–124.
11. Goodell, M. A., Rosenzweig, M., Kim, H., et al. (1997). Dye efflux studies suggest that hematopoietic stem cells expressing low or undetectable levels of CD34 antigen exist in multiple species. Nat Med 3, 1337–1345.
12. Gage, F. H., Ray, J., and Fisher, L. J. (1995). Isolation, characterization, and use of stem cells from the CNS. Annu Rev Neurosci 18, 159–192.
13. Ray, J., Raymon, H. K., and Gage, F. H. (1995). Generation and culturing of precursor cells and neuroblasts from embryonic and adult central nervous system. Methods Enzymol 254, 20–37.

14. Toma, J. G., Akhavan, H., Fernandes, K. S., et al. (2001). Isolation of multipotent adult stem cells from the dermis of mammalian skin. Nat Cell Biol 3, 778–784.
15. Gussoni, E., Soneoka, Y., Strickland, C. D., et al. (1999). Dystrophin expression in the mdx mouse restored by stem cell transplantation. Nature 401, 390–394.
16. Jackson, K. A., Mi, T., and Goodell, M. A. (1999). Hematopoietic potential of stem cells isolated from murine skeletal muscle. Proc Natl Acad Sci U S A 96, 14,482–14,486.
17. Seale, P., Asakura, A., and Rudnicki, M. A. (2001). The potential of muscle stem cells. Dev Cell 1, 333–342.
18. Megeney, L. A. and Rudnicki, M. A. (1995). Determination vs differentiation and the MyoD family of transcription factors. Biochem Cell Biol 73, 723–732.
19. Cornelison, D. D. and Wold, B. J. (1997). Single-cell analysis of regulatory gene expression in quiescent and activated mouse skeletal muscle satellite cells. Dev Biol 191, 270–283.
20. Sabourin, L. A. and Rudnicki, M. A. (2000). The molecular regulation of myogenesis. Clin Genet 57, 16–25.
21. Rudnicki, M. A., Braun, T., Hinuma, S., and Jachisch, R. (1992). Inactivation of MyoD in mice leads to up-regulation of the myogenic HLH gene Myf-5 and results in apparently normal muscle development. Cell 71, 383–390.
22. Megeney, L. A., Kablar, B., Garrett, K., Anderson, J. E., and Rudnicki, M. A. (1996). MyoD is required for myogenic stem cell function in adult skeletal muscle. Genes Dev 10, 1173–1183.
23. Bailey, P., Holowacz, T., and Lassar, A. B. (2001). The origin of skeletal muscle stem cells in the embryo and the adult. Curr Opin Cell Biol 13, 679–689.
24. Borycki, A. G., Mendham, L., and Emerson, C. P., Jr. (1998). Control of somite patterning by Sonic hedgehog and its downstream signal response genes. Development 125, 777–790.
25. Borycki, A. G., Brunk, B., Tajbakhsh, S., et al. (1999). Sonic hedgehog controls epaxial muscle determination through Myf5 activation. Development 126, 4053–4063.
26. Kruger, M., Mennerich, D., Fees, S., et al. (2001). Sonic hedgehog is a survival factor for hypaxial muscles during mouse development. Development 128, 743–752.
27. Cossu, G., and Borello, U. (1999). Wnt signaling and the activation of myogenesis in mammals. Embo J 18, 6867–6872.
28. De Angelis, L., Berghella, L., Colletta, M., et al. (1999). Skeletal myogenic progenitors originating from embryonic dorsal aorta coexpress endothelial and myogenic markers and contribute to postnatal muscle growth and regeneration. J Cell Biol 147, 869–878.
29. Amthor, H., Christ, B., and Patel, K. (1999). A molecular mechanism enabling continuous embryonic muscle growth—a balance between proliferation and differentiation. Development 126, 1041–1053.
30. Macleod, K. F., Sherry N., Hannon, G., et al. (1995). p53-dependent and inde-

pendent expression of p21 during cell growth, differentiation, and DNA damage. Genes Dev 9, 935–944.

31. Borycki, A. G., Li, J., Jin, F., Emerson, C. P., and Epstein, J. A. (1999). Pax3 functions in cell survival and in pax7 regulation. Development 126, 1665–1674.

32. Seale, P., Sabourin, L. A., Girgis-Gabardo, A., et al. (2000). Pax7 is required for the specification of myogenic satellite cells. Cell 102, 777–786.

33. Ferrari, G., Cusella-De Angelis, G., Coletta, M., et al. (1998). Muscle regeneration by bone marrow-derived myogenic progenitors. Science 279, 1528–1530.

34. Galli, R., Borello, U., Gritti, A., et al. (2000). Skeletal myogenic potential of human and mouse neural stem cells. Nat Neurosci 3, 986–991.

35. Rietze, R. L., Valcanis, H., Brooker, G. F., et al. (2001). Purification of a pluripotent neural stem cell from the adult mouse brain. Nature 412, 736–739.

36. Goodell, M. A., Brose, K., Paradis, G., Conner, A. S., and Mulligan, C. (1996). Isolation and functional properties of murine hematopoietic stem cells that are replicating in vivo. J Exp Med 183, 1797–1806.

37. Petersen, B. E., Bouen, W. C., Patrene, K. D., et al. (1999). Bone marrow as a potential source of hepatic oval cells. Science 284, 1168–1170.

38. Bjornson, C. R., Rietze, R. L., Reynolds, B. A., Magli, M. C., and Vescovi, A. L., et al. (1999). Turning brain into blood: a hematopoietic fate adopted by adult neural stem cells in vivo. Science 283, 534–537.

39. Morshead, C. M., Benveniste, P., Iscove, N. N., and van Der Kooy, D., et al. (2002). Hematopoietic competence is a rare property of neural stem cells that may depend on genetic and epigenetic alterations. Nat Med 8, 268–273.

40. Uchida, N., Jerabek, L., and Weissman, I. L. (1996). Searching for hematopoietic stem cells. II. The heterogeneity of Thy-1.1(lo)Lin(–/lo)Sca-1+ mouse hematopoietic stem cells separated by counterflow centrifugal elutriation. Exp Hematol 24, 649–659.

41. Orkin, S. H. and Zon, L. I. (2002). Hematopoiesis and stem cells: plasticity vs developmental heterogeneity. Nat Immunol 3, 323–328.

42. Cossu, G., De Angelis, L., Borello, U., et al. (2000). Determination, diversification and multipotency of mammalian myogenic cells. Int J Dev Biol 44, 699–706.

43. Sata, M., Saiura, A., Kunisato, A., et al. (2002). Hematopoietic stem cells differentiate into vascular cells that participate in the pathogenesis of atherosclerosis. Nat Med 8, 403–409.

44. Sabourin, L. A., Girgis-Gabardo, A., Seale, P., Asakura, A., and Rudnicki, M. A. (1999). Reduced differentiation potential of primary MyoD–/– myogenic cells derived from adult skeletal muscle. J Cell Biol 144, 631–643.

45. Odelberg, S. J., Kollhoff, A., and Keating, M. T. (2000). Dedifferentiation of mammalian myotubes induced by msx1. Cell 103, 1099–1109.

46. McGann, C. J., Odelberg, S. J., and Keating, M. T. (2001). Mammalian myotube dedifferentiation induced by newt regeneration extract. Proc Natl Acad Sci U S A 98, 13,699–13,704.

47. Borycki, A. G. and Emerson, C. P. (1997). Muscle determination: another key player in myogenesis? Curr Biol 7, R620–R623.

48. Borycki, A. G., Strunk, K. E., Savary, R., and Emerson, C. P. Jr. (1997). Distinct signal/response mechanisms regulate pax1 and QmyoD activation in sclerotomal and myotomal lineages of quail somites. Dev Biol 185, 185–200.

49. Ozawa, E., Noguchi, S., Mizuno, Y., Hagiwara, Y., and Yoshida, M., et al. (1998). From dystrophinopathy to sarcoglycanopathy: evolution of a concept of muscular dystrophy. Muscle Nerve 21, 421–438.

50. Partridge, T. (2000). The current status of myoblast transfer. Neurol Sci 21, S939–S942.

51. Lee, J. Y., Qu-Petersen, Z., Cao, B., et al. (2000). Clonal isolation of muscle-derived cells capable of enhancing muscle regeneration and bone healing. J Cell Biol 150, 1085–1100.

52. Torrente, Y., Tremblay, J. P., Pisati, F., et al. (2001). Intraarterial injection of muscle-derived CD34(+)Sca-1(+) stem cells restores dystrophin in mdx mice. J Cell Biol 152, 335–348.

53. Bhardwaj, G., Murdoch, B., Wu, D., et al. (2001). Sonic hedgehog induces the proliferation of primitive human hematopoietic cells via BMP regulation. Nat Immunol 2, 172–180.

54. Ross, S. E., Hemati, N., Longo, K. A., et al. (2000). Inhibition of adipogenesis by Wnt signaling. Science 289, 950–953.

4

Gene Therapy Using
Muscle-Derived Stem Cells

Christopher Chermansky, Johnny Huard, and Michael B. Chancellor

1. INTRODUCTION

Novel molecular techniques such as ex vivo gene therapy and tissue engineering have only recently been introduced to the field of urology. Stem cell-based tissue engineering can be a platform for ex vivo gene therapy, the aim of which is to replace, repair, or enhance the function of damaged tissues or organs. The ex vivo process involves harvesting cells from donors, manipulating the cells in vitro to express genes that would enhance the therapeutic potential of the harvested cells, and injecting the cells back into the donor.

Investigations have delivered strong evidence that myogenic cells isolated from skeletal muscle are heterogeneic, and a small proportion of these cells displays stem cell characteristics *(1)*. These muscle-derived stem cells (MDSCs) survive transplantation, are pluripotent, and are characteristically distinguishable from myoblasts *(2)*. The most obvious diseases for the use of the MDSC are the muscular dystrophies. Yet, based on the capacity of somatic stem cells for multilineage differentiation, skeletal muscle is viewed as an accessible source from which stem cells can be isolated noninvasively for eventual use in cell replacement therapies in many medical disciplines *(3)*. This chapter discusses the MDSC as a stem cell, viral vectors for gene therapy, strategies for gene transfer into muscle, and potential applications of these technologies in urology and other fields of medicine.

2. CHARACTERIZING THE MDSC AS A STEM CELL

Adult stem cells are defined by two major functions: multilineage differentiation and self-renewal. Satellite cells, which many refer to as muscle stem cells, are myogenic precursors capable of regenerating muscle *(4)*. They are named *satellite cells* because of their location in the basal lamina

From: *Adult Stem Cells*
Edited by: K. Turksen © Humana Press Inc., Totowa, NJ

adjacent to muscle fibers. When transplanted into host muscle, satellite cells can either fuse with host myofibers or fuse together into myotubes, which will consequently differentiate into muscle fibers; however, they are considered committed to the myogenic lineage, and the majority of transplanted satellite cells die rapidly following injection *(5)*.

The MDSC, which may represent a predecessor of the satellite cell, is considered distinct in that it may not be restricted to the myogenic lineage or even to mesenchymal tissues *(6)*. The lack of specific markers to identify myogenic cells at very early stages of development makes the characterization of MDSC difficult. The CD34 antigen, a transmembrane cell surface glycoprotein, has been identified on MDSC *(7)*. It is characterized as a hematopoietic stem cell (HSC) marker; however, studies suggest that the expression of CD34 may be reversible and too unreliable to identify pluripotent populations *(8)*. Stem cell antigen 1 (Sca-1) is a protein expressed in hematopoietic stem cells *(7)*. It has been consistently identified on the putative MDSC. A proposed mechanism of skeletal muscle precursor cell differentiation in the myogenic lineage, including the MDSC, is presented in Fig. 1.

In addition to markers, HSCs and MDSCs share common morphological and adherence characteristics. MDSCs from a primary isolation are relatively small, round, and nonadherent *(9)*. Isolation techniques, such as the preplate technique, have utilized these features to purify the slowly adherent cells from which the MDSC are isolated. Furthermore, a fluorescence-activated cell sorting method, used to isolate a bone marrow side population with hematopoietic reconstitution capacity, has been applied to skeletal muscle to isolate a side population *(10)*. This side population of muscle cells was capable of reconstituting the hematopoietic lineage of irradiated hosts.

In summary, the marker studies and functional assays indicate that myogenic cells are heterogenic. Furthermore, a minority of myogenic cells seems to harbor defining characteristics of true stem cells. For this reason, much recent research has focused on the isolation and purification of MDSC.

3. VIRAL VECTORS FOR GENE THERAPY

Gene therapy, the transferring of a functional gene into a particular tissue to alleviate a biochemical deficiency, has emerged as a novel and exciting form of molecular medicine. At its inception, gene therapy focused on the treatment of such inherited diseases as cystic fibrosis, hemophilia, and Duchenne muscular (DMD) dystrophy *(11)*. Progress in the field was made when gene therapy was introduced for such acquired diseases as cancer and acquired immunodeficiency syndrome (AIDS) *(12)*. Evans and Robbins ini-

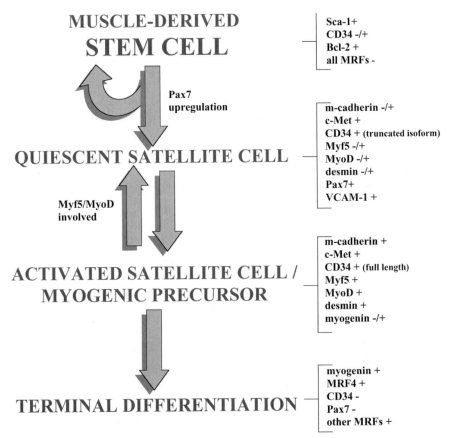

Fig. 1. A proposed mechanism of skeletal muscle precursor cell differentiation in the myogenic lineage, including the MDSC. *Abbr:* Bcl-2, apoptosis inhibiting protein; MRFs, muscle regulatory factors; c-Met, cell–cell signaling receptor required for myogenic migration; Mfy-5 and MyoD, muscle regulatory factors expressed early in myoblast maturation; Pax 7, paired box transcription factor; VCAM, vascular cell adhesion molecule.

tiated a human clinical trial to study gene therapy as a novel drug delivery approach for the treatment of rheumatoid arthritis *(13)*.

Because of a number of properties, muscle tissue has been suggested as a promising target for gene therapy. First, because skeletal muscle is composed of multinucleated, postmitotic fibers, it may facilitate high- and long-term expression of introduced genes. Second, the ability of myogenic cells to fuse with, or into, myofibers in vivo has established them as promising gene delivery vehicles. Finally, the high level of vascularization in muscle

Table 1
Viral Vectors for Gene Therapy

Vector	Integrating	Muscle Cells Infection				General
		In vitro		In vivo		
		Myoblasts	Myotubes	New-born muscle fibers	Adult muscle fibers	
Plasmid DNA and liposomes	No	+	+	+	+	Very low efficiency of gene delivery Low immunogenicity/ cytotoxicity
Retrovirus	Yes	+++	–	–	–	Low toxicity/immunogenicity Infects mitotically active cells Low gene insert capacity
Adenovirus	No	+++	–	+++	–	Infects mitotic/ postmitotic cells Low cytotoxicity Immune rejection problems New-generation vectors (less immunogenic; under investigation)
Herpes simplex virus (type1)	No	+++	+++	+++	–	Large insert capacity Infects mitotic/ postmitotic cells Immune rejection problems New-generation vectors (less immunogenic; under investigation)
Adeno-associated virus	Yes	+++	+++	+++	+++	Low immunogenicity/cytotoxicity High persistence of gene transfer Low gene insert capacity

may facilitate the systemic delivery of potentially therapeutic muscle and nonmuscle proteins, such as growth factors, factor IX, or erythropoietin. The following sections describe five different vectors used for gene transfer to skeletal muscle *(3,14)*. Table 1 summarizes these vectors. Of note, plasmid DNA (deoxyribonucleic acid) is a nonviral vector, yet it is included to complete the discussion of gene delivery.

3.1. Plasmid DNA/Liposomes

Naked, functional DNA can be taken into cells by a number of physical methods, including coprecipitation with calcium phosphate, conjugation with a polycation or lipid, or encapsulation into liposomes *(15)*. The advantages of the intramuscular injection of plasmid DNA include low toxicity and immunogenicity; however, the major disadvantage is the low transfection efficiency *(16)*. Improved transfection efficiencies in muscle can be obtained by pretreating with a myonecrotic agent or hypertonic sucrose *(17)*. Injecting the plasmid DNA into the myotendinous junction can also improve transfection efficiency. Several studies have shown improved plasmid transfection efficiencies in muscle using nontargeted liposomes *(18)*. In particular, the recent development of ligand-directed DNA–liposome complexes capable of transducing myogenic cells in a receptor-dependent manner is promising *(19)*.

3.2. Retrovirus

Retrovirus vectors are safe and can infect dividing myoblasts with high efficiency *(20)*. Furthermore, the ability of retroviruses to become stably integrated into the host genome can provide long-term, stable expression of the delivered gene; however, retroviruses are incapable of infecting postmitotic cells *(21)*. Because muscle cells become postmitotic early in their development, muscle shows a significant loss of retroviral transduction during maturation. Finally, retroviruses have a small gene insert capacity, and there is a risk of mutagenesis with gene insertion.

3.3. Adenovirus

Adenovirus vectors can infect both mitotic myoblasts and postmitotic myofibers; however, the stability and long-term expression of transgenes delivered to muscle using first-generation adenovirus vectors have been limited by immune rejection and maturation-dependent adenoviral transduction *(22)*. Fortunately, the low gene insert capacity of the first-generation adenovirus vectors has recently been overcome by the development of new mutant adenovirus vectors that lack all viral genes and have an expanded gene insert capacity *(23)*. In addition, these newer-generation adenovirus vectors promise to reduce immunogenicity dramatically.

3.4. Herpes Simplex Virus

Viral vectors derived from herpes simplex virus (HSV) type 1 are naturally capable of carrying large DNA fragments, such as the 14-kb dystrophin cDNA (cyclic DNA), and they have been studied for their ability to transduce muscle cells *(24)*. HSV vectors, which persist in the host cell in a nonintegrated state, have shown efficient transduction of myoblasts, myotubes, and immature myofibers. Yet, the HSV vectors are highly cytotoxic and immunogenic, both of which hamper long-term transgene expression. The deletion of viral immediate-early genes from mutant HSV vectors has been shown to reduce cytotoxicity in many cell types, including muscle *(25)*.

3.5. Adeno-Associated Virus

Recombinant adeno-associated virus vectors have been used as gene delivery vehicles for muscle cells. Although long-term gene expression and highly efficient transduction in all muscle cell types has been observed, the application of these vectors for gene therapy purposes is limited by their restrictive gene insert capacity *(26)*. The genes of certain growth factors, such as basic fibroblast growth factors and insulinlike growth factors, are small enough to make the adeno-associated virus vectors potential delivery vehicles for the musculoskeletal system.

4. STRATEGIES FOR GENE TRANSFER INTO MUSCLE

Various approaches can be used to achieve gene transfer to the musculoskeletal system, including cell therapy (myoblast transplantation), gene therapy (based on viral and nonviral vectors), and a combination of both techniques. A description of the various approaches used to deliver growth factors into injured skeletal muscle follows. It is hoped that, through cell and gene therapy, substances such as growth factors can be delivered to promote efficient healing and complete functional recovery following injury.

4.1. Cell Therapy

MDSC transplantation is the implantation of MDSCs into damaged muscle for muscle regeneration. The transplantation of MDSCs into dystrophin-deficient muscle to create a reservoir of dystrophin-producing cells has been studied extensively in mdx mice (an animal model for DMD) and in patients with DMD *(27)*. MDSC transplantation is capable of delivering dystrophin, the missing protein in DMD, and increasing muscle strength in DMD muscles. Yet, immune rejection as well as the poor survival and spread of injected MDSC posttransplantation have greatly limited the success of MDSC transplantation *(28)*. As such, recent research has focused on using viral or nonviral vectors to deliver genes to skeletal muscle, an approach called ex vivo gene transfer.

4.2. Direct Gene Therapy

Direct gene therapy is another approach to deliver genes to skeletal muscle. Muscle cells have been successfully transduced in vitro and in vivo using the intramuscular inoculation of replicative-deficient adenovirus, retrovirus, and HSV carrying the -galactosidase (LacZ) reporter gene; however, a major limitation of using these viral vectors alone is the differential transducibility observed throughout skeletal muscle development *(22,29)*.

The direct gene transfer of recombinant adenovirus carrying the LacZ reporter gene is capable of highly transducing injured muscle. Many LacZ-expressing myofibers have been found in the injured site of contused, lacerated, and strained muscles at 5 d following the direct gene transfer approach *(30)*. Although the transient expression of the transgene is likely an immune reaction, the use of the new-generation adenovirus vectors and the adeno-associated virus vectors may reduce the immune response and allow persistent expression of the transgene into the injected muscle *(23,31)*.

4.3. MDSC Transplantation Using Ex Vivo Gene Transfer

The ex vivo approach combines MDSC transplantation and gene therapy using an autologous MDSC transfer to deliver genes to skeletal muscle. This approach involves establishing a primary MDSC culture from dystrophic and injured muscles, which is then engineered by adequate transfection or transduction to produce dystrophin and growth factors in vitro. The engineered MDSCs are then injected into the same host to avoid immune rejection against the injured myoblasts. This method has been performed using adenovirus, retrovirus, and HSV-1-carrying reporter genes *(20,32,33)*. The transduced MDSCs fuse and reintroduce the reporter genes into the injected muscle. The ex vivo approach was used to deliver dystrophin into dystrophic muscle, and the efficiency of viral transduction using the ex vivo approach is greater than that of direct injection of the same amount of virus *(34)*.

Also, the ex vivo approach has been used to deliver genes into injured muscle. Primary MDSCs have been isolated, engineered following transduction with an adenovirus carrying the LacZ reporter gene, and injected

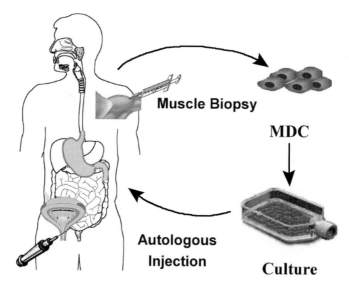

Fig. 2. MDSC-based tissue engineering in the lower urinary tract.

into injured muscle. The LacZ-transduced myofibers persisted in the injured muscle for at least 35 d postinjection. This suggests that the ex vivo gene transfer of autologous MDSCs is feasible and leads to persistent expression of marker gene in injured muscle *(30,35)*.

In summary, MDSC transplantation using ex vivo gene transfer may be advantageous for repairing muscle because the cells can be used as a reservoir for secreting growth factors. In addition, the engineered MDSCs can serve as a source of exogenous cells capable of participating in the healing process. The use of pluripotent MDSC in an ex vivo approach may become very attractive for the healing of damaged muscle.

5. GENE THERAPY APPLICATIONS IN THE LOWER URINARY TRACT

The lower urinary tract is ideally suited for minimally invasive therapy. All of the lower urinary tract can be reached either percutaneously or through endoscopy. Using MDSCs to deliver growth factor genes, an ex viv approach, could treat such disabling and prevalent conditions as urinary incontinence, interstitial cystitis (IC), and erectile dysfunction (ED) and limit the risk of systemic side effects. Figure 2 depicts MDSC-based tissue engineering in the lower urinary tract. This section focuses on gene therapy strategies that use both viral and nonviral approaches in the lower urinary tract.

5.1. Urinary Incontinence

Urinary incontinence is a serious and prevalent condition worldwide. The pharmaceutical industry has recently realized the significance of this disability and the potential market size that urinary incontinence represents. In the United States alone, an estimated 17 million men and women suffer from bladder control problems *(36)*. The Agency for Health Care Policy and Research and the World Health Organization have estimated the total economic costs of urinary incontinence as $26 billion a year *(37)*. As the population within developed countries continues to age, the economic costs will continue to soar. Thus, the economic impact of urinary incontinence is staggering.

The three main types of urinary incontinence are stress, urge, and overflow incontinence *(38)*. Stress and urge incontinence each account for 45% of all incontinence cases, whereas overflow incontinence accounts for approx 5% of cases. *Stress incontinence* occurs when the urethral sphincter muscle is weak and cannot prevent urine leakage during activities that put stress on the abdomen, such as coughing, sneezing, or jumping. *Urge incontinence* is a condition characterized by urinary urgency and frequency associated with uncontrollable urine leakage. *Overflow incontinence* is a devastating condition in which patients cannot urinate because of damage to the nerves innervating the bladder. One of the most common causes of overflow incontinence is diabetes mellitus, which produces bladder neuropathy.

5.1.1. Stress Incontinence

There are three approaches to the treatment of stress incontinence: exercise, bladder suspension surgery, or injection of bulking agents *(38)*. Intrinsic sphincteric deficiency (ISD) is the most severe type of stress incontinence. It is usually the result of prior surgery that damaged the sphincter muscle or pudendal nerve. The injection of bulking agents, such as collagen, into the urethra at the level of the urinary sphincter serves to produce a functional obstruction that will help to correct urine leakage. The injection of collagen into the urinary sphincter is a quick outpatient procedure that gives the patient little pain or risk; however, the long-term success of this procedure is limited by several disadvantages *(39)*. Collagen is often reabsorbed, which adversely affects a successful outcome and requires repeat injections in the majority of patients. An average of three collagen injections is needed to achieve partial or total improvement. Also, 5% of patients are allergic to bovine collagen, and in most injected with collagen, antibodies develop to bovine antigens.

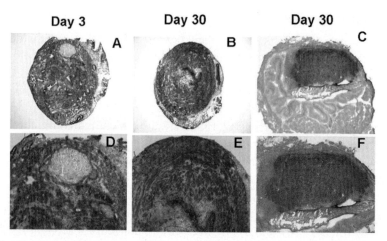

Fig. 3. A histologic comparison of collagen and MDSC injected into the rat urethra. A and D, d 3 after collagen injection. B and E, d 30 after collagen injection. C and F, d 30 after MDSC injection. A, B, and C reduced from ×40. D, E, and F reduced from ×100. The following were noted: (1) greater persistence of injected MDSCs vs bovine collagen; (2) autologous MDSC infection may be preferred for stress incontinence (nonallergenic and long-term persistence). (*See* color plate 4 in the insert following page 82.)

In one study, the feasibility of injecting MDSCs into the urethra was studied in rats and compared to bovine collagen injection *(40)*. A large number of the cells, transduced with the LacZ reporter gene, expressed -galactosidase 3 and 30 d after autologous MDSC injection. The persistence of the MDSCs at 3 d was similar to that of collagen; however, at 30 d, 88% of the cells survived, compared to only a scant amount of collagen. Figure 3 shows the histologic comparison between collagen and MDSCs in this study. In addition, the injection of the MDSCs into the urethral wall had no adverse effects. Thus, autologous MDSC injection may be more desirable than collagen injection for treating stress incontinence because the MDSCs are nonallergenic, they persist long term, and they directly correct the underlying pathophysiology by regenerating damaged urethral muscle. Also, it can be speculated that the cost of MDSC injection would be much less than the cost of many collagen injections because the MDSCs are obtained from the host patient.

In another study, stem cell tissue engineering was used to restore deficient urethral sphincter muscle in a rat model *(41)*. In the study, MDSCs were isolated from normal rats, transduced with LacZ for labeling, and injected into the proximal urethra of rats with denervated urinary sphincters.

After 2 wk, strips of urethra were tested from normal, denervated, and denervated-MDSC-injected rats. Fast twitch muscle contractions were recorded after electrical field stimulation. The amplitude of fast twitch muscle contractions was decreased in denervated sphincters and was improved by 88% in denervated sphincters injected with MDSCs. Histological examination revealed the formation of new skeletal muscle fibers at the injection sites of the urethral sphincter. This study lays the foundation for further investigation into the use of stem cells to treat urinary incontinence.

5.1.2. Urge Incontinence

The standard therapy for urge incontinence involves anticholinergic drugs, which work to reduce the involuntary bladder contractions, and behavioral therapy *(42)*. Gene therapy strategies for overactive bladders might include suppression of bladder muscle activity or neural pathways that trigger the micturition reflex. Christ et al. studied K^+ channel gene therapy as a treatment for urge incontinence in rats *(43)*. The intravesical inoculation of naked hSlo/pcDNA suppressed bladder hyperactivity in rats with partial urethral obstruction. They postulated that the overexpression of K^+ channels in the bladder might inhibit the overactive bladder.

5.1.3. Overflow Incontinence

Preliminary work has been carried out using HSV to treat overflow incontinence caused by diabetic neurogenic bladder dysfunction. Diabetic animals exhibit a decrease in nerve growth factor (NGF) production in target tissues *(44)*. There is evidence that the neuronal gene targets for NGF are understimulated *(45)*. Apfel et al. reported that exogenously administered NGF is capable of preventing the behavioral and biochemical manifestations of diabetic sensory neuropathy in the streptozotocin (STZ)-induced diabetic rat model *(46)*. In one study, HSV-1 vectors carrying NGF were injected into the bladder walls of rats 6 wk after the induction of diabetes by STZ *(47)*. Bladder function was compared between STZ-induced diabetic rats injected with the NGF expressed vector or rats injected with the control vector at 4 wk following HSV injection. The NGF-treated diabetic rats exhibited a 50% decrease in volume each void without significant changes in total urine output when compared with control diabetic rats. These results indicate that NGF expression via HSV vectors in the bladder afferent pathways improved bladder function in diabetic rats.

5.2. Interstitial Cystitis

Interstitial cystitis (IC) is a voiding dysfunction that affects nearly a million people in the United States *(48)*. IC is characterized by chronic pelvic pain associated with bladder symptoms of urinary frequency and urgency. A

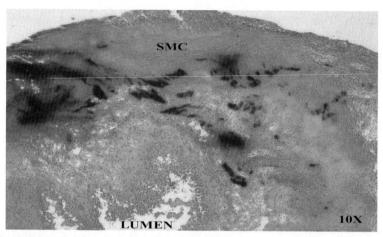

Fig. 4. Extensive β-galactosidase staining in the rat bladder. (*See* color plate 5 in the insert following page 82.)

possible gene therapy approach involves the delivery of preproenkephalin to the peripheral nerves of the bladder. This method delivers low, but therapeutic, quantities of enkephalin only to sensory nerves that innervate the organ in pain, but not to the whole animal. In one study, the preproenkephalin gene was transferred and maintained in the bladders and bladder afferent nerves of rats using the HSV-1 vector *(49)*. Also, this study concluded that the increased expression of enkephalin in bladder afferent pathways suppressed nociceptive responses induced by bladder irritation. Figure 4 depicts gene therapy for such lower urinary tract dysfunction as overflow incontinence and interstitial cystitis. This technique of gene transfer may be useful for treating IC and other types of visceral pain syndromes.

5.3. Erectile Dysfunction

Several studies of nitric oxide synthase (NOS) gene therapy for erectile dysfunction have been reported. Champion et al. demonstrated the feasibility of gene transfer of endothelial NOS (eNOS) augmenting erectile responses in the aged rat *(50)*. They administered a recombinant adenovirus containing the eNOS gene into the corpora cavernosa of the aged rat. An increase in cavernosal pressure with cavernosal nerve stimulation was enhanced in animals transfected with eNOS. In another study, Tirney et al. assessed inducible NOS (iNOS) gene therapy into the corpus cavernosum of adult rats *(51)*. They compared injections of plasmid, adenovirus, or adenovirus-transduced muscle cells. Muscle cell-mediated gene therapy was more successful for delivering iNOS into the corpus cavernosum than direct adenovirus or plasmid transfection methods. Gene therapy of NOS, using MDSCs, could open new avenues of treatment for erectile dysfunction. Control of NOS expression would be necessary to prevent prolonged erection. Finally, Wessells and Williams demonstrated endothelial cell-based ex vivo gene therapy for erectile dysfunction in rats *(52)*.

6. CONCLUSION

In summary, gene therapy using the ex vivo approach has the potential to revolutionize the treatments of many urological diseases. Although the safety and efficiency of this approach must be determined in clinical trials, many clinicians and scientists are excited about the relief that millions of patients will have with this revolutionary technology. However, two problems with gene therapy are the transient expression in transduced cells and the incomplete knowledge regarding the regulation of gene expression. Efforts of MDSC researchers continue to focus on the origin and identity of the population responsible for stem cell-like capabilities, the isolation of these cells from human tissues, the expansion to clinically relevant cell numbers, and controlling the differentiation and signaling processes that regulate progression within a lineage.

REFERENCES

1. Deasy, B. M., Jankowski, R. J., and Huard, J. (2001). Muscle-derived stem cells: characterization and potential for cell-mediated therapy. Blood Cells Mol Dis 27, 924–933.
2. Qu, Z., Balkir, L., van Deutekom, J. C., Robbins, P. D., Pruchnic, R., and Huard, J. (1998). Development of approaches to improve cell survival in myoblast transfer therapy. J. Cell Biol 14, 1257–1267.
3. Blau, H. M., and Springer M. L. (1995). Muscle-mediated gene therapy. N Engl J Med 333, 1554–1556.
4. Mauro, A. (1961). Satellite cells of skeletal muscle fibers. J Biochem Biophys Cytol 9, 493–498.
5. Beauchamp, J. R., Pagel, C. N., and Partridge, T. A. (1997). A dual-marker system for quantitative studies of myoblast transplantation in the mouse. Transplantation 63, 1794–1797.
6. Lee, J., Qu-Peterson, Z., Cao, B., et al. (2000). Clonal isolation of muscle-derived cells capable of enhancing muscle regeneration and bone healing. J Cell Biol 150, 1085–1099.
7. Gussoni, E., Soneoka, Y., Strickland, C. D., et al. (1999). Dystrophin expression in the mdx mouse restored by stem cell transplantation. Nature 401, 390–394.
8. Goodell, M. A. (1999). CD34+ or CD34−: does it really matter? Blood 94, 2545–2547.
9. Qu-Peterson, Z., Deasy, B. M., Jankowski, R. J., et al. (2002). Identification of

a novel population of muscle stem cells in mice: potential for muscle regeneration. J Cell Biol 157, 851–864.

10. Jackson, K. A., Mi, T., and Goodell, M. A. (1999). Hematopoietic potential of stem cells isolated from murine skeletal muscle. Proc Natl Acad Sci U S A 96, 14,482–14,486.

11. Huard, J., Bouchard, J. P., Roy, R., et al. (1992). Human myoblast transplantation: preliminary results of four cases. Muscle Nerve 15, 550–560.

12. Provinciali, M., Argentati, K., and Tibaldi, A. (2000). Efficacy of cancer gene therapy in aging: adenocarcinoma cells engineered to release IL-2 are rejected but do not induce tumor specific immune memory in old mice. Gene Ther 7, 624–632.

13. Evans, C. H., and Robbins, P. D. (1995). Possible orthopedic applications of gene therapy. J Bone Joint Surg Am 77, 1103–1114.

14. Svensson, E. C., Tripathy, S. K., and Leiden, J. M. (1996). Muscle based gene therapy: realistic possibilities for the future. Mol Med Today 2, 166–172.

15. Wolff, J. A., Ludtke, J. J., Assadi, G., Williams, P., and Jani, A. (1992). Long term persistence of plasmid DNA and foreign gene expression in mouse muscle. Hum Mol Genet 1, 363–369.

16. Danko, I., Fritz, J. D., Latendresse, J. J., Herweijer, H., Schultz, E., and Wolff, J. A. (1993). Dystrophin expression improves myofiber survival in mdx muscle following intramuscular plasmid DNA injection. Hum Mol Genet 2, 2055–2061.

17. Davis, H. L., Whalen, R. G., and Demeneix, B. A. (1993). Direct gene transfer into skeletal muscle in vivo: factors affecting efficiency of transfer and stability of expression. Hum Gene Ther 4, 151–159.

18. Vitiello, L., Chonn, A., Wasserman, J. D., Duff, C., and Worton, R. G. (1996). Condensation of plasmid DNA with polylysine improves liposome mediated gene transfer into established and primary muscle cells. Gene Ther 3, 396–404.

19. Feero, W. G., Li, S., Rosenblatt, J. D., et al. (1997). Selection and use of ligands for receptor-medicated gene delivery to myogenic cells. Gene Ther 4, 664–674.

20. Salvatori, G., Ferrari, G., Messogiorno, A., et al. (1993). Retroviral vector mediated gene transfer into human primary myogenic cells leads to expression in muscle fibers in vivo. Hum Gene Ther 4, 713–723.

21. Miller, D. G., Adam, M. A., and Miller, A. D. (1990). Gene transfer by retrovirus vectors occurs only in cells that are actively replicating at the time of infection. Mol Cell Biol 10, 4239–4242.

22. Van Deutekom, J. C., Floyd, S. S., Booth, D. K., et al. (1998). The development of approaches to improve viral gene delivery to mature skeletal muscle. Neuromuscul Disord 8, 135–148.

23. Kochanek, S., Clemens, P. R., Mitani, K., Chen, H. H., Chan, S., and Caskey, C. T. (1996). A new adenoviral vector: replacement of all viral coding sequences with 28kb of DNA independently expressing both full length dystrophin and beta-galactosidase. Proc Natl Acad Sci U S A 93, 5731–5736.

24. Huard, J., Krisky, D., Oligino, T., et al. (1997). Gene transfer to muscle using herpes simplex virus-based vectors. Neuromuscul Disord 7, 299–313.

25. Marconi, P., Krisky, D., Oligino, T., et al. (1996). Replication defective HSV vectors for gene transfer in vivo. Proc Natl Acad Sci U S A 93, 11,319–11,320.
26. Fisher, K. J., Jooss, K., Alston, J., et al. (1997). Recombinant adeno-associated virus for muscle directed gene therapy. Nat Med 3, 306–312.
27. Partridge, T. A., Morgan, J. E., Coulton, G. R., Hoffman, E. P., and Kunkel, L. M. (1989). Conversion of mdx myofibers from dystrophin negative to positive by injection of normal myoblasts. Nature 337, 176–179.
28. Beauchamps, J. R., Morgan, J. E., Pagel, C. N., and Partridge, T. A. (1994). Quantitative studies of the efficacy of myoblast transplantation. Muscle Nerve 18, S261.
29. Van Deutekom, J. C., Hoffman, E. P., and Huard, J. (1998). Muscle maturation: implications for gene therapy. Mol Med Today 4, 214–220.
30. Kasemkijwattana, C., Menetrey, J., Somogyi, G., et al. (1998). Development of approaches to improve the healing following muscle contusion. Cell Transplant 7, 585–598.
31. Haecker, S. E., Stedman, H. H., Balice-Gordon, R. J., et al. (1996). In vivo expression of full length human dystrophin from adenoviral vectors deleted of all viral genes. Hum Gene Ther 7, 1907–1914.
32. Huard, J., Acsadi, G., Jani, A., Massie, B., and Karpati, G. (1994). Gene transfer into skeletal muscles by isogenic myoblasts. Hum Gene Ther 5, 949–958.
33. Booth, D. K., Floyd, S. S., Day, C. S., Glorioso, J. C., Kovesdi, I., and Huard, J. (1997). Myoblast mediated ex vivo gene transfer to mature muscle. J Tissue Eng 3, 125–133.
34. Floyd, S. S., Clemens, P. R., Ontell, M. R., et al. (1998). Ex vivo gene transfer using adenovirus mediated full length dystrophin delivery to dystrophic muscles. Gene Ther 5, 19–30.
35. Kasemkijwattana, C., Menetrey, J., Day, C. S., et al. (1998). Biologic intervention in muscle healing and regeneration. Sports Med Arthrosc Rev 6, 95–102.
36. Abrams, P., Khoury, S., and Wein, A. (Eds.). (1999). WHO First International Consultation in Incontinence. Plymouth, UK: Plymouth Distributors.
37. Fantyl, J. A., Newman, D. K., Colling, J., et al. (1996). Urinary incontinence in adults: acute and chronic management. In: Clinical Practice Guideline, No. 2 Update. Rockville, MD: Department of Health and Human Services, Public Health Service, Agency for Health Care Policy and Research, AHCPR publication no. 96–0682.
38. Blaivas, J. G., and Groutz, A. (2002). Urinary incontinence: pathophysiology, evaluation, and management overview. In: Walsh, P. C., ed., Campbell's Urology, 8th ed. St. Louis, MO: Saunders, pp. 1027–1052.
39. Appell, R. A. (1994). Collagen injection therapy for urinary incontinence. Urol Clin North Am 21, 177–182.
40. Yokoyama, T., Yoshimura, N., Dhir, R., et al. (2001). Persistence and survival of autologous muscle derived cells vs bovine collagen as potential treatment of stress urinary incontinence. J Urol 165, 271–276.
41. Lee, J. Y., Cannon, T. W., Pruchnic, R., Fraser, M. O., Huard, J., and Chancellor, M. B. (2002). The effect of preiurethral muscle derived stem cell injection

on leak point pressure in a female rat model of stress urinary incontinence. J Urol 167, A799, 198.

42. Payne, C. K. (2002). Urinary incontinence: nonsurgical management. In: Walsh, P. C., ed., Campbell's Urology, 8th ed. St. Louis, MO: Saunders, pp. 1069–1091.
43. Christ, G. J., Day, N. S., Day, M., et al. (2001). Bladder injection of "naked" hSlo/pcDNA3 ameliorates detrusor hyperactivity in obstructed rats in vivo. Am J Physiol Regul Integr Comp Physiol 281, R1699–R1709.
44. Hellweg, R., Raovocj, G., Hartung, H. D., Hock, C., and Kreutzberg, G. W. (1994). Axonal transport of endogenous nerve growth factor (NGF) and NGF receptor in experimental diabetic neuropathy. Exp Neurol 130, 24–30.
45. Diemel, L. T., Stevens, E. J., Willars, G. B., and Tomlinson, D. R. (1992). Depletion of substance P and calcitonin gene-related peptide in sciatic nerve of rats with experimental diabetes; effects of insulin and aldose reductase inhibition. Neurosci Lett 137, 253–256.
46. Apfel, S. C., Arezzo, J. C., Brownlee, M., Rederoff, H., and Kessler, J. A. (1994). Nerve growth factor administration protects against experimental diabetic sensory neuropathy. Brain Res 634, 7–12.
47. Goins, W. F., Yoshimura, N., Phelan, M. W., et al. (2001). Herpes simplex virus mediated nerve growth factor expression in bladder and afferent neurons: potential treatment for diabetic bladder dysfunction. J Urol 165, 1748–1754.
48. Rivas, D. A., Chancellor, M. B., and Blaivas, J. G. (1994). Interstitial cystitis, current status of diagnosis and management. In: McGuire, E. J., ed., Advances in Urology, Vol. 7. New York: Churchill Livingstone, pp. 229–265.
49. Yoshimura, N., Franks, M. E., Sasaki, K., et al. (2001). Gene therapy of bladder pain with herpes simplex virus (HSV). Vectors expressing preproenkephalin (PPE). Urology 57, 116.
50. Champion, H. C., Bivalacqua, T. J., Hyman, A. L., Ignarro, L. J., Hellstrom, W. J., and Kadowitz, P. J. (1999). Gene transfer of endothelial nitric oxide synthase to the penis augments erectile responses in the aged rat. Proc Natl Acad Sci U S A 96, 11,648–11,652.
51. Tirney, S., Mattes, C. E., Yoshimura, N., et al. (2001). Nitric oxide synthase gene therapy for erectile dysfunction: comparison of plasmid, adenovirus, and adenovirus-transduced myoblast vectors. Mol Urol 5, 37–43.
52. Wessells, S., and Williams, S. K. (1999). Endothelial cell transplantation into the corpus cavernosum: moving towards cell-based gene therapy. J Urol 162, 2162–2164.

5

Human Dental Pulp Stem Cells
Characterization and Developmental Potential

Stan Gronthos, Natasha Cherman, Pamela Gehron Robey, and Songtao Shi

1. INTRODUCTION

The ongoing debate concerning the ethics of human embryonic stem cells has brought to light the clinical potential of stem cell-based therapies for treating various human afflictions, ranging from cancer, Alzheimer's disease, heart disease, cirrhosis of the liver, and diabetes. Although the future of stem cell research holds much promise, tissue bioengineering utilizing embryonic stem cell technology is still at an early stage of development *(1–3)*. The routine use of postnatal stem cells in clinical applications has been successfully demonstrated in certain circumstances, such as the rescue of hematopoietic bone marrow using mobilized peripheral blood stem cells in cancer patients *(4,5)*. These results have fueled investigations into the potential of other stem cell populations found to exist in a wider range of adult tissues than originally anticipated. One important development to emerge from these studies is the concept of *plasticity*, the ability of stem cell populations to develop into multiple cell lineages from diverse tissues, therefore greatly expanding the potential of stem cells beyond the constraints of their tissue of origin. In this chapter, we describe the stem cell-like qualities of putative postnatal dental pulp stem cells and speculate on the future clinical benefits that may arise from their further characterization.

During development, tooth morphogenesis progresses through mutually inductive signaling between interacting oral epithelial and mesenchymal cells of neural ectodermal origin *(6–8)*. This results in the formation of the hard external enamel layer of the crown, a highly mineralized acellular matrix produced by specialized epithelial cells known as ameloblasts, which undergo apoptosis during tooth eruption and are absent in mature teeth.

The synthesis of the initial inner layer of supportive mineralized dentin (primary) is because of the activities of another cell type, called *odonto-*

From: *Adult Stem Cells*
Edited by: K. Turksen © Humana Press Inc., Totowa, NJ

blasts, derived from the developing dental papilla. These cells align on the outer edge of the central chamber, which is comprised of a vascularized soft connective tissue called *pulp*, and continue to form dentin (secondary) throughout the life of the tooth. Odontogenic cytoplasmic extensions span the mineralized dentin matrix through fine canals or tubules. Both blood vessels and sensory nerve fibers infiltrate the pulp tissue via the apical foraman.

The whole tooth structure is held in place with the surrounding bone by a fibrocellular stratum of peridontal ligaments that link the aveolar bone with a thin outer layer of cementum covering the root dentin. The complexity of the different structures and cell types that constitute permanent teeth have evolved to withstand the forces of mastication beyond the average human life span without the need for multiple successive teeth or continuous eruption.

In spite of the extensive knowledge concerning the development and pathology of teeth, restoration of pulpal tissue and dentin regeneration damaged by either carious lesions or mechanical trauma has at best been limited using conventional treatments *(9–11)*. Under steady-state conditions, human teeth do not undergo remodeling, unlike many other mineralized tissues, such as bone, which remodels, albeit slowly, throughout postnatal life. However, following tooth eruption, dentinal damage caused by mechanical trauma, exposure to chemicals, or infectious processes induces the formation of reparative (tertiary) dentin, a more poorly organized mineralized matrix, compared to primary and secondary dentin, that serves as a protective barrier to the dental pulp *(11–15)*.

Once the odontoblast layer has been breached, it is presumed that preodontoblasts are recruited from somewhere in the pulpal tissue *(16,17)* because functional odontoblasts are postmitotic cells and often are destroyed by the damage. This is supported by animal studies demonstrating the regenerative capacity of developing neonatal rodent and fetal bovine dental papilla to form mineralized dentin–pulp complexes following transplantation into ectopic sites in vivo *(18–22)*.

We recently identified candidate dental pulp stem cells (DPSCs) derived from human adult third molars *(23)*. To confirm the stem cell-like properties of human DPSCs, studies were designed to satisfy several basic criteria characteristic of other stem cell populations, including the ability to form clonogenic cell clusters, a high proliferative potential, the ability to self-renew, and the capacity to differentiate into multiple functional cell types *(1,3)*.

2. CHARACTERIZATION OF HUMAN DENTAL PULP STEM CELLS

2.1. Proliferation Capacity of DPSCs In Vitro

We first identified putative human DPSCs by their ability to generate clonogenic cell colonies in vitro, a common feature displayed by various stem cell populations previously isolated from other tissues. To determine the colony-forming efficiency of DPSCs from whole pulp tissue, single-cell suspensions were prepared by collagenase–dispase treatment of pulp fragments, followed by filtration through a fine mesh to remove cell aggregates prior to seeding the cells at low plating densities *(23)*. The resulting colonies consisted of adherent, fibroblasticlike cells, analogous to colony-forming units-fibroblastic (CFU-F), which form in vitro by osteogenic precursors known as bone marrow stromal stem cells (BMSSCs) or mesenchymal stem cells *(24–26)*. An average incidence of approx 40 dental pulp-derived CFU-F were generated per 10,000 cells plated *(23)*.

Recent cloning experiments have indicated that the majority (67%) of the individual colonies failed to proliferate beyond 20 population doublings. Therefore, as primary DPSC cultures are expanded over successive cell passages, progeny arising from the minor subfraction of highly proliferative colonies constitute the bulk of the cell population. This mirrors the growth patterns observed for individual CFU-F derived from primary BMSSCs following ex vivo expansion. Multicolony-derived DPSCs were consistent in their capacity to proliferate in vitro at an average 30% greater rate compared to that observed for BMSSCs.

To account for the rapid rate of proliferation demonstrated by DPSCs, we employed cyclic deoxyribonucleic acid (cDNA) microarray analysis to identify differences in gene expression profiles between DPSCs and BMSSCs *(27)*. The high incidence of DPSCs undergoing S phase was recently correlated with high expression levels of the cell cycle activator, cyclin-dependent kinase 6 (CDK6) *(27)*. The activation of CDK6 is mediated by D-type cyclins to promote the progression of cells through G1 to the start of DNA synthesis *(28,29)*. In turn, D-cyclins are activated by various growth factors, such as mitogen IGF-2 (insulin-like growth factor 2), which was also found to be highly expressed by DPSCs compared to BMSSCs. Conversely, DPSCs expressed lower levels of insulin-like growth factor binding protein 7 (IGFBP-7) than BMSSCs *(27)*, a factor known to bind to IGF-1 and IGF-2 that induces inhibition of cell growth *(30)*.

The consequences of these and other differentially expressed genes regarding the growth and development of mineralized dentin and bone are currently under investigation. Studies thus far indicate that DPSCs maintain their high rate of proliferation even after extensive subculture beyond 40 population doublings. Taken together with their clonogenic nature, highly proliferative DPSCs satisfy two characteristics of postnatal somatic stem cells *(1,3)*.

2.2. Phenotypic Analysis of DPSCs

Elucidating the gene and protein expression patterns of primitive DPSC populations and functional odontoblasts is pivotal for understanding the process of odontogenesis. Mineralized dentin is composed of a complex scaffold of extracellular matrix, consisting mainly of collagen type I and some noncollagenous glycoproteins (dentin matrix protein 1, collagen type I, osteonectin, osteopontin, bone sialoprotein, and osteocalcin) also commonly found in the matrix of bone *(27,31–37)*.

These similarities are intriguing considering that, during embryogenesis, odontoblasts are derived from neuroectodermal mesenchyme, in contrast to osteoblasts of the axial and peripheral skeleton, which arise from mesodermal mesenchyme. However, two matrix proteins, dentin sialoprotein (DSP) *(38)* and dentin phosphoprotein (DPP), are thought to be expressed uniquely in dentin *(39)*. Both DSP and DPP have been shown to be encoded by a single gene, known as dentin sialophosphoprotein (DSPP) *(40)*, which is expressed following formation of the collagen-rich predentin matrix, prior to mineralization *(41)*.

Importantly, primary DPSC cultures that had not been induced to differentiate were negative for the odontoblast specific marker DSPP, initially by *in situ* hybridization *(23)* and more recently by Western blot analysis using a human-specific DSPP polyclonal antibody developed by Dr. Larry Fisher (Fig. 1A). Immunohistological studies identified dentin-specific staining during early stages of ectopic mineralization by DPSCs in xenogeneic transplants (Fig. 1B). In sections of human pulp tissue, the DSPP antibody only reacted with cells in the mature odontoblast layer *(61)*. These data suggest that the clonogenic dental pulp–derived cells represent an undifferentiated preodontogenic phenotype. This was also supported in animal studies that failed to detect DSPP messenger RNA (mRNA) expression in cultured dental papilla cells derived from rat incisors using reverse transcriptase polymerase chain reaction (RT-PCR) in contrast to the high expression of DSPP detected from freshly isolated odontoblast–pulp tissue *(42,43)*.

To date, the precise anatomical location of DPSCs is largely unknown because of a lack of markers specific to the preodontogenic population. Cir-

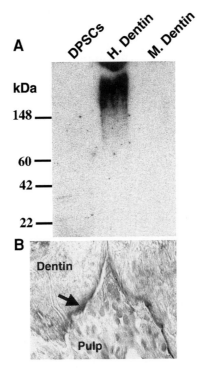

Fig. 1. DSPP expression by DPSCs. (**A**) Western blot analysis of cultured DPSCs, human dentin, and mouse dentin with a rabbit polyclonal antihuman DSPP antibody. Ex vivo expanded DPSCs were transplanted with HA/TCP subcutaneously into immunocompromised mice. (**B**) Immunoreactivity of DSPP antibody *(arrow)* with the dentin–pulp interface is shown in recovered 8-wk-old DPSC transplants.

cumstantial evidence suggests that preodontoblasts may originate from pericytes migrating from the pulpal endothelium *(44)*. Extensive immunophenotyping of ex vivo expanded DPSCs demonstrated their expression of various markers associated with endothelial or smooth muscle cells such as vascular cell adhesion molecule-1(VCAM-1), CD146 (MUC-18), and α-smooth muscle actin *(23)*. In addition, α-smooth muscle actin-positive cells have also been detected close to mineralized deposits in human dental pulp cultures *(45)*. The expression of these perivascular markers implicates a possible niche for DPSCs in association with blood vessel walls. It is hoped further characterization of DPSCs using current molecular technology will provide novel markers useful in their identification *in situ* and isolation and purification ex vivo.

3. DIFFERENTIATION POTENTIAL OF DPSCS IN VITRO AND IN VIVO

3.1. Regenerating a Dentin–Pulp Complex In Vivo

Mineralization within the papal chamber is a frequent event that usually manifests as small calcified pulp stones because of caries, aging, trauma, and systemic conditions *(46)*. Previous studies have established animal cell cultures from dental pulp tissue using a variety of culture methods and noted the ability of such cultures to form mineralized nodules in vitro *(43,47–50)*. In analogy, human-derived dental pulp can also be cultivated in vitro and possesses the capacity to form mineralized deposits in the presence of inductive media containing ascorbic acid, dexamethasone, and an excess of phosphate *(23,51,52)*. The use of methodology such as infrared microspectroscopic examination and X-ray diffraction electron microscopic (EM) analysis has confirmed the dentinlike nature of the crystalline structures that comprise the mineralized nodules in vitro, which are distinct from the crystal structures of mineralized enamel and bone in vivo *(47,49,51,52)*.

To determine the capacity of ex vivo expanded DPSCs to generate a functional dentin–pulplike tissue in vivo, we utilized an established transplantation system previously optimized for the formation of ectopic bone by cultured BMSSCs *(53,54)*. Until recently, the ability to evoke ectopic dentin formation in vivo was only demonstrated successfully in animal models that utilized rodent or bovine developing papilla tissue *(18–21)*. Similar studies using human intact developing dental papilla or adult dental pulp tissue failed to generate a mineralized dentin matrix or odontoblastlike cells following transplantation into immunocompromised mice *(55,56)*. Previous reports showed that, unlike rodent-derived bone marrow stromal and dental pulp cells, human equivalents require a suitable conductive carrier, such as hydroxyapatite/tricalcium phosphate (HA/TCP) particles, to induce the formation of bone and dentin in vivo *(19,54)*. HA/TCP and other biomaterials have also been used, with partial success, in the clinic to stimulate a pupal proliferation response to aid in the repair of damaged dentin *(9,11)*.

We previously demonstrated that cultured adult human dental pulp cells are capable of generating a dentin–pulplike complex in vivo in conjunction with HA/TCP as a carrier vehicle *(23)*. Typical DPSC transplants developed areas of vascularized pulp tissue surrounded by a well-defined layer of odontoblastlike cells, aligned around mineralized dentin with their processes extending into tubular structures. The odontoblastlike cells and fibrous pulp tissue in the transplants were shown to be donor in origin by their reactivity to the human-specific, alu cDNA probe *(23)*. In addition, orientation of the collagen fibers in the dentin was characteristic of ordered primary dentin, perpendicular to the odontoblast layer. Backscatter EM analysis demonstrated that the dentinlike material formed in the transplants had a globular appearance consistent with the structure of dentin *in situ* (unpublished observations). Moreover, the presence of human DSPP detected in the transplants confirmed the ability of DPSCs to regenerate a human dentin–pulp microenvironment in vivo.

Studies explored whether DPSCs possess the ability to self-renew. To answer this question, we harvested primary DPSC implants at 2 mo posttransplantation and liberated the cells by enzymatic digestion for subsequent expansion in vitro. Donor human cells were isolated from the cultures by fluorescence activated cell sorting (FACS) using a human β_1-integrin-specific monoclonal antibody, then retransplanted into immunodeficient mice for 2 mo. Recovered secondary transplants yielded the same dentin–pulplike structures as observed in the primary transplants. Human DSPP protein was found in the dentin matrix by immunohistochemical staining, and *in situ* hybridization studies confirmed the human origin of the odontoblast–pulp cells contained in the secondary DPSC transplants (Fig. 2). Efforts are now under way to determine whether the self-renewing stem cell compartment is localized in the fibrous pulp tissue of the primary transplants.

The developmental potential of individual ex vivo expanded DSPC colonies were also assessed. Of the clones from the initial primary cultures, 25% demonstrated a reduced capacity to form ectopic dentin in vivo; 30% showed an increased capacity when compared to parental multicolony-derived cells. These data are suggestive of a hierarchy of pulp cell differentiation that corresponds to the variations seen in the proliferation rates and developmental potential between individual DSPC clones. Therefore, pulp tissue seems to harbor a rare population of high-proliferating cells with the ability to regenerate a dentin–pulp structure in vivo and the capacity for self-renewal.

3.2. Adipogenic Potential of DPSCs In Vitro

To determine whether DPSCs represent multipotent stem cells, we cultured the cells in various inductive media previously shown to promote the differentiation of adipocytes. The development of fat is not a feature of dental pulp, as opposed to the abundance of fat cells in bone marrow. Analogous to this, the dentin–pulp structures observed in DPSC transplants failed to support either a hematopoietic marrow or any fat cell development, commonly detected in BMSSC transplants following significant bone formation *(23)*. In vitro studies also failed to induce adipogenesis in long-term DPSC cultures grown in the presence of the glucocorticoid dexamethasone, in con-

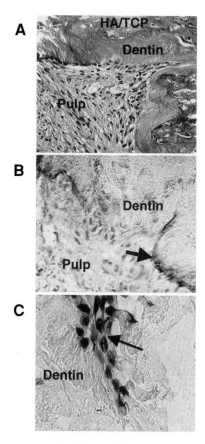

Fig. 2. Self-renewal capacity of DPSCs. Cell cultures were established from 3-mo-old DPSC primary transplants following collagenase/dispase treatment. Ex vivo expanded human cells were selected by FACS, then retransplanted into immunocompromised mice with HA/TCP. (**A**) Secondary transplants developed a dentin–pulp complex in vivo. (**B**) The dentin–pulp interface stained positive *(arrow)* for human DSPP protein. (**C**) Fibrous pulp tissue was positive *(arrow)* for the human-specific alu repetitive element by *in situ* hybridization.

trast to the abundant clusters of lipid-laden adipocytes observed in corresponding BMSSC cultures. More recent studies using a potent cocktail of adipogenenic inductive agents (0.5 m*M* methylisobutylxanthine, 0.5 μ*M* hydrocortisone, 60 μ*M* indomethacin) *(57)* have demonstrated the presence of Oil red O-positive fat-containing adipocytes in DPSC cultures following several weeks of induction (Fig. 3A). This was also correlated with an upregulation of the early adipogenic master regulatory gene peroxisome

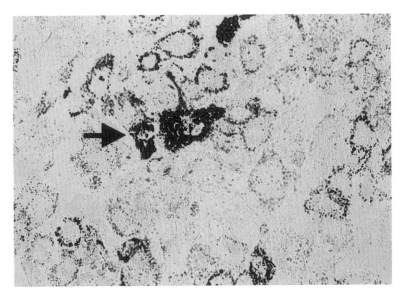

Fig. 3. Fat development in vitro. Histochemical staining of oil red O-positive *(arrow)* lipid-laden adipocytes in DPSC cultures following 5 wk of induction with 0.5 m*M* methylisobutylxanthine, 0.5 µ*M* hydrocortisone, and 60 µ*M* indomethacin.

proliferator-activated receptor r2 (PPAR2) and the mature adipocyte marker lipoprotein lipase using RT-PCR *(62)*. These observations highlight the plasticity of the DPSC population to develop into functional stromal cell types not normally associated with dental pulp tissue.

3.3. Neuronal Potential of DPSC

Dental pulp contains prominent nerve fibers that penetrate through the tubules alongside the odontogenic cellular processes and act as a protective system in response to degradation of the dentin layer *(6)*. This system of nerve fibers in the dentin matrix allows teeth to receive external stimulation that acts through pain receptors. During development, dental nerve tissue and odontoblasts are both presumed to originate from migratory neural crest cells *(6–8)*. Recent investigations have explored the possibility that DPSCs have the potential to differentiate into neural-like cells. Ex vivo expanded DPSCs were constitutively expressed nestin, an early marker of neural precursor cells, and glial fibrillary acidic protein (GFAP), an antigen characteristic of glial cells (Fig. 4A).

In accord with these findings, other investigators have identified the same markers in dental pulp tissue *in situ (58,59)*. When DPSCs were cultured

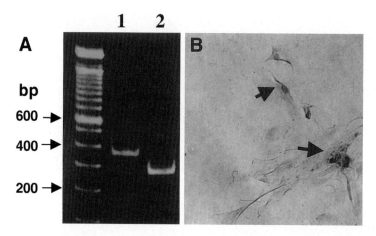

Fig. 4. Neuronal differentiation in vitro. (**A**) Basal mRNA expression levels of *(1)* GFAP and *(2)* nestin transcripts in DPSCs cultured under normal conditions. (**B**) NeuN protein expression *(arrows)* following 2 wk of culture in neoronal inductive conditions: Neuroblast A medium (Invitrogen/GIBCO), 5% horse serum, 1% fetal bovine serum, transferrin 100 µg/mL, insulin 25 µg/mL, retinoic acid 0.5 µ*M*, and BDNF (brain-derived neurotrophic factor) 10 ng/mL.

under defined neural inductive conditions, there was enhanced expression of both nestin and GFAP. Morphological assessment of induced DPSCs identified long cytoplasmic processes protruding from rounded cell bodies, in contrast to their usual bipolar fibroblasticlike appearance. Moreover, DPSCs cultured under neural inductive conditions *(60)* were found to express the neuron-specific marker neuronal nuclei (NeuN) by immunohistochemical staining (Fig. 4B).

These preliminary studies provided the first experimental evidence that adult human DPSCs may possess the potential to differentiate into neural-like cells with expression of nestin, GFAP, and NeuN in vitro. Transplantation studies are now under way to determine the capacity of human DPSCs to form functional neuronal tissue following their transplantation into different brain sites in immunocompromised mice.

4. SUMMARY

Although the loss of teeth is not a life-threatening event, edentulism is a major problem in the aging population. The study of DPSCs becomes a quality-of-life issue for many individuals afflicted with severe peridontal disease and tooth loss or for those patients who have undergone radical reconstructive craniofacial surgery. Human DPSCs, in association with biocompatible materials, may be an ideal candidate for future cell-based therapies for treating dental disorders; this is because of their capacity to regenerate a dentin–pulplike complex in vivo. However, it is not known whether DPSCs have the potential to differentiate into other cell lineages such as seen with bone marrow stromal stem cells. Preliminary evidence suggests that DPSCs are not merely committed preodontoblastlike cells, but appear to have a more primitive multipotential phenotype capable of developing into other stromal type cells, such as functional adipocytes, normally absent in dental pulp tissue. Moreover, it is their potential to develop into neuronal-like tissue that may offer enormous possibilities to treat different neurological disorders given the easy accessibility of this noncontroversial stem cell population.

REFERENCES

1. Bianco, P., and Robey, P. G. (2001). Stem cells in tissue engineering. Nature 414, 118–121.
2. Weissman, I. L., Anderson, D. J., and Gage, F. (2001). Stem and progenitor cells: origins, phenotypes, lineage commitments, and transdifferentiations. Annu Rev Cell Dev Biol 17, 387–403.
3. Fuchs, E., and Segre, J. A. (2000). Stem cells: a new lease on life. Cell 100, 143–155.
4. Bacigalupo, A., Frassoni, F., and Van Lint, M. T. (2000). Bone marrow or peripheral blood as a source of stem cells for allogeneic transplants. Curr Opin Hematol 7, 343–347.
5. Buckner, C. D. (1999). Autologous bone marrow transplants to hematopoietic stem cell support with peripheral blood stem cells: a historical perspective. J Hematother 8, 233–236.
6. Orchardson, R., and Cadden, S. W. (2001). An update on the physiology of the dentine–pulp complex. Dent Update 28, 200–206, 208, 209.
7. Peters, H., and Balling, R. (1999). Teeth. Where and how to make them. Trends Genet 15, 59–65.
8. Thesleff, I., and Aberg, T. (1999). Molecular regulation of tooth development. Bone 25, 123–125.
9. Kaigler, D., and Mooney, D. (2001). Tissue engineering's impact on dentistry. J Dent Educ 65, 456–462.
10. Baum, B. J., and Mooney, D. J. (2000). The impact of tissue engineering on dentistry. J Am Dent Assoc 131, 309–318.
11. Levin, L. G. (1998). Pulpal regeneration. Pract Periodontics Aesthet Dent 10, 621–624.
12. About, I., Murray, P. E., Franquin, J. C., Remusat, M., and Smith, A. J. (2001). Pulpal inflammatory responses following non-carious class V restorations. Oper Dent 26, 336–342.

13. About, I., Murray, P. E., Franquin, J. C., Remusat, M., and Smith, A. J. (2001). The effect of cavity restoration variables on odontoblast cell numbers and dental repair. J Dent 29, 109–117.

14. Murray, P. E., About, I., Franquin, J. C., Remusat, M., and Smith, A. J. (2001). Restorative pulpal and repair responses. J Am Dent Assoc 132, 482–491.

15. Murray, P. E., About, I., Lumley, P. J., Smith, G., Franquin, J. C., and Smith, A. J. (2000). Postoperative pulpal and repair responses. J Am Dent Assoc 131, 321–329.

16. Murray, P. E., About, I., Lumley, P. J., Franquin, J. C., Remusat, M., and Smith, A. J. (2000). Human odontoblast cell numbers after dental injury. J Dent 28, 277–285.

17. Ruch, J. V. (1998) Odontoblast commitment and differentiation. Biochem Cell Biol 76, 923–938.

18. Lyaruu, D. M., van Croonenburg, E. J., van Duin, M. A., Bervoets, T. J., Woltgens, J. H., and de Blieck-Hogervorst, J. M. (1999). Development of transplanted pulp tissue containing epithelial sheath into a tooth-like structure. J Oral Pathol Med 28, 293–296.

19. Holtgrave, E. A., and Donath, K. (1995). Response of odontoblast-like cells to hydroxyapatite ceramic granules. Biomaterials 16, 155–159.

20. Ishizeki, K., Nawa, T., and Sugawara, M. (1990). Calcification capacity of dental papilla mesenchymal cells transplanted in the isogenic mouse spleen. Anat Rec 226, 279–287.

21. Prime, S. S., and Reade, P. C. (1980). Xenografts of recombined bovine odontogenic tissues and cultured cells to hypothymic mice. Transplantation 30, 149–152.

22. Bartlett, P. F., Sim, F. R., Reade, P. C., and Prime, S. S. (1978). Transplantation of bovine odontogenic tissues and dissociated odontogenic cells to hypothymic mice. Transplantation 25, 126–130.

23. Gronthos, S., Mankani, M., Brahim, J., Robey, P. G., and Shi, S. (2000). Postnatal human dental pulp stem cells (DPSCs) in vitro and in vivo. Proc Natl Acad Sci U S A 97, 13,625–13,630.

24. Pittenger, M. F., Mackay, A. M., Beck, S. C., et al. (1999). Multilineage potential of adult human mesenchymal stem cells. Science 284, 143–147.

25. Castro-Malaspina, H., Gay, R. E., Resnick, G., et al. (1980). Characterization of human bone marrow fibroblast colony-forming cells (CFU-F) and their progeny. Blood 56, 289–301.

26. Friedenstein, A. J. (1976). Precursor cells of mechanocytes. Int Rev Cytol 47, 327–359.

27. Shi, S., Gehron Robey, P., and Gronthos, S. (2001). Comparison of gene expression profiles for human, dental pulp and bone marrow stromal stem cells by cDNA microarray analysis. Bone 29, 532–539.

28. Ekholm, S. V., and Reed, S. I. (2000). Regulation of G(1) cyclin-dependent kinases in the mammalian cell cycle. Curr Opin Cell Biol 12, 676–684.

29. Grossel, M. J., Baker, G. L., and Hinds, P. W. (1999). cdk6 can shorten G(1) phase dependent upon the N-terminal INK4 interaction domain. J Biol Chem 274, 29,960–29,967.

30. Kato, M. V., Sato, H., Tsukada, T., Ikawa, Y., Aizawa, S., and Nagayoshi, M. (1996). A follistatin-like gene, mac25, may act as a growth suppressor of osteosarcoma cells. Oncogene 12, 1361–1364.

31. Buchaille, R., Couble, M. L., Magloire, H., and Bleicher, F. (2000). A substractive PCR-based cDNA library from human odontoblast cells: identification of novel genes expressed in tooth forming cells. Matrix Biol 19, 421–430.

32. Shiba, H., Fujita, T., Doi, N., et al. (1998). Differential effects of various growth factors and cytokines on the syntheses of DNA, type I collagen, laminin, fibronectin, osteonectin/secreted protein, acidic and rich in cysteine (SPARC), and alkaline phosphatase by human pulp cells in culture. J Cell Physiol 174, 194–205.

33. D'Souza RN, Cavender A, Sunavala G, et al. (1997). Gene expression patterns of murine dentin matrix protein 1 (Dmp1) and dentin sialophosphoprotein (DSPP) suggest distinct developmental functions in vivo. J Bone Miner Res 12, 2040–2049.

34. Butler, W. T., Ritchie, H. H., and Bronckers, A. L. (1997). Extracellular matrix proteins of dentine. Ciba Found Symp 205, 107–115.

35. Butler, W. T., and Ritchie, H. (1995). The nature and functional significance of dentin extracellular matrix proteins. Int J Dev Biol 39, 169–179.

36. Nakashima, M., Nagasawa, H., Yamada, Y., and Reddi, A. H. (1994). Regulatory role of transforming growth factor-beta, bone morphogenetic protein-2, and protein-4 on gene expression of extracellular matrix proteins and differentiation of dental pulp cells. Dev Biol 162, 18–28.

37. Kuo, M. Y., Lan, W. H., Lin, S. K., Tsai, K. S., and Hahn, L. J. (1992). Collagen gene expression in human dental pulp cell cultures. Arch Oral Biol 37, 945–952.

38. Ritchie, H. H., Hou, H., Veis, A., and Butler, W. T. (1994). Cloning and sequence determination of rat dentin sialoprotein, a novel dentin protein. J Biol Chem 269, 3698–3702.

39. Gorter de Vries, I., Quartier, E., Van Steirteghem, A., Boute, P., Coomans, D., and Wisse, E. (1986). Characterization and immunocytochemical localization of dentine phosphoprotein in rat and bovine teeth. Arch Oral Biol 31, 57–66.

40. MacDougall, M., Simmons, D., Luan, X., Nydegger, J., Feng, J., and Gu, T. T. (1997). Dentin phosphoprotein and dentin sialoprotein are cleavage products expressed from a single transcript coded by a gene on human chromosome 4. Dentin phosphoprotein DNA sequence determination. J Biol Chem 272, 835–842.

41. Feng, J. Q., Luan, X., Wallace, J., et al. (1998). Genomic organization, chromosomal mapping, and promoter analysis of the mouse dentin sialophosphoprotein (DSPP) gene, which codes for both dentin sialoprotein and dentin phosphoprotein. J Biol Chem 273, 9457–9464.

42. Dey, R., Son, H. H., and Cho, M. I. (2001). Isolation and partial sequencing of potentially odontoblast-specific/enriched rat cDNA clones obtained by suppression subtractive hybridization. Arch Oral Biol 46, 249–260.

43. Ueno, A., Kitase, Y., Moriyama, K., and Inoue, H. (2001). MC3T3-E1-conditioned medium-induced mineralization by clonal rat dental pulp cells. Matrix Biol 20, 347–355.

44. Carlile, M. J., Sturrock, M. G., Chisholm, D. M., Ogden, G. R., and Schor, A. M. (2000). The presence of pericytes and transitional cells in the vasculature of the human dental pulp: an ultrastructural study. Histochem J 32, 239–245.

45. Alliot-Licht, B., Hurtrel, D., and Gregoire, M. (2001). Characterization of α-smooth muscle actin positive cells in mineralized human dental pulp cultures. Arch Oral Biol 46, 221–228.

46. Holan, G. (1998). Tube-like mineralization in the dental pulp of traumatized primary incisors. Endod Dent Traumatol 14, 279–284.

47. Unda, F. J., Martin, A., Hilario, E., Begue-Kirn, C., Ruch, J. V., and Arechaga, J. (2000). Dissection of the odontoblast differentiation process in vitro by a combination of FGF1, FGF2, and TGFβ1. Dev Dyn 218, 480–489.

48. Yokose, S., Kadokura, H., Tajima, Y., et al. (2000). Establishment and characterization of a culture system for enzymatically released rat dental pulp cells. Calcif Tissue Int 66, 139–144.

49. Hanks, C. T., Sun, Z. L., Fang, D. N., et al. (1998). Cloned 3T6 cell line from CD-1 mouse fetal molar dental papillae. Connect Tissue Res 37, 233–249.

50. Tsukamoto, Y., Fukutani, S., Shin-Ike, T., et al. (1992). Mineralized nodule formation by cultures of human dental pulp-derived fibroblasts. Arch Oral Biol 37, 1045–1055.

51. About, I., Bottero, M. J., de Denato, P., Camps, J., Franquin, J. C., and Mitsiadis, T. A. (2000). Human dentin production in vitro. Exp Cell Res 258, 33–41.

52. Couble, M. L., Farges, J. C., Bleicher, F., Perrat-Mabillon, B., Boudeulle, M., and Magloire, H. (2000). Odontoblast differentiation of human dental pulp cells in explant cultures. Calcif Tissue Int 66, 129–138.

53. Kuznetsov, S. A., Krebsbach, P. H., Satomura, K., et al. (1997). Single-colony derived strains of human marrow stromal fibroblasts form bone after transplantation in vivo. J Bone Miner Res 12, 1335–1347.

54. Krebsbach, P. H., Kuznetsov, S. A., Satomura, K., Emmons, R. V., Rowe, D. W., and Robey, P. G. (1997). Bone formation in vivo: comparison of osteogenesis by transplanted mouse and human marrow stromal fibroblasts. Transplantation 63, 1059–1069.

55. Buurma, B., Gu, K., and Rutherford, R. B. (1999). Transplantation of human pulpal and gingival fibroblasts attached to synthetic scaffolds. Eur J Oral Sci 107, 282–289.

56. Prime, S. S., Sim, F. R., and Reade, P. C. (1982). Xenografts of human ameloblastoma tissue and odontogenic mesenchyme to hypothymic mice. Transplantation 33, 561–562.

57. Gimble, J. M., Morgan, C., Kelly, K., et al. (1995). Bone morphogenetic proteins inhibit adipocyte differentiation by bone marrow stromal cells. J Cell Biochem 58, 393–402.

58. Hainfellner, J. A., Voigtlander, T., Strobel, T., et al. (2001). Fibroblasts can

express glial fibrillary acidic protein (GFAP) in vivo. J Neuropathol Exp Neurol 60, 449–461.

59. About, I., Laurent-Maquin, D., Lendahl, U., and Mitsiadis, T. A. (2000). Nestin expression in embryonic and adult human teeth under normal and pathological conditions. Am J Pathol 157, 287–295.

60. Sanchez-Ramos, J., Song, S., Cardozo-Pelaez, F., et al. (2000). Adult bone marrow stromal cells differentiate into neural cells in vitro. Exp Neurol 164, 247–256.

61. Shi, S. and Gronthos, S. (2003) Perivascular niche of postnatal mesenchymal stem cells in human bone marrow and dental pulp. J Bone Miner Res 18, 696–704.

62. Gronthos, S., Brahim, J., Li, W., et al. (2002) Stem cell properties of human dental pulp stem cells. J Dent Res 81, 531–535.

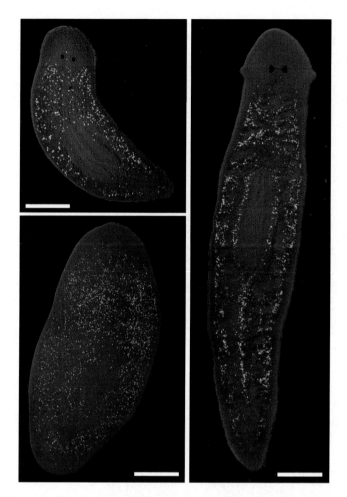

Color Plate 1, Fig. 1. (*See* full caption and discussion in Chapter 1, p. 3.) Bromodeoxyuridine labeling of regenerative stem cells in planarians *Phagocata* sp. *(upper left)*; *Girardia dorotocephala (lower left)*; and *Schmidtea mediterranea (right)*...

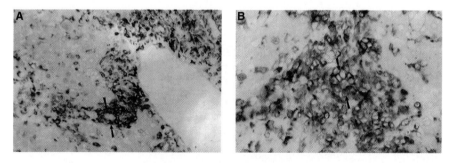

Color Plate 2, Fig. 2. (*See* full caption and discussion in Chapter 1, p. 7.) Hepatic reconstitution of bone marrow-derived cells. Brown Norway L21-negative liver was transplanted into a Lewis animal (L21 positive)...

Color Plate 3, Fig. 4. Functional blood vessels derived from highly purified, bone marrow-derived cells. (Courtesy Edward Scott.)

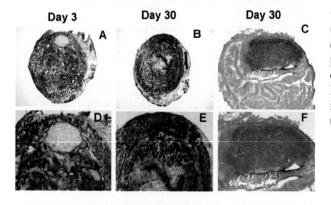

Day 3   Day 30   Day 30

A   B   C

D   E   F

Color Plate 4, Fig. 3. (*See* full caption and discussion in Chapter 4, p. 60.) A histologic comparison of collagen and MDSC injected into the rat urethra...

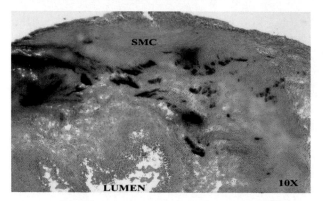

SMC

LUMEN   10X

Color Plate 5, Fig. 4. Extensive β-galactosidase staining in the rat bladder.

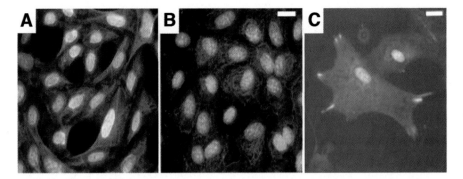

Color Plate 6, Fig. 3. (*See* full caption and discussion in Chapter 14, p. 273.) The DC condition induced cells in clone UB/OC-2 cultures to express fimbrin. Immunofluorescence images of cell line UB/OC-2 derived from E13 auditory sensory epithelia...

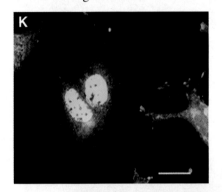

Color Plate 7, Fig. 4K. (*See* full caption and discussion in Chapter 14, pp. 274–275.) The DC condition induced cells in clonal cell line UB/UE-1 to differentiate into either hair cell or support cell phenotypes...

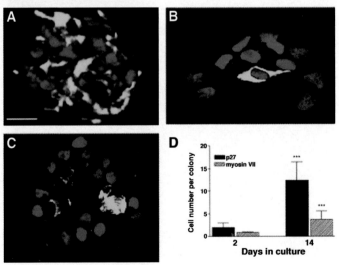

Color Plate 8, Fig. 5. (*See* full caption and discussion in Chapter 14, p. 277.) Nestin (+) cells isolated from the P0 rat organ of Corti (1) proliferate, (2) form otospheres, (3) generate hair cell and support cell phenotypes in response to EGF...

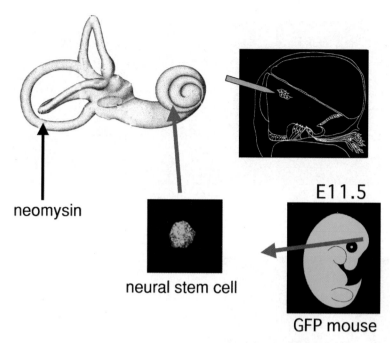

neomysin

E11.5

neural stem cell

GFP mouse

Color Plate 9, Fig. 7. (*See* full caption and discussion in Chapter 14, p. 283.) Donor embryonic neural stem cells can be transplanted into the scala media of oto-toxic-damaged cochleae of neonatal rats...

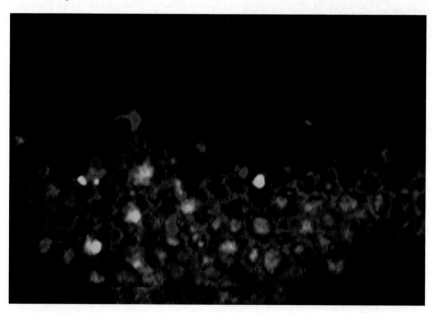

Color Plate 10, Fig. 8. (*See* full caption and discussion in Chapter 14, p. 284.) Donor neural stem cells integrate into the inner ear sensory receptor epithelium of the recipient's inner ear...

6
Epithelial Stem/Progenitor Cells in Thymus Organogenesis

Hans-Reimer Rodewald

1. INTRODUCTION

The thymus provides a unique three-dimensional hematopoietic environment that is essential for the development of T (thymus-derived) lymphocytes. T-cell progenitors do not arise in the thymus itself, but migrate to the thymus from central hematopoietic organs such as the liver and bone marrow during fetal and adult life, respectively (1–3). Despite the fact that prethymic T lineage-committed progenitors have been identified (e.g., in fetal blood) (4), adult thymus-colonizing pro-T cells have remained elusive. Moreover, the molecular and cellular basis underlying thymus homing has not been resolved. Intrathymic stages of T-cell development have been studied in great detail both at the cellular and the molecular levels (reviewed in refs. 5–7). The intrathymic developmental sequence can be ordered into three major stages:

1. Growth factor–driven proliferation and protection from apoptosis of rare pro-T cells (*growth factor expansion phase*; reviewed in refs. 8 and 9).
2. Rearrangement of β, γ, and δ T-cell antigen receptor (TCR) genes, followed by thymocyte selection for further development based on the expression of productive TCR β-chains. This process, termed *β-selection*, depends on the assembly of the pre-TCR complex (reviewed in ref. 10). The β-selected thymocytes undergo massive proliferation and become CD4$^+$CD8$^+$, still immature, thymocytes.
3. Following TCR α-chain rearrangements, the pre-TCR is replaced by the αβ-TCR complex at the CD4$^+$CD8$^+$ stage. Cells that succeed to assemble a complete αβ-TCR can be selected based on TCR-MHC (major histocompatibility complex) interactions (*αβ-selection*). Thymocytes undergo further differentiation as CD4$^+$CD8$^-$ (helper-type) or CD4$^-$CD8$^+$ (cytotoxic-type) single-positive cells before they populate as MHC-restricted (*positive selection*) and self-tolerant (*negative selection*) antigen-reactive T cells the circulation and the peripheral lymphoid organs (reviewed in ref. 5).

From: *Adult Stem Cells*
Edited by: K. Turksen © Humana Press Inc., Totowa, NJ

The thymus is unique in its property to support T-cell development, as shown by the fact that impaired or abrogated thymus organogenesis (see below) or neonatal thymectomy *(11)* cause severe T-cell immunodeficiencies. This is apparent in athymic mutants such as the well-known nude mouse, a widely used animal deficient in T cells. This mouse harbors a natural mutation in the Forkhead box n 1 gene (*Foxn1*, formerly termed winged-helix-nude [*Whn*]) *(12,13)* (for the nomenclature, *see* ref. *14*). Humans affected by DiGeorge syndrome carry deletions on chromosome 22q11. Children with this syndrome lack, partially or completely, a thymus.

It has been shown that null mutations in the T box gene 1 (*Tbx1*) (a gene that is, among other genes, located to the DiGeorge region) result in an athymic phenotype in mice. Thus, mutations in *Tbx1* in mice resemble the DiGeorge syndrome in humans *(15,16)*. These natural mutations demonstrate that formation of a thymus is essential for T-cell immunity. Hence, the interest in the development of a functional thymus in ontogeny. Moreover, because functional thymus tissue deteriorates with age, a phenomenon known as *thymus involution* (reviewed in ref. *17*), the idea to "rejuvenate" an old thymus draws considerable interest.

Thymocytes develop in tight contact with thymic epithelium (reviewed in refs. *18–20*). Thymus epithelium is a major component of the thymic "stroma." For the purpose of this chapter, I define stromal cells as those cells and tissues that, regardless of origin, support thymocyte development. Such supporting cells can be of diverse origin: thymic epithelial cells (TECs) (i.e., thymus stroma "proper" forming cortical and medullary zones), mesenchymal cells such as fibroblasts, hematopoietic cells such as dendritic cells, and macrophages. Like other organs, the thymus also harbors endothelial cell-lined vasculature *(21)* and neuronal cells *(22,23)*.

One of the better-understood examples of the molecular function of thymic epithelium is the production of specific growth factors. Thymic epithelial cells can support T-cell development by providing soluble factors or by providing cell–cell contact. Critical growth factors expressed in thymic epithelium are interleukin 7 (IL-7) and stem cell factor (SCF), which bind to the IL-7 receptor, and the tyrosine kinase receptor c-Kit (reviewed in refs. *8* and *9*). Both of these receptors are expressed on pro-T cells, and their synergistic stimulation is essential for T-cell development at birth *(24)*. During adult life, these cytokines are nonredundant. Adult mice lacking c-Kit *(25)* or IL-7 *(26)* show severe defects in the maintenance of lymphopoiesis.

For reasons still unclear, T-cell development requires a three-dimensional epithelial stroma cell architecture. Such permissive environment cannot be provided in "simple" two-dimensional stromal cell cultures, which have been used successfully to analyze, for instance, B-cell development in vitro (reviewed in ref. *27*). With very few exceptions, such as a particular T-cell sublineage in the bone marrow *(28)*, it has proven difficult to recapitulate the physiology of intrathymic T-cell development in monolayer cultures of thymic stromal cells.

Therefore, techniques to assemble three-dimensional epithelial stromal cultures (reaggregate fetal thymic organ cultures [RFTOCs]) have greatly improved experimental access to stromal cell–thymocyte interactions. In RFTOCs, T-cell development and stromal cell requirements have been studied extensively in vitro (reviewed in refs. *19* and *20*). Prior to the introduction of RFTOC, fetal thymic organ cultures (FTOCs), depleted of endogenous progenitors and repopulated by exogenous progenitors, have been widely used to study intrathymic development (reviewed in ref. *29*). Expression of critical TEC genes (e.g., MHC class II and whn) is rapidly lost in monolayer cultures, but is maintained in three-dimensional cultures such as FTOCs and RFTOCs, suggesting that the stromal cell architecture is critical for the maintenance of thymic epithelial cell "identity" *(30,31)*.

In vivo, thymus organogenesis has been studied mostly by detailed observations of cell and cell sheet movements in the area of the third pharyngeal pouch during ontogeny in various species *(32–34)*. These classical reports concluded that the thymus architecture and its typical organization into medullary and cortical compartments result from the invagination of an endodermal into an ectodermal epithelial sheet at the third pharyngeal pouch and cleft, respectively. Recently, Manley staged thymus organogenesis according to the "generic" order of events in "general" organogenesis: positioning, induction, and outgrowth, leading to a thymus rudiment in the proper location in the embryo, followed by patterning and differentiation to give rise to a fully developed, functional thymus *(35)*. Several mutants have been identified in which thymus organogenesis is affected or abrogated, and these mutants have been placed according to the order described above (reviewed in ref. *35*).

Former models of thymus organogenesis *(34)* relied on epithelial sheets rather than epithelial stem/progenitor cells. It has been speculated that epithelial stem cells may exist in the thymus *(36–38)*, but compelling evidence for the existence of such cells, and for their role in thymus development has only been obtained very recently. This chapter discusses these recent findings in the context of thymus organogenesis.

2. THYMUS ORGANOGENESIS

2.1. Third Pharyngeal Pouch and Arch Epithelium and Neural Crest-Derived Mesenchyme

Steps of thymus organogenesis and some of the known genes involved in this process are outlined in Fig. 1. The thymus originates from the third pharyngeal pouch and the third and fourth pharyngeal arches. On embryonic d 10, at this location, endodermal epithelium is situated next to neural crest–derived mesenchymal cells. Several studies have shown that mesenchyme plays a critical role for thymus organogenesis and function *(39–42)*. Mesenchymal–epithelial interactions are essential for thymus morphogenesis (e.g., capsule formation, lobulation). The role of the mesenchyme can be substituted in vitro by epidermal growth factor (EGF) or transforming growth factor-α (TGF-α) *(40)*. *Pax3* plays a role in the migration of neural crest-derived mesenchyme toward the thymus *(43)*.

Following positioning of the thymus anlage, endodermal epithelium buds and grows to form a more circular structure. According to classical models of thymus organogenesis, rapid circumferential growth of ectodermal epithelium, derived from the ectodermal cervical vesicle, occurs close to the endodermal pouch at this stage. These ectodermal cells supposedly surround the endodermal tissue with a later annotation of the endodermal part to form the inner, medullary, epithelium, and the ectodermal cells to build the outer, cortical, epithelium. Current models of thymus organogenesis have abandoned an involvement of ectoderm (e.g., *see* refs. *35* and *44*).

Patterning and further differentiation of the thymus take place beginning on d 13 of embryonic development, and these steps involve lobulation, segmentation, and formation of medulla–cortex architecture. Medulla–cortex architecture may be influenced by colonization of pro-T cells, a process termed "cross-talk" between thymic epithelium and thymocytes *(18,45,46)*. The term "cross-talk" is more commonly used for mutual interactions of intracellular signaling pathways, and has remained an ill-defined phenomenon pointing vaguely at thymocyte–stroma interactions.

Fig. 1. Thymus organogenesis. (**A**) Stages of thymus development are depicted schematically closely based on the model proposed by Manley *(35)*. (**B**) Genes likely to play a role in thymus organogenesis are shown without close timing to the stages shown in A. (**C**) The exact role of thymus epithelial stem/progenitor cells in thymus organogenesis is not known, but such cells can definitively contribute to medulla–cortex organization and thus to patterning.

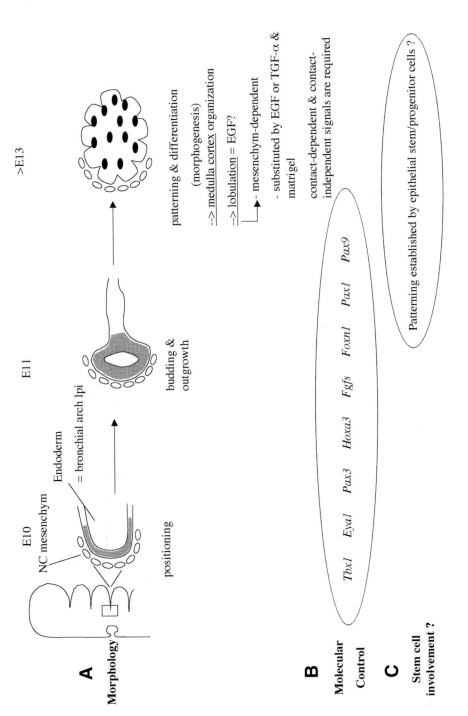

A Morphology

E10   E11   >E13

NC mesenchym

Endoderm
= bronchial arch lpi

positioning

budding &
outgrowth

patterning & differentiation
(morphogenesis)
--> medulla cortex organization

--> lobulation = EGF?

└── mesenchym-dependent

- substituted by EGF or TGF-α &
matrigel

contact-dependent & contact-
independent signals are required

B Molecular Control

Tbx1   Eya1   Pax3   Hoxa3   Fgfs   Foxn1   Pax1   Pax9

C Stem cell involvement ?

Patterning established by epithelial stem/progenitor cells ?

2.2. Genes Involved in Thymus Organogenesis

Mutations impinging on pharyngeal arch and pouch development abrogate or perturb thymus development. For instance, third pouch development is abrogated in *Tbx1*-deficient mice *(15,16)*, and null mutations in the eyes absent gene 1 *(Eya1)* result in morphogenetic defects in thymus, parathyroid, and thyroid development *(47)*. Neural crest migration is controlled by *Pax3*.

Transcription factors of the *hox* family are key molecules in the direction of the development of pharyngeal arch-derived organs (i.e., thymus, parathyroid gland, ultimobranchial body). A key gene is *Hoxa3 (48)*, which is expressed in endodermal cells in the third pouch. *Hoxa3* appears to be positioned "upstream" of *Pax1 (49–51)* and *Pax9 (44,52)*, both of which are also critical for thymus development, ventral migration of the two lobes, or thymocyte development.

The transcription factor *Foxn1 (12)* is essential for the development of the thymic epithelium at the stage of the thymus anlage past the initial induction and outgrowth. *Foxn1*, allelic with the nude gene, continues to be expressed in both cortical and medullary epithelium during adult life *(13)*, but the functional significance of this expression is not known. The block in thymus development in nude mice places the *Foxn1* gene between *Hoxa3* and *Pax9 (44)*. Bleul and Boehm identified several target genes of *Foxn1 (53)*, and have proposed that misguided chemokine expression may contribute to the lack of pro-T cell homing to the nude thymus *(54)*. An identification of those target genes of the *Foxn1* gene that direct thymus development may be the key to deepen the understanding of thymus epithelial differentiation. For a detailed discussion of other genes involved in thymus organogenesis, see the review by Manley *(35)*.

3. DEFINING COMPONENTS OF THE THYMIC EPITHELIAL ENVIRONMENT USING MONOCLONAL ANTIBODIES

Several groups generated comprehensive panels of monoclonal antibodies (mAbs) recognizing distinct elements of thymic epithelium (*see* refs. *18* and *55–57*). According to their reactivity, these antibodies have been grouped into markers recognizing pan-epithelium, subcapsular and septum structures, fibroblasts, cortical epithelium, medullary epithelium, Hassall's corpuscles (enigmatic structures in the medullary zones), endothelium, and "miscellaneous" structures. Most of these reagents are reactive to intracellular antigens and are therefore useful to describe thymus morphology under various physiological and pathological conditions, but are of limited use to purify viable subsets by cell surface phenotypes.

Some of these markers have been used to observe changes in thymus architecture in T-cell developmental mutants, and attempts have been made to correlate a block in T-cell development with a block in TEC development *(46)*. It should be noted, however, that such correlation may be very indirect, and these systems are likely to be too complex to provide definitive and conclusive results.

3.1. In Vivo Reconstruction of a Functional Thymus Environment From Purified Epithelial Cells: Functional Properties of Thymic Epithelial Cell Grafts

Thymic epithelium has become experimentally accessible with the introduction of methods to dissociate and reassemble thymic epithelium in vitro *(39)*. Using this approach, the requirements for the formation of a functional thymus could be examined in vitro. In this type of experiment, input of either stromal cells or progenitor cell types can be varied (e.g., *see* refs. *19, 30, 58,* and *59*). Both RFTOCs and FTOCs cannot be cultured for periods extending longer than approx 12 d. Thus, such thymus cultures represent only a "transient thymus"; consequently, the architecture of RFTOCs maintained in vitro does not resemble normal thymic structures *(31,60)*.

Transplantation of cultured thymic epithelial fragments *(61)* or fused thymic tissue fragments derived from the third pharyngeal pouch *(62)* was reported some time ago. In both cases, a functional thymus structure was restored in vivo. Specifically, these grafts were colonized by lymphocyte progenitors from the host that developed into functional T cells. In these experiments, however, the input of thymic epithelial cells could not be varied qualitatively because the tissue fragments could not be manipulated at the level of a cell suspension. Such manipulation is only possible if thymic epithelium is first dissociated, then reassembled, and subsequently grafted into a recipient mouse.

The methodology and the initial results from RFTOC grafting experiments were first reported in 1996 *(63)* and were specified in 2000 *(60)*. In these experiments, alymphoid RFTOCs were grafted under the kidney capsules of host mice. The recipients were either immunocompromised mice, and thus incompetent to reject allogeneic grafts (e.g., nude or recombination-activating genes [RAG]-deficient hosts), or histocompatible mice bearing a congenic marker. The congenic marker is required to identify the origin of thymocytes in the graft (host bone marrow origin vs "carryover" within the graft). The structural and functional in vivo properties after transplantation of reaggregates of thymic epithelium have yielded the following novel results:

1. When RFTOCs, assembled in vitro from purified fetal thymic epithelium, are cultured as alymphoid thymic "organoids" for several days, these structures develop the rigidity required for handling and subsequent transplantation.
2. RFTOC grafts can attract host (bone marrow)-derived T-cell progenitors from the circulation.
3. T-cell development proceeds along the well-defined stages also found in an endogenous thymus.
4. The fidelity of negative and positive selection is normal when "provoked" in TCR-transgenic recipient mice, in which negative and positive selection depend on MHC molecules absent from the host, but present exclusively on thymic epithelium in the graft.

Thus, RFTOC transplantation uncovered that purified thymic epithelium, starting from a single-cell suspension, has the remarkable capacity to self-reorganize into a structurally and functionally competent microenvironment promoting T-cell development in vivo *(31,60,63–65)*.

3.2. Phenotype of Thymic Epithelial Cells With the Potential to Generate a Functional Thymus In Vivo

In vivo reconstruction of a functional thymus environment was originally performed using aggregates of either CD45⁻ MHC class II⁺ epithelial cells *(60)* or all CD45⁻ cells *(31)*. CD45⁻ MHC class II⁻ epithelial cells were not sufficient. The fact that thymus epithelium formation potential resides in the CD45⁻ MHC class II⁺ thymic epithelium *(31,60,63)* has been confirmed and extended in recent reports *(64,65)*.

Gill and colleagues *(64)* showed that a functional thymus can be generated from a major subset of CD45⁻ MHC class II⁺ epithelial cells, defined as MTS24⁺. In a similar study, Bennett and colleagues *(65)* defined thymus-forming cells as MTS20⁺MTS24⁺. It should be noted that CD45⁻ MHC class II⁺ MTS24⁺ and MTS20⁺MTS24⁺ cells are essentially identical populations *(64)*. Moreover, given that around 50% of CD45⁻ MHC class II⁺ are MTS24⁺, and the other half of MHC class II⁺ are MTS24⁻ *(64)*, this population is enriched only by a factor of approx 2 compared to CD45⁻ MHC class II⁺ epithelium.

Despite the fact that no clonal assays were performed, these two reports claimed evidence for an identification and characterization of thymic epithelial progenitor cells *(64,65)*. Although it is not impossible that CD45⁻ MHC class II⁺MTS24⁺ cells contain some epithelial progenitors, this conclusion is clearly not proven by the data shown in these reports. The grafts were assembled from bulk populations, and without single cell readouts, bulk experiments cannot yield information on precursor activities. Moreover, ratios of cell numbers (input vs output) have not been determined, probably owing to the difficulties in retrieving epithelial cells quantitatively from the grafts. Therefore, no attempts were made by Gill et al. or Bennett et al. to demonstrate an increase in epithelial cellularity, which is expected from precursor activity *(64,65)*.

3.3. Reorganization of Thymic Epithelium in the Graft

RFTOC maintained in vitro lack recognizable medulla–cortex architecture. Indeed, they appear to be randomly organized. In marked contrast, histological examination of RFTOC grafts reveals a striking reappearance of proper medulla–cortex organization in vivo *(31,60)*. Thus, in vivo, but not in vitro, RFTOCs can "self-reorganize" into a functional thymic architecture with clear medulla–cortex boundaries.

This finding raised the interesting question of how the thymus can reestablish its key morphological pattern (i.e., the division into distinct epithelia characterized as medulla and cortex) once the original pattern is destroyed by enzymatic digestion. Given that in vitro reaggregation occurred from a single-cell suspension, initially yielding a random structure in vitro, medulla–cortex organization could take place via segregation and clustering of preexisting medullary epithelial cells ("sorting out"). Alternatively, growth of single progenitors or stem cells might contribute to formation of distinct thymic compartments, such as cortex and medulla. These possibilities have been raised and experimentally addressed by Rodewald and colleagues *(31)*.

In principle, two types of experiments, summarized below, were performed to distinguish between these possibilities. In one set of experiments, reaggregates were assembled from mixtures of thymic epithelium isolated from two mouse strains differing in their MHC class II haplotypes (C57BL/6 [I-A^b]) and BALB/c [I-A^d]). Such mixed reaggregates were transplanted, and analysis of MHC class II expression could trace the origin of the epithelium to either of the two donors. In the other set of experiments, chimeric mice were generated by injection of embryonic stem (ES) cells into blastocysts using a combination of ES cells and blastocysts that, again, differed in MHC class II. The results from both types of experiments are summarized below.

3.3.1. Evidence for Epithelial Stem/Progenitor Cells in Thymus Organogenesis From Mixed Thymic Epithelial Cell Grafts

The principles of the generation, transplantation, and analysis of mixed reaggregates of thymic epithelium are shown in Fig. 2. Fetal thymus was enzymatically digested, and thymus epithelium was purified by cell sorting or by depletion of CD45$^+$ (hematopoietic cells) from the single-cell suspension. At the stage of isolation (fetal d 15 or 16), medullary and cortical epi-

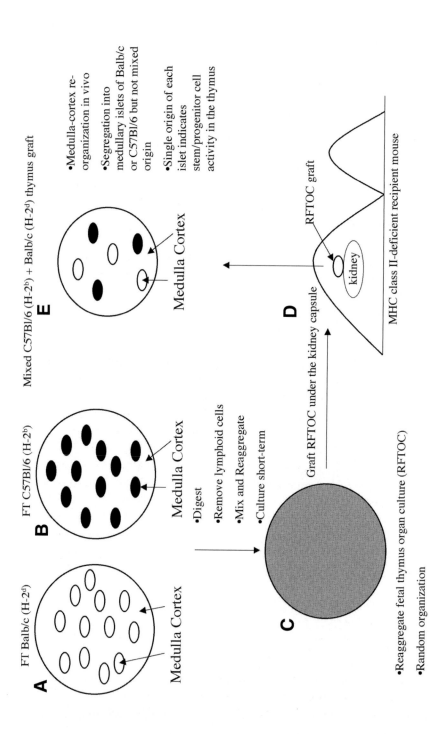

Fig. 2. Experimental strategy leading to the identification of epithelial stem/progenitor cell activity in the thymus (*31*). Epithelium from fetal thymi from (**A**) Balb/c mice and (**B**) C57Bl/6 mice was reaggregated to form (**C**) MHC-mismatched

thelium were already separated into clearly defined areas, as schematically depicted in Fig. 2A,B. The donor mouse strains used were C57BL/6 (I-A^b) and BALB/c (I-A^d). Mixed reaggregates showed a random distribution of I-A^{b+} (C57BL/6-derived) and I-A^{d+} (BALB/c-derived) epithelium, as determined by staining with antibodies specific for I-A^b or I-A^d. Moreover, medulla-specific (MTS10) and cortex-specific (MTS44) antibodies *(56)* revealed that RFTOCs kept in vitro contained individual cells stained with either of these reagents, but no separable cortex or medulla pattern (schematically depicted in Fig. 2C).

Mixed reaggregates were transplanted under the kidney capsule of MHC class II-deficient C57BL/6 mice *(66)* (Fig. 2D). MHC class II-deficient recipients were chosen because class II molecules, in addition to thymic epithelium, are also expressed on bone marrow–derived dendritic cells, and their presence in the thymus might "complicate" an analysis of the epithelial composition and origin of the grafts. Specifically, in the absence of MHC class II expression on hematopoietic cells, all class II expression is exclusively confined to thymic epithelium of the grafted type.

In vivo, RFTOC grafts were colonized by host bone marrow-derived pro-T cells. Intrathymic T-cell development in such class II$^+$ grafts included the generation of mature CD4$^+$CD8$^-$ thymocytes. This population, owing to lack of class II expression in the host, failed to develop in the endogenous host thymus *(66–68)*. Ex vivo histological analyses of tissue sections from RFTOC grafts showed that medullary areas reappeared. Double staining for the donor origin using antibodies specific for anti-I-A^b and anti-I-A^d demonstrated, surprisingly, that distinct medullary areas (islets) were derived from mutually exclusive donor epithelia; that is, these epithelial structures were of either I-A^b or I-A^d origin, but not mixed (depicted in Fig. 2E). Furthermore, immunofluorescence analyses using cytokeratin-specific antibodies proved that the lineage of these MHC class II$^+$ areas was epithelial. These experiments led to the intriguing conclusions that (1) medullary areas can be reestablished in thymic epithelial grafts, and (2) these areas are derived from either of the two, but not both, donor origins. The latter observation strongly

---

Fig. 2. *(continued)* chimeric thymic epithelium. After culture, mixed reaggregated thymus grafts were implanted into MHC class II-deficient mice. Grafts developed in vivo into a fully functional ectopic thymus as shown by the generation of all thymocyte subsets and the MHC class II-dependent CD4$^+$CD8$^-$ single-positive thymocytes. Remarkably, medulla–cortex reorganization occurred in vivo, and medullary zones segregated into islets of Balb/c or C57Bl/6, but not mixed, origin. This single origin of each islet indicates stem/progenitor cell activity in the thymus.

suggests that medullary epithelial structures can arise from single stem/progenitor cells by formation of medullary islets *(31)*.

3.3.2. Evidence for Epithelial Stem/Progenitor Cells in Thymus Organogenesis From MHC-Chimeric Mice In Vivo

The data obtained from RFTOC transplantation experiments uncovered the remarkable property of thymic epithelium to reestablish proper architecture starting from an initially disrupted, randomly arranged composition formed in vitro. Another set of experiments has been reported in which aspects of thymus epithelial morphogenesis were studied more directly in vivo. The applied strategy was based on the fact that each cell in an organ of chimeric mice, generated by injection of ES cells into genetically distinct blastocysts, can originate from either the ES cell or the blastocyst. The quantitative contribution of cells derived from the ES cell or blastocyst to an individual tissue in an individual mouse can vary from animal to animal.

To identify mice in which the proportions of ES- and blastocyst-derived tissues were approximately comparable (balanced mice), the contribution of tissues derived from ES cells vs tissues derived from blastocysts was analyzed in an ectodermal (skin), a mesodermal (muscle), and an endodermal (liver) tissue. Mice were typed in these tissues using microsatellite markers specific for ES or blastocyst genomic deoxyribonucleic acid (DNA). Large numbers of chimeras were generated by injection of ES cells into MHC-mismatched blastocysts. The ES cells used were either from CBA (I-A^k) or BALB/c (I-A^d) mice, and the blastocysts were from C57BL/6 (I-A^b) mice. Using this approach, mice were identified in which epithelium derived from both ES cells and blastocysts contributed comparably to thymus formation *(31)*.

Chimeric thymi were analyzed in detail by three-color histology using antibodies specific for the two donor epithelia (anti-I-A^k and anti-I-A^b) and specific for pan-thymic epithelial cytokeratin. Thymi were also examined by staining with the antibody MTS10, which recognizes medullary zones. Interestingly, these studies revealed that, in a physiologically developed thymus, medullary areas are composed of individual epithelial islets, each derived from either the ES cell or the blastocyst.

Histological measurements led to the calculation that these epithelial islets vary in diameter from a minimum of 60 × 40 to a maximum of 170 × 170 μm. Each cell cluster harbors between 5 and 45 epithelial cells in a two-dimensional lattice. Serial sectioning of an entire mouse thymus lobe demonstrated that, in the mouse, one thymus lobe harbors about 300 medullary areas. Each area can include several islets. Therefore, it was estimated that one lobe contains approx 900 islets *(31)*. The morphological "isletlike" character of medullary epithelium was noted in an earlier study *(21)*. The islet character of medullary epithelium is most apparent in the juvenile thymus. In contrast, medullary epithelial islets tend to be larger in the adult thymus; here, the medulla can appear more confluent. Collectively, these experiments provided the first evidence that at least part of the thymic epithelium is composed of individual islets.

As summarized above, medulla–cortex compartmentalization has been thought for a long time to occur via invagination of an endodermal into an ectodermal epithelial sheet at the third pharyngeal pouch and cleft, respectively. Despite the fact that epithelial stem or progenitor cells have been invoked in thymus development, based on marker studies, no experimental evidence for such cells had been obtained. Data from chimeric mice, as well as data from RFTOC grafts, have provided the first evidence for an involvement of epithelial stem or progenitor cells in thymus morphogenesis. However, it should be noted that these experiments provide the first evidence for stem/progenitor cells in thymus organogenesis, but such cells were not physically purified by phenotype. Isolation of highly enriched thymic epithelial stem/progenitor cells will be required before their prospective developmental behavior can be studied.

4. EVIDENCE FOR THYMIC EPITHELIAL STEM CELLS

Although these experiments employing RFTOC grafts and mixed chimeras did not identify thymic stem/progenitor cells directly, they provided the first compelling experimental evidence for the existence of such long-postulated cells *(31)*. It has been speculated previously, based on marker studies, that the thymus may harbor common stem cells for both cortex and medulla. Like a blueprint of the developmental pattern of immature double-positive $CD4^+CD8^+$ thymocytes giving rise to mature single-positive $CD4^+CD8^-$ and $CD4^-CD8^+$ thymocytes, Ritter and Boyd *(37)* and Ropke and colleagues *(38)* speculated that epithelial cells bearing markers of both cortex and medulla may represent putative thymic epithelial progenitors. Direct experimental evidence for such cells, or any epithelial progenitor activity in the thymus, was not provided in these reports *(37,38)*; therefore, the described mixed RFTOC grafts and MHC chimeric mice studies uncovered the first evidence for a role of stem/progenitor cells in thymus organogenesis *(31)*.

In this context, it should be pointed out that the first assay for hematopoietic stem cell (HSC) activity was the colony-forming unit spleen (CFU-S) assay, discovered in 1961 by Till and McCullouch *(69)*. Macroscopically visible spleen colonies were detected when nonirradiated bone marrow cells were injected into lethally irradiated mice. In this assay, which preceded the physical enrichment of HSC by several decades *(70)*, the *activity* of stem cells was evident, and this assay has been instrumental as a readout for stem cell enrichment procedures. By analogy, an assay is now at hand to test candidate thymic epithelial cells for stem or progenitor cell activity. As is true for all experiments on stem or progenitor cells, clonogenicity must be the gold standard.

5. IMPLICATIONS

A better understanding of the cellular components and the molecular mechanisms underlying thymus organogenesis and maintenance of a functional thymus will be of interest both for basic and, potentially, for clinical immunology. Thymus function deteriorates with age or under certain thymotoxic conditions, such as irradiation, steroid treatment, chemotherapy, or human immunodeficiency virus (HIV) infection. It would be highly desirable to improve thymus function (i.e., *de novo* T-cell production) under these conditions and in the elderly. Recent research has provided novel insights into thymus organogenesis, primarily by the observations that a functional thymus can be regenerated from purified thymic epithelium, and that formation of a functional thymus architecture is a result of thymus epithelial stem or progenitor cell activity, both in normal development and following transplantation of thymus epithelial reaggregate grafts.

ACKNOWLEDGMENTS

I thank my colleagues Drs. Claudia Waskow and Hans Joerg Fehling (Ulm) and Thomas Boehm (Freiburg) for comments on the manuscript. Support was received from grants from the Deutsche Forschungsgemeinschaft (Sonderforschungsbereich 497-B5), the Landesforschungsschwerpunkt Baden-Württemberg, and the Interdisciplinary Center for Clinical Research (IZKF) in Ulm.

REFERENCES

1. Moore, M. A. S., and Owen, J. J. T. (1967). Stem-cell migration in developing myeloid and lymphoid systems. Lancet 2, 658–659.
2. Suniara, R. K., Jenkinson, E. J., and Owen, J. J. (1999). Studies on the phenotype of migrant thymic stem cells. Eur J Immunol 29, 75–80.
3. Foss, D. L., Donskoy, E., and Goldschneider, I. (2001). The importation of hematogenous precursors by the thymus is a gated phenomenon in normal adult mice. J Exp Med 193, 365–374.
4. Rodewald, H.-R., Kretzschmar, K., Takeda, S., Hohl, C., and Dessing, M. (1994). Identification of pro-thymocytes in murine fetal blood: T lineage commitment can precede thymus colonization. EMBO J 13, 4229–4240.

5. Kisielow, P., and von Boehmer, H. (1995). Development and selection of T cells: facts and puzzles. Adv Immunol 58, 87–209.
6. Shortman, K., and Wu, L. (1996). Early T lymphocyte progenitors. Annu Rev Immunol 14, 29–47.
7. Rodewald, H.-R., and Fehling, H. J. (1998). Molecular and cellular events in early thymocyte development. Adv Immunol 69, 1–112.
8. Akashi, K., Kondo, M., and Weissman, I. L. (1998). Role of interleukin-7 in T-cell development from hematopoietic stem cells. Immunol Rev 165, 13–28.
9. DiSanto, J. P., and Rodewald, H.-R. (1998). In vivo roles of receptor tyrosine kinases and cytokine receptors in thymocyte development. Curr Opin Immunol 10, 196–207.
10. von Boehmer, H., and Fehling, H. J. (1997). Structure and function of the pre-T cell receptor. Annu Rev Immunol 15, 433–452.
11. Miller, J. F. A. P. (1961). Immunological function of the thymus. Lancet 2, 748–749.
12. Nehls, M., Pfeifer, D., Schorpp, M., Hedrich, H., and Boehm, T. (1994). New member of the winged-helix protein family disrupted in mouse and rat nude mutations. Nature 372, 103–107.
13. Nehls, M., Kyewski, B., Messerle, M., et al. (1996). Two genetically separable steps in the differentiation of thymic epithelium. Science 272, 886–889.
14. Kaestner, K. H., Knochel, W., and Martinez, D. E. (2000). Unified nomenclature for the winged helix/forkhead transcription factors. Genes Dev 14, 142–146.
15. Lindsay, E. A., Vitelli, F., Su, H., et al. (2001). Tbx1 haploinsufficieny in the DiGeorge syndrome region causes aortic arch defects in mice. Nature 410, 97–101.
16. Jerome, L. A., and Papaioannou, V. E. (2001). DiGeorge syndrome phenotype in mice mutant for the T-box gene, Tbx1. Nat Genet 27, 286–291.
17. Rodewald, H. R. (1998). The thymus in the age of retirement [comment]. Nature 396, 630–631.
18. van Ewijk, W. (1991). T-cell differentiation is influenced by thymic microenvironments. Annu Rev Immunol 9, 591–615.
19. Anderson, G., Moore, N. C., Owen, J. J. T., and Jenkinson, E. J. (1996). Cellular interactions in thymocyte development. Annu Rev Immunol 14, 73–99.
20. Anderson, G., and Jenkinson, E. J. (2001). Lymphostromal interactions in thymic development and function. Nature Rev Immunol 1, 31–40.
21. Anderson, M., Anderson, S. K., and Farr, A. G. (2000). Thymic vasculature: organizer of the medullary epithelial compartment? Int Immunol 12, 1105–1110.
22. von Gaudecker, B. (1991). Functional histology of the human thymus. Anat Embryol (Berl) 183, 1–15.
23. Kranz, A., Kendall, M. D., and von Gaudecker, B. (1997). Studies on rat and human thymus to demonstrate immunoreactivity of calcitonin gene-related peptide, tyrosine hydroxylase and neuropeptide Y. J Anat 191, 441–450.
24. Rodewald, H.-R., Ogawa, M., Haller, C., Waskow, C., and DiSanto, J. P. (1997). Pro-thymocyte expansion by c-Kit and the common cytokine receptor γ chain is essential for repertoire formation. Immunity 6, 265–272.
25. Waskow, C., Paul, S., Haller, C., Gassmann, M., and Rodewald, H. R. (2002).

Viable c-kit(W/W) mutants reveal pivotal role for c-Kit in the maintenance of lymphopoiesis. Immunity 17, 277–288.

26. Carvalho, T. L., Mota-Santos, T., Cumano, A., Demengeot, J., and Vieira, P. (2001). Arrested B lymphopoiesis and persistence of activated B cells in adult interleukin 7(–/–) mice. J Exp Med 194, 1141–1150.

27. Rolink, A., Ghia, P., Grawunder, U., et al. (1995). In-vitro analyses of mechanisms of B-cell development. Semin Immunol 7, 155–167.

28. Dejbakhsh-Jones, S., and Strober, S. (1999). Identification of an early T cell progenitor for a pathway of T cell maturation in the bone marrow. Proc Natl Acad Sci U S A 96, 14,493–14,498.

29. Jenkinson, E. J., and Owen, J. J. (1990). T-cell differentiation in thymus organ cultures. Semin Immunol 2, 51–58.

30. Anderson, K. L., Moore, N. C., McLoughlin, D. E., Jenkinson, E. J., and Owen, J. J. (1998). Studies on thymic epithelial cells in vitro. Dev Comp Immunol 22, 367–377.

31. Rodewald, H. R., Paul, S., Haller, C., Bluethmann, H., and Blum, C. (2001). Thymus medulla consisting of epithelial islets each derived from a single progenitor. Nature 414, 763–768.

32. Crisan, C. (1935). Die Entwicklung des thyreo-parathyreo-thymischen Systems der weissen Maus. Z Anat Entwicklungsgesch 104, 327–358.

33. Norris, E. H. (1938). The morphogenesis and histogenesis of the thymus gland in man: in which the origin of the Hassall's corpuscles of the human thymus is discovered. Contrib Embryol 166, 191–221.

34. Cordier, A. C., and Haumont, S. M. (1980). Development of thymus, parathyroids, and ultimo-branchial bodies in NMRI and nude mice. Am J Anat 157, 227–263.

35. Manley, N. R. (2000). Thymus organogenesis and molecular mechanisms of thymic epithelial cell differentiation. Semin Immunol 12, 421–428.

36. Lampert, I. A., and Ritter, M. A. (1988). The origin of the diverse epithelial cells of the thymus: is there a common stem cell? In: Kendall, R. M., ed., Thymus Update. London: Harwood, pp. 5–25.

37. Ritter, M. A., and Boyd, R. L. (1993). Development in the thymus: it takes two to tango. Immunol Today 14, 462–469.

38. Ropke, C., Van Soest, P., Platenburg, P. P., and Van Ewijk, W. (1995). A common stem cell for murine cortical and medullary thymic epithelial cells? Dev Immunol 4, 149–156.

39. Anderson, G., Jenkinson, E. J., Moore, N. C., and Owen, J. J. (1993). MHC class II–positive epithelium and mesenchyme cells are both required for T-cell development in the thymus. Nature 362, 70–73.

40. Shinohara, T., and Honjo, T. (1996). Epidermal growth factor can replace thymic mesenchyme in induction of embryonic thymus morphogenesis in vitro. Eur J Immunol 26, 747–752.

41. Shinohara, T., and Honjo, T. (1997). Studies in vitro on the mechanism of the epithelial/mesenchymal interaction in the early fetal thymus. Eur J Immunol 27, 522–529.

42. Suniara, R. K., Jenkinson, E. J., and Owen, J. J. (2000). An essential role for thymic mesenchyme in early T cell development. J Exp Med 191, 1051–1056.

43. Conway, S. J., Henderson, D. J., and Copp, A. J. (1997). Pax3 is required for cardiac neural crest migration in the mouse: evidence from the splotch (Sp2H) mutant. Development 124, 505–514.

44. Hetzer-Egger, C., Schorpp, M., Haas-Assenbaum, A., Balling, R., Peters, H., and Boehm, T. (2002). Thymopoiesis requires Pax9 function in thymic epithelial cells. Eur J Immunol 32, 1175–1181.

45. van Ewijk, W., Shores, E. W., and Singer, A. (1994). Crosstalk in the mouse thymus. Immunol Today 15, 214–217.

46. van Ewijk, W., Hollander, G., Terhorst, C., and Wang, B. (2000). Stepwise development of thymic microenvironments in vivo is regulated by thymocyte subsets. Development 127, 1583–1591.

47. Xu, P. X., Zheng, W., Laclef, C., et al. (2002). Eya1 is required for the morphogenesis of mammalian thymus, parathyroid and thyroid. Development 129, 3033–3044.

48. Manley, N. R., and Capecchi, M. R. (1995). The role of Hoxa-3 in mouse thymus and thyroid development. Development 121, 1989–2003.

49. Dietrich, S., and Gruss, P. (1995). Undulated phenotypes suggest a role of Pax-1 for the development of vertebral and extravertebral structures. Dev Biol 167, 529–548.

50. Su, D. M., and Manley, N. R. (2000). Hoxa3 and Pax1 transcription factors regulate the ability of fetal thymic epithelial cells to promote thymocyte development. J Immunol 164, 5753–5760.

51. Su, D., Ellis, S., Napier, A., Lee, K., and Manley, N. R. (2001). Hoxa3 and Pax1 regulate epithelial cell death and proliferation during thymus and parathyroid organogenesis. Dev Biol 236, 316–329.

52. Peters, H., Neubuser, A., Kratochwil, K., and Balling, R. (1998). Pax9-deficient mice lack pharyngeal pouch derivatives and teeth and exhibit craniofacial and limb abnormalities. Genes Dev 12, 2735–2747.

53. Bleul, C. C., and Boehm, T. (2001). Laser capture microdissection-based expression profiling identifies PD1-ligand as a target of the nude locus gene product. Eur J Immunol 31, 2497–2503.

54. Bleul, C. C., and Boehm, T. (2000). Chemokines define distinct microenvironments in the developing thymus. Eur J Immunol 30, 3371–3379.

55. Rouse, R. V., Bolin, L. M., Bender, J. R., and Kyewski, B. A. (1988). Monoclonal antibodies reactive with subsets of mouse and human thymic epithelial cells. J Histochem Cytochem 36, 1511–1517.

56. Boyd, R. L., Tucek, C. L., Godfrey, D. I., et al. (1993). The thymic microenvironment. Immunol Today 14, 445–459.

57. Gray, D. H., Chidgey, A. P., and Boyd, R. L. (2002). Analysis of thymic stromal cell populations using flow cytometry. J Immunol Methods 260, 15–28.

58. Oosterwegel, M. A., Haks, M. C., Jeffry, U., Murray, R., and Kruisbeek, A. M. (1997). Induction of TCR gene rearrangements in uncommitted stem cells by a

subset of IL-7 producing, class II–expressing thymic stromal cells. Immunity 6, 351–360.

59. Muller, K. M., Luedecker, C. J., Udey, M. C., and Farr, A. G. (1997). Involvement of E-cadherin in thymus organogenesis and thymocyte maturation. Immunity 6, 257–264.

60. Rodewald, H.-R. (2000). Thymus epithelial cell reaggregate grafts. Curr Top Microbiol Immunol 251, 101–108.

61. Kendall, M. D., Schuurman, H. J., Fenton, J., Broekhuizen, R., and Kampinga, J. (1988). Implantation of cultured thymic fragments in congenitally athymic (nude) rats. Cell Tissue Res 254, 283–294.

62. Pyke, K. W., Bartlett, P. F., and Mandel, T. E. (1983). The in vitro production of chimeric murine thymus from non-lymphoid embryonic precursors. J Immunol Methods 58, 243–254.

63. Rodewald, H. R. (1996). Reconstitution of selective hematopoietic lineages and hematopoietic environments in vivo. In: (Drews, J. and Ryser, S., eds.), Human Disease—From Genetic Cause to Biochemical Effects. Proceedings of the Symposium "The Genetic Basis of Human Disease." Basel, Switzerland: Blackwell Science, pp. 51–57.

64. Gill, J., Malin, M., Hollander, G. A., and Boyd, R. (2002). Generation of a complete thymic microenvironment by MTS24(+) thymic epithelial cells. Nat Immunol 3, 635–642.

65. Bennett, A. R., Farley, A., Blair, N. F., Gordon, J., Sharp, L., and Blackburn, C. C. (2002). Identification and characterization of thymic epithelial progenitor cells. Immunity 16, 803–814.

66. Kontgen, F., Suss, G., Stewart, C., Steinmetz, M., and Bluethmann, H. (1993). Targeted disruption of the MHC class II Aa gene in C57BL/6 mice. Int Immunol 5, 957–964.

67. Grusby, M. J., Johnson, R. S., Papaioannou, V. E., and Glimcher, L. H. (1991). Depletion of CD4+ T cells in major histocompatibility complex class II–deficient mice. Science 253, 1417–1420.

68. Gosgrove, D., Gray, D., Dierich, A., et al. (1991). Mice lacking MHC class II molecules. Cell 66, 1051–1066.

69. Till, J. E., and McCullouch, E. A. (1961). A direct measurement of the radiation sensitivity of normal mouse bone marrow cells. Radiat Res 14, 213–222.

70. Spangrude, G. J., Heimfeld, S., and Weissman, I. L. (1988). Purification and characterization of mouse hematopoietic stem cells. Science 242, 58–62.

7
Adult Liver Stem Cells

William B. Coleman, Joe W. Grisham, and Nadia N. Malouf

1. INTRODUCTION

The search for liver stem cells has been ongoing for decades. Until very recently, the existence of liver cells with stemlike potential was critically questioned and not generally accepted *(1,2)*. In contrast, it has long been known that tissues with high cellular turnover, like epidermis, intestine, and bone marrow, contain stem cells that function to maintain tissue homeostasis through continuous renewal of the cell lineage *(3–5)*. Evidence for liver stem cells emerged from studies of liver injury, regeneration, and carcinogenesis in rodent models; evidence for human liver stem cells has also been mustered. Investigations into the roles that these cells play in response to hepatic injury and carcinogenesis have escalated. In addition, there is interest in pursuing the potential application of stem cells to the treatment of liver disease through gene therapy and cell transplantation approaches *(6–11)*.

New understanding of cell lineage generation and of the relationships of cells composing a lineage in adult organisms have modified the traditional thinking that stem cells are necessarily undifferentiated cells with a limited degree of potency *(12)*. However, modification of the stem cell paradigm has not resulted in a universally accepted definition of what a stem cell represents *(13)*. Rather, there has been a reevaluation of the contributions of various cell types, both differentiated and undifferentiated, to normal lineage renewal, response to injury, and tissue regeneration. In the liver, these changing concepts have given rise to a recognition that there are multiple liver epithelial cell types that have the potential to originate new cell populations *(14,15)*. In addition, a number of intriguing new observations have been made that suggest that stem cells may yield progeny that evince a high degree of phenotypic plasticity *(16,17)*. Several reports have demonstrated that adult stem cells from one tissue can be induced to differentiate into

From: *Adult Stem Cells*
Edited by: K. Turksen © Humana Press Inc., Totowa, NJ

parenchymal cells of other tissues when transplanted into appropriate sites; this differentiation includes the derivation in the liver of hepatocytes by transplantation of stem cells from extrahepatic sources *(18)*. These results suggest that stem cells of adult tissues may not be intrinsically restricted in their differentiation commitment, or that they are capable of responding to different tissue microenvironments with an alternative cellular differentiation that is appropriate for the new site *(19)*.

In this chapter, we review the evidence for liver stem cells, including (1) the types of cells in the liver that exhibit stemlike potential; (2) sources of liver stem cells from extrahepatic tissues; (3) evidence for human liver stem cells; (4) isolation, culture, and characterization of liver stem cells from rats; (5) evidence for the differentiation potential of cultured rat liver stem cells; and (6) evidence for the multipotentiality of cultured rat liver stem cells. Given the scope of this undertaking and the large number of published studies on these topics, we do not attempt to review the literature comprehensively. Therefore, in the last sections of this chapter, we focus our review on studies of the well-characterized WB-F344 rat liver epithelial stem cell line *(20)*. Furthermore, if possible, we recommend excellent reviews of these subjects.

2. A SHORT HISTORY OF LIVER STEM CELL BIOLOGY

The idea that the liver of adult rodents contains stem cells that can give rise to the epithelial cell types of the liver (hepatocytes and biliary epithelial cells) has developed from evidence that has accumulated from nearly 100 yr of investigation *(15,21)*. Early in the last century, potential lineage relationships among biliary epithelial cells, transitional cells, and hepatocytes were recognized in several studies of liver regeneration *(22–25)*. Studies on hepatocarcinogenesis in experimental animals produced additional evidence that hepatocytes could be generated through the formation, proliferation, and differentiation of transitional cells possessing features of both ductular cells and hepatocytes *(26,27)*. Subsequently, Wilson and Leduc *(28)* suggested that cells contained in the cholangioles or terminal bile ductules constitute a compartment of stem cells that can proliferate and generate hepatocytes in some forms of liver injury. Since then, the concept of the liver stem cell has generated considerable controversy, argument, and discussion *(1,2,14,21,29–33)*.

The debate surrounding the liver stem cell is fueled by the fact that stem cells and their phenotypic characteristics are largely intuitive concepts. As is true for other tissues, putative stem cells of the liver have not been identified microscopically *in situ* and have not been prospectively isolated from the normal liver in pure form. In addition, hepatocytes and biliary epithelial cells of the normal liver demonstrate very little cellular turnover and do not operate as a typical stem cell-fed lineage system *(32,34)* like those of other self-renewing tissues, such as intestine *(35,36)*, epidermis *(37–39)*, and bone marrow *(40,41)*.

These observations and the fact that the adult liver retains the capacity for complete and rapid renewal of cell numbers by the amplification of fully differentiated cells (both hepatocytes and biliary epithelial cells) in response to cell loss seem to argue against the need for an epithelial stem cell in the adult liver tissue. However, these observations do not eliminate the possibility that cells possessing a broader differentiation potential are present in the adult liver as a "reserve" *(29)* or "facultative" *(42)* stem cell compartment, or that cells with stem properties may serve a physiological function other than (or in addition to) lineage renewal. It has been suggested that participation of stemlike cells in liver growth processes may depend on several factors, including the presence or absence of liver injury, the type and extent of injury, and the capacity of hepatocytes to respond to growth stimuli *(32)*, consistent with the concept of the facultative liver stem cell *(42)*.

Despite the controversy surrounding the concept of the liver stem cell, considerable evidence has accumulated that supports the notion that the adult rodent liver contains cells with stemlike properties that can serve as progenitor cells for both hepatocytes and biliary epithelial cells under certain pathophysiological circumstances *(15)*. Three major sources of evidence support the existence of liver stem cells: (1) the founding of lineages of hepatocytes and biliary epithelial cells from hepatoblasts during embryonic development of the liver and the expression of differentiated cell-specific traits in cultured hepatoblasts; (2) the reestablishment of epithelial lineages following the proliferation of simple epithelial cells (oval cells) in livers subjected to carcinogenic or noncarcinogenic liver injury; (3) the isolation and propagation from the livers of adult rodents of simple epithelial cells that demonstrate the ability to differentiate into hepatocytes or biliary epithelial cells when transplanted into appropriate sites in vivo or when cultured under specific conditions ex vivo. These sources of evidence have been comprehensively discussed in several reviews *(11,15,21,43–45)*.

3. ESSENTIAL PROPERTIES OF STEM CELLS

The major properties thought to characterize stem cells have been inferred from investigations of classic stem cell-fed lineage renewal systems, including bone marrow, intestinal epithelium, and epidermis. Essential properties expected of stem cells include the capacity to (1) proliferate repeatedly, (2) renew the stem cell population, and (3) generate sufficient differentiated progeny to maintain or regenerate the functional capacity of a tissue *(36,39)*. Classic stem cells are thought to exhibit undifferentiated cellular phenotypes, to express variable differentiation potentials, and to be able to proliferate continuously (actual stem cells) or to be proliferationally quiescent until needed (potential or facultative stem cells) *(36,39)*. Although classic stem cell-fed lineage systems have been used to infer the properties of stem cells, evidence now suggests the existence of stem cells in many tissues that do not contain active stem cell-fed lineages *(46)*, such as the central nervous system *(47–50)* and liver *(15,21)*. These newly discovered stem cells appear to have properties that differ from those proposed for classic stem cells.

4. STEM CELLS OF THE ADULT LIVER

Unlike rapidly renewing epithelial tissues (such as the intestinal mucosa or epidermis), in which an active stem cell lineage system continually initiates replacement of differentiated cells that are shed *(4,5)*, the liver is normally a quiescent organ with minimal or slow rates of cell turnover in the adult *(51)*. Nonetheless, the liver possesses an extraordinary capacity for the regeneration of tissue mass following loss of normal hepatocyte numbers because of partial tissue loss (surgical resection) or hepatotoxic injury (necrosis).

A number of different cell types can be activated to repair or regenerate the liver depending on the nature and extent of injury or tissue deficit *(15)*. In an otherwise healthy liver, the replacement of hepatocytes (and tissue mass) lost to surgical resection or toxic injury is achieved through the proliferation of differentiated, normally quiescent hepatocytes contained in the residual (viable) tissue (Fig. 1). However, certain forms of liver injury impair the capacity of the remaining differentiated hepatocytes to proliferate in response to liver tissue deficit. When this occurs, a reserve or facultative stem cell compartment is activated to proliferate and replace the lost hepatocytes. Evidence suggests that there are at least two distinct cell populations that can be activated to generate new hepatocytes or cholangiocytes (Fig. 1). Rodent livers contain a population of normally quiescent (facultative), undifferentiated stem cells that reside in or around the biliary ductules of the portal tracts, which can be activated under certain pathological conditions to reestablish a proliferating–differentiating lineage (the oval cell reaction), capable of generating hepatocytes and some other cell types *(52,53)*. In addition, the adult liver contains a population of incompletely differentiated small hepatocytelike progenitor cells (SHPC) that can be activated to replace lost hepatocytes in some forms of tissue injury *(54)*.

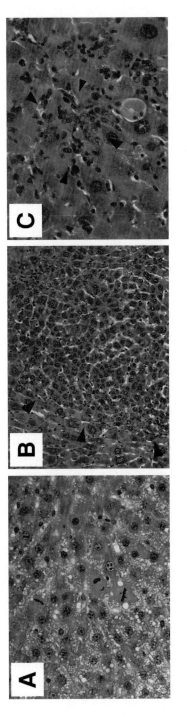

Fig. 1. Cellular responses to signals for liver regeneration. (**A**) Liver regeneration after surgical partial hepatectomy involves proliferation of differentiated hepatocytes. However, when differentiated hepatocytes cannot proliferate in response to the regenerative stimulus (such as in retrorsine-exposed rats), other liver progenitor cell populations are activated, such as (**B**) small hepatocyte progenitor cells or (**C**) oval cells.

These observations combine to suggest that there are at least three distinct populations of cells with stemlike potential in the adult rat liver (Fig. 1). It is conceivable that all these stem cells of the adult liver are derived from the same primordial stem cell population of the developing liver. In the following sections, the evidence for the existence of each of these stem cell populations in liver is reviewed.

4.1. Unipotential Liver Progenitor Cells: Differentiated Hepatocytes and Biliary Epithelial Cells

The liver possesses an enormous capacity to replace cells lost to surgical resection or necrosis (55,56). Activation of undifferentiated stem cells does not occur after cell loss when mature hepatocytes and biliary epithelial cells are capable of proliferating to restore the normal liver mass and structure (57,58). In rats subjected to surgical partial hepatectomy, the residual (viable) hepatocytes undergo a rapid burst of proliferation that ultimately restores the normal hepatocyte number (51,59–62). Likewise, biliary epithelial cells proliferate after partial hepatectomy to form expansions of the intrahepatic duct system (51,62–64). Irrespective of their location in the parenchyma (periportal to pericentral), virtually all hepatocytes proliferate and divide at least once during restoration of the hepatocyte number (65,66).

The ability of quiescent hepatocytes to reenter the cell cycle and proliferate in response to liver deficit has fascinated investigators throughout history. More recently, the extensive growth potential and enormous proliferative (replicative) capacity of the mature hepatocyte has become evident. In rats, hepatocytes proliferated and divided at least 8 to 12 times during the prolonged process of liver growth following five consecutive partial hepatectomies (67). In transgenic mice that express the urokinase gene in the liver under the direction of the albumin promoter–enhancer, the majority of hepatocytes succumbed to the toxic transgene product (68). In this model, the toxic transgene became inactivated in random hepatocytes, enabling them to proliferate, undergoing 10–12 cycles of cell division to yield discrete nodular aggregates (clones) that repopulated the liver parenchyma (69).

In a similar experimental system, transplanted normal hepatocytes repopulated the livers of transgenic mice that lack fumarylacetoacetate hydrolase (FAH$^{-/-}$) enzyme activity (70) because of the targeted disruption of exon 5 of the *Fah* gene (71). In this model, the transplanted FAH-expressing hepatocytes exhibited a selective growth advantage over host FAH-deficient hepatocytes, allowing the former to repopulate the livers of mutant mice. In these studies, it was estimated that transplanted hepatocytes proliferated through at least 15 cell divisions during repopulation of mutant livers *(70)*.

In other studies, wild-type male hepatocytes were serially transplanted at limiting dilution through the livers of female FAH$^{-/-}$ mice *(72)*. Complete repopulation of the diseased liver was accomplished in each round. The complete replacement of host liver by the progeny of transplanted hepatocytes through seven rounds of transplantation suggests that the transplanted hepatocytes were capable of at least 100 population doublings *(72)*.

Together, these studies demonstrate the incredible capacity for cell proliferation by differentiated hepatocytes, consistent with the suggestion that these cells represent a unipotential progenitor cell population of the adult liver.

4.2. Unipotential Liver Progenitor Cells: Small Hepatocyte Progenitor Cells

In several experimental models, hepatocytes were rendered incapable of proliferation through treatment with mito-inhibitory compounds, facilitating the outgrowth of stem cells in response to liver deficit. We recently described the cellular responses and time course for liver regeneration after surgical partial hepatectomy (PH) in rats with retrorsine-induced hepatocellular injury *(54)*. Similar to other models of chemical liver injury *(15,21)*, systemic exposure to retrorsine, a member of the pyrrolizidine alkaloid (PA) family, resulted in severe inhibition of the replicative capacity of fully differentiated hepatocytes *(54,73–75)*. When confronted with a strong proliferative stimulus such as PH *(54,73,74)* or hepatocellular necrosis *(76)*, retrorsine-injured hepatocytes synthesized deoxyribonucleic acid (DNA), but were unable to complete mitosis and arrested as nonproliferative giant cells (megalocytes). In this model, neither retrorsine-injured, fully differentiated hepatocytes nor oval cells proliferated sufficiently to contribute significantly to the restoration of liver mass after PH. Instead, the entire liver mass was reconstituted after PH through a novel cellular response that was mediated by the emergence and rapid expansion of a population of SHPCs, which share some phenotypic traits with fetal hepatoblasts, oval cells, and fully differentiated hepatocytes, but are morphologically and phenotypically distinct from each *(54)*. SHPCs emerged in all regions of the liver lobule after PH and were not solely associated with modest oval cell outgrowth in periportal regions, suggesting that SHPCs represent a novel cell population *(54)*.

The SHPCs morphologically most closely resemble differentiated (but small) hepatocytes at early time points after PH, perhaps suggesting that

SHPCs are a subset of retrorsine-resistant hepatocytes and not a novel progenitor cell population. However, the phenotype of SHPCs indicates that they are in fact distinct from fully differentiated hepatocytes because a subset of SHPCs expresses the oval cell/bile duct/fetal liver markers OC.2 and OC.5 through 5 d post-PH *(54)*. Coexpression of hepatocyte markers and oval cell markers by early-appearing SHPCs suggests that these cells are not fully differentiated, and that they display a phenotype similar to that expected for a cell type transitional between the bipotential hepatoblast (E14) and a fetal hepatocyte (E18–E20).

Retrorsine-exposed rats were able to regenerate their liver mass completely after PH, as evidenced by liver weights and liver:body weight ratios *(54)*. At 30 d post-PH, liver weights and liver:body weight ratios do not differ significantly after either retrorsine/PH or control/PH *(54)*. By this time, the progeny of SHPCs occupied virtually the entire (87% by area) parenchyma in retrorsine/PH rats. However, comparison of the time course for liver regeneration in control and retrorsine-exposed rats after partial hepatectomy showed that liver regeneration through activation and expansion of SHPCs is a much more protracted process. Complete regeneration of the liver mass in retrorsine-exposed animals required nearly 30 d, compared to about 10 d in control rats *(54)*.

Using a combined approach involving gene expression analysis of tissues isolated using laser capture microdissection and *in situ* immunohistochemistry, the expression patterns of select mRNAs and proteins were examined in the earliest (least-differentiated) SHPCs that emerged after PH in retrorsine-exposed rat livers *(77)*. The results showed that early-appearing SHPCs (at 3–7 d post-PH) expressed messenger RNA (mRNA) or protein for all of the major liver-enriched transcription factors (hepatic nuclear factor 1α [HNF1α], HNF1β, HNF3α, HNF3β, HNF3γ, HNF4, HNF6, C/EBPα, C/EBPβ, and C/EBPγ), WT1, α-fetoprotein, and P-glycoprotein *(77)*.

Compared to surrounding hepatocytes, early-appearing SHPCs lack (or have significantly reduced) expression of mRNA for hepatocyte differentiation markers tyrosine aminotransferase and α1-antitrypsin *(77)*. Likewise, SHPCs that emerge and proliferate during the early phase of liver regeneration lack (or have reduced expression of) several hepatic CYP (cytochrome P450) proteins known to be induced in rat livers after retrorsine exposure (CYP2E1, CYP1A2, and CYP3A1). However, by 30 d post-PH, expression patterns for all markers expressed by SHPCs mirrored that expected for fully differentiated hepatocytes. Both α-fetoprotein and WT1 protein are uniquely expressed by SHPCs during the early phase of liver regeneration, suggesting that these markers may be used to identify the earliest progenitors of these cells *(77)*. These results suggest that SHPCs represent a unique paren-chymal (less differentiated than mature hepatocytes) progenitor cell popula-tion of adult rodent liver *(54,77)*.

4.3. Multipotential Liver Progenitor Cells: Oval Cells

Oval cells, which proliferate in several hepatocarcinogenesis models *(78–80)* and in some forms of noncarcinogenic liver damage *(81–85)*, may be related to liver stem cells. A number of different experimental models elicit the proliferation of oval cells *(52,81,86–89)*. All of these models are characterized by concurrent stimulation of liver growth and inhibition of normal mechanisms for liver tissue restoration (i.e., blockade of the prolif-eration of hepatocytes). The stimulus for liver growth can be satisfied through several different methods, including surgical resection, nutritional stress, or chemically induced necrosis. Blockade of hepatocyte proliferation is frequently achieved using chemicals (such as 2-acetylaminofluorene) that impede or prevent mitotic division of mature hepatocytes *(52)*. The cellular response common to each of these models involves the proliferation of small cells with scant cytoplasm and ovoid nuclei that are morphologically described as *oval cells (90)*. Although most of the models of oval cell prolif-eration involve rats, similar models have been developed using carcinogen-treated mice *(91–94)*, transgenic mice that express viral oncogenes *(95,96)*, or other transgenes *(97)*.

The timing of cellular events differs, sometimes dramatically, among the various models of oval cell proliferation *(15)*. However, the majority of oval cell proliferation models share a common sequence of events: (1) prolifera-tion of oval cells in or around the portal spaces, (2) invasion of the lobular parenchyma by the proliferating oval cells, (3) appearance of transitional cell types and immature hepatocytes, and (4) maturation of hepatocytes and restoration of normal liver structure. Oval cells are initially seen in the por-tal zones of the liver lobule in the regions of terminal bile ductules or cholangioles *(29,98)*. Proliferating oval cells are recognized as representing a collection of phenotypically distinct cells that compose a heterogeneous cell population or *compartment (32,99,100)*. Morphologically, the typical oval cell possesses cellular characteristics similar to those of cells of termi-nal bile ductules *(101–103)*. However, the oval cell compartment also con-tains transitional cells that display morphologic features intermediate between oval cells and hepatocytes *(86,103)*.

Proliferating oval cells form irregular ductlike structures connected to preexisting bile ducts *(103–107)*. As they proliferate, oval cells migrate from the portal regions into the lobular parenchyma, sometimes occupying a large percentage of the liver mass. Groups of small basophilic hepatocytes appear among oval cells; these immature hepatocytes proliferate and differentiate as the oval cells gradually disappear and the normal liver structure is restored *(83,93,103,108–110)*. The possibility that oval cells might possess stemlike properties and give rise to hepatocytes or biliary epithelial cells has been recognized for some time *(26,28,90,111)*.

Several studies have attempted to document the fate of oval cells that proliferated in various hepatocarcinogenesis models and after noncarcinogenic liver injury, producing evidence that oval cells are precursors of hepatocytes *(83,108–110,112,113)*. Using the modified Solt–Farber model of oval cell proliferation, Evarts et al. *(109)* demonstrated unequivocally that oval cells radiolabeled with ^{3}H-thymidine could give rise directly to tagged basophilic hepatocytes. More recently, Alison and colleagues examined the proliferation and fate of oval cells in rats using the modified Solt–Farber model with various doses of 2-acetylaminofluorene *(114)*.

Proliferation of oval cells also has been described in the chronic injury produced in mouse liver by transgenic expression of both hepatitis B virus *(95)* and SV40 T antigen *(96)*. Bennoun and colleagues *(96)* demonstrated a transition between proliferating oval cells in SV40 T antigen transgenic mice and newly formed hepatocytes. Likewise, in mice treated with diethylnitrosamine, oval cells proliferated and subsequently differentiated into hepatocytes *(94)*.

Employing the D-galactosamine model of oval cell proliferation in rats, Lemire et al. *(110)* and Dabeva and Shafritz *(83)* also demonstrated the transfer of radiolabel from oval cells to small hepatocytes. In both of these studies, transition from oval cells to hepatocytes was accompanied by a shift from the biliary epithelial/oval cell phenotype (expression of α-fetoprotein, γ-glutamyltranspeptidase, and biliary epithelial-type cytokeratins) to a cellular phenotype characteristic of hepatocyte differentiation (expression of albumin, glucose-6-phosphatase, and other hepatocyte markers; reduction of α-fetoprotein expression) *(83,110)*.

4.4. Liver Progenitor Cells From Extrahepatic Tissues

In addition to the progress characterizing stem cell responses in liver and the various populations of liver cells with stemlike potential, advances have also been made in the identification of multipotent adult stem cells with broad differentiation potential that includes liver. A number of studies have identified extrahepatic sources of stemlike cells that can colonize the liver or give rise to hepatocytes in vivo or in culture. Most recently, several investigators have reported that progenitor cells in bone marrow or peripheral blood can give rise to cells of the liver. In somewhat older studies, the ability of pancreatic cells to give rise to hepatocytes has been described. In addition to these two extrahepatic sources of stem cells for liver epithelial cells, several other sources have also been suggested, including neural stem cells. Evidence that extrahepatic stem cells can give rise to liver is summarized in the sections that follow.

4.4.1. Liver From Progenitor Cells of the Bone Marrow

Bone marrow contains several different cell types with stemlike potential, including hematopoietic *(115,116)*, stromal *(117)*, and mesenchymal stem cells *(118–120)*. In addition to these progenitor cell types, it has been suggested that bone marrow contains a multipotent adult progenitor cell that expresses a broader tissue differentiation potential. However, whether this multipotent progenitor cell compartment of the bone marrow represents a single cell type with broad differentiation capacity or whether it represents an admixture of several tissue stem cell types has not been resolved *(3)*. In fact, the multipotent progenitor cell of bone marrow may be related (or identical) to one of these other cell types (hematopoietic or mesenchymal stem cells) of the bone marrow. Bone marrow transplants generate cell lineages of the blood and have now been suggested to give rise to a number of other cell types, including cardiac muscle *(121)*, skeletal muscle *(122,123)*, neurons *(124–126)*, lung epithelium *(127)*, oval cells *(128)*, hepatocytes *(129–133)*, and biliary epithelial cells *(127,130)*. Bone marrow progenitor cell–derived hepatocytes have been demonstrated in rats *(128)*, mice *(129,132,133)*, and humans *(130,131)*.

Transdifferentiation of bone marrow progenitor cells into hepatocytes has yielded variable replacement of liver, possibly related to the animal model employed. Transplantation of unfractionated bone marrow from male donors into lethally irradiated syngeneic female mice (B6D2F1) resulted in efficient reconstitution of the host hematopoietic system and generation of donor-derived hepatocytes in the host livers *(129)*. The bone marrow-derived hepatocytes were identified in the hepatic plates of recipient livers using Y chromosome *in situ* hybridization *(129)*. Quantitative analysis suggested that 1–2% of hepatocytes were bone marrow derived *(129)*. These results suggested that bone marrow-derived stem cells could engraft and give rise to hepatocytes, albeit at low frequency, in normal liver.

Grompe and colleagues employed the murine model of hereditary tyrosinemia type I *(70,71)* to investigate the potential for bone marrow-derived stem cells to repopulate diseased liver *(132,133)*. Transplantation of unfractionated bone marrow into FAH$^{-/-}$ mice resulted in replacement of 30–50% of the liver mass after a period of selection *(132)*. Furthermore, when highly purified KTLS (c-kit^{high}Thy^{low}Lin$^-$Sca-1$^+$) hematopoietic stem cells from male ROSA26/BA mice were transplanted into lethally irradiated female FAH$^{-/-}$ mice, liver engraftment and hepatocytic differentiation of transplanted cells was observed *(132)*. The hepatocyte phenotype of engrafted cells was confirmed by expression of albumin and bile canalicular dipeptidylpeptidase IV *(132)*. In addition, the engrafted cells expressed the FAH protein and were positive for β-galactosidase and the Y chromosome *(132)*. It was also suggested that the c-kit$^-$ and Lin$^+$ fractions of the bone marrow do not contain significant numbers of progenitor cells that can give rise to hepatocytes *(132)*. Although substantial liver repopulation by bone marrow-derived hepatocytes was observed when selective conditions were employed *(132,133)*, negligible hepatocyte replacement was observed in the absence of selective pressure *(133)*.

In a similar study, unfractionated bone marrow from transgenic mice expressing Bcl2 under the control of the liver pyruvate kinase gene promoter was transplanted into normal mice, some of which were subjected to lethal irradiation *(134)*. In this study, the frequency of hepatocyte differentiation from transplanted bone marrow progenitor cells was rare. However, when selection pressure was applied through the administration of anti-Fas antibodies, the small number of bone marrow-derived hepatocytes present in the livers of recipient mice expanded 6-fold to 20-fold, ultimately occupying approx 1% of the liver mass *(134)*.

These studies showed that positive selection pressure can result in a higher degree of replacement of liver by bone marrow-derived hepatocytes. Other investigators have failed to detect donor-derived hepatocytes in undamaged livers after bone marrow transplant despite reconstitution of the hematopoietic system and replacement of the liver endothelium *(135,136)*. These observations suggest that generation of hepatocytes from bone marrow stem cells is uncommon in the absence of strong selective pressure, such as that in the FAH$^{-/-}$ mouse model.

A few studies have examined the differentiation potential of bone marrow–derived progenitor cells in vitro. Reyes and colleagues isolated and established cultured multipotent adult progenitor cells from bone marrow of humans, mice, and rats *(119,120)*. These multipotent adult progenitor cells copurified with the mesenchymal stem cell fraction of the bone marrow *(119,120)*. When these cells are propagated on Matrigel in the presence of fibroblast growth factor 4 (FGF-4) and human growth factor (HGF), hepatocytelike cells expressing albumin, cytokeratin 18 (CK18), and HNF3β resulted after 14 d of culture *(137)*. Furthermore, these multipotent adult progenitor cell-derived hepatocytes expressed several functional characteristics of mature hepatocytes, including secretion of urea and albumin, expression of phenobarbital-inducible cytochrome P450, the capacity to store glycogen, and the ability to take up low-density lipoprotein (LDL) *(137)*, suggesting that the bone marrow of adult mammals contains progenitor cells with the potential to give rise to hepatocytes, and that these cells can be propagated in culture without loss of potency *(137)*.

4.4.2. Liver From Stem Cells of the Pancreas

Several experimental models have been developed in which large eosinophilic cells that morphologically and phenotypically resemble hepatocytes are induced in the pancreas of rats and hamsters following severe pancreatic injury *(138–141)*. Rats maintained on a copper-deficient diet for 8–10 wk showed widespread injury to exocrine elements of the pancreas *(139,140)*. When the copper-deficient diet was replaced with a normal diet, hepatocytelike cells developed during the regeneration of the pancreatic tissue *(139,140)*. In this model, cells that resembled hepatic oval cells were thought to represent the progenitor cells for pancreatic hepatocytes *(140)*.

These observations coupled with those of other studies led to the suggestion that liver and pancreas may share a common stem cell *(142)*. To examine the possibility that pancreatic oval cells could serve as liver progenitor cells, proliferating pancreatic oval cells were isolated and introduced into the livers of dipeptidylpeptidase IV-deficient rats via transplantation into the spleen *(143)*. Following transplantation, hepatocytelike cells that express dipeptidylpeptidase IV activity were observed in the livers of recipient dipeptidylpeptidase IV-deficient animals, suggesting that oval cells proliferating in response to pancreatic injury caused by the copper-deficient diet can serve as hepatocyte progenitor cells *(143)*. In a similar study, suspensions of pancreatic cells from normal adult mice were transplanted into FAH$^{-/-}$ mice to examine the possibility that the normal pancreas contains a population of hepatocyte progenitor cells *(144)*. When selection pressure was applied, extensive liver repopulation (>50% replacement of liver) was observed in a subset of recipient mice, and another subset showed nodules of donor-derived hepatocytes *(144)*.

Chen et al. *(145)* examined the fate of normal rat pancreatic ductal epithelial cells following implantation into the abdominal subcutaneous tissue or intraperitoneal cavity of adult syngeneic rats. In these studies, RP-2 pancreatic duct epithelial cells *(146)* were embedded in a gel composed of extracellular matrix (collagen I and Matrigel) prior to implantation *(145)*. Eight weeks following subcutaneous implantation, nests of eosinophilic epithelioid cells and rare duct structures were observed in the recovered extracellular matrix gel *(145)*. Six weeks following intraperitoneal implantation, trabeculae and clusters of large polygonal epithelioid cells with granular eosinophilic cytoplasms, resembling mature hepatocytes of the adult liver, were observed *(145)*. These hepatocytelike cells expressed high levels of tyrosine aminotransferase, albumin, and transferrin and stained positively with HES6 monoclonal antibodies *(145)*.

These studies demonstrated that normal pancreatic duct epithelial cells can differentiate into functional hepatocytes following implantation into an appropriate host microenvironment. They provide additional support for the suggestion that liver and pancreas share a similar stem cell with differentiation options that are determined by the tissue microenvironment.

4.4.3. Liver From Neural Stem Cell Cultures

Cultured neural stem cells isolated from mice *(147)* have been shown to give rise to neurons and glia following transplantation into brain tissue of host animals *(148–150)*. Such neural stem cells have been suggested to have a differentiative plasticity when transplanted into various tissue microenvironments other than the brain. For instance, neural stem cells derived from adult donor tissue differentiate into hematopoietic lineages when engrafted into the bone marrow *(151)*. When introduced into developing mouse blastocysts, neural stem cells contribute to cells of various germ layers and tissues of chimeric embryos, including the liver *(152)*. Additional studies will be required to demonstrate whether neural stem cells can give rise to differentiated hepatocytes in the adult liver.

4.5. Human Liver Stem Cells

4.5.1. Evidence for Liver Stem Cells in Humans

In recent years, numerous investigators have attempted to identify and isolate human liver stem cells or have made observations in pathological human livers that suggest the existence of these cells. In many instances, investigators have attempted to determine if activation and proliferation of oval cells occurs in humans in a fashion similar to the oval cell reaction observed in rodents *(153)*. Using morphologic criteria or immunohistochemical staining, several reports suggested the presence of cells resembling oval cells in several different human liver diseases *(154–169)*. In some studies, cells exhibiting a phenotype consistent with oval cells were observed in normal human liver *(169,170)*.

4.5.2. Liver From Bone Marrow in Humans

In rodents, bone marrow has been suggested to contain stem cells with hepatocytic differentiation potential *(129,132)*. Likewise, several studies have suggested that human bone marrow contains stemlike progenitor cells that can differentiate into hepatocyte progeny under various conditions *(130,131)*. The first evidence that bone marrow-derived progenitor cells give rise to hepatocytes in humans emerged from a study of archival autopsy or biopsy liver specimens from gender-mismatched transplant patients *(130)*.

Liver tissue from two female patients who received therapeutic bone marrow transplants from male donors and liver tissue from four male patients who received orthotopic liver transplants from female donors was studied *(130)*. Variable numbers of Y chromosome–positive hepatocytes and biliary epithelial cells were detected using *in situ* hybridization *(130)*. In this study, the interval from transplant to liver sampling was 1 mo to 2 yr, and the numbers of Y chromosome-positive hepatocytes observed varied from 1 to 8% among the patients *(130)*. However, when the investigators adjusted the data to account for their assessment of the insensitivity of Y chromosome *in situ* hybridization, the corrected results suggested that 5–40% of hepatocytes observed were derived from circulating progenitor cells, probably of bone marrow origin *(130)*.

In a similar study, the livers of 9 female patients who received bone marrow transplant from male donors and the livers of 11 male patients who received orthotopic liver transplants from female donors were evaluated for Y chromosome-positive hepatocytes *(131)*. Among these two groups of patients, 0.5–2% of hepatocytes were derived from extrahepatic progenitor cells, and clusters of Y chromosome-positive hepatocytes were observed in several instances, suggesting that hepatocyte progeny had clonally expanded after colonization of the liver by circulating progenitor cells *(131)*.

In a third study, patients with hematologic or breast malignancies received high-dose chemotherapy and transplants of allogeneic peripheral blood stem cells from donors pretreated with granulocyte colony-stimulating factor *(171)*. Y chromosome-positive hepatocytes were detected as early as 13 d posttransplant, and 4–7% of hepatocytes examined were donor derived *(171)*.

In all of these studies, few hepatocytes derived from bone marrow (or other circulating progenitor cells) were detected, and the interval from transplant to sampling of the liver was short (less than a year in most cases). It has been suggested that the engraftment of bone marrow-derived stem cells in the liver may represent an early feature in liver transplants, but that hepatocytes derived from bone marrow progenitor cells may not persist as a long-term feature of grafted livers *(172)*. In a study of gender-mismatched liver transplant patients with long interval between liver transplant and biopsy (1.2–12 yr), no host-derived hepatocytes were detected when the Y chromosome was used in *in situ* hybridization experiments *(172)*.

4.6. Isolation and Culture of Adult Liver Stem Cells

Several types of epithelial cells can be isolated from rodent livers and established in primary or propagable cultures *(173)*. Differentiated hepatocytes and bile duct cells can be maintained in primary culture for short periods, but generally these cell types exhibit limited propagability and lifespan in culture *(174–176)*. In contrast, simple (undifferentiated) liver epithelial cells can be readily established and propagated in culture *(177)*. These simple liver epithelial cell types possess some stemlike properties, suggesting that they may represent the cultured counterpart of epithelial stem cells in the adult liver *(178)*.

4.6.1. Early Studies of Propagable Liver Epithelial Cells

Early long-term rat liver epithelial cell cultures were established from cell outgrowths in liver tissue explant cultures *(179)*. Development of enzymatic techniques for the preparation of viable single-cell suspensions of liver cells made possible the selective culture of several liver epithelial cell types *(180,181)*. Brief digestion of liver tissue with collagenase produces liver cell suspensions enriched for hepatocytes *(181,182)*, whereas enrichment of nonparenchymal epithelial cell types can be accomplished by the selective removal of hepatocytes from collagenase-dispersed liver using various strong proteases, such as Pronase or trypsin *(183–185)*. The nonparenchymal cells remaining following protease treatment of liver cell suspensions include macrophages (Kupffer cells), endothelial cells, bile ductular cells, Ito cells, and various hematopoietic cells *(183,186)*. Also present in dispersed liver cell suspensions are simple epithelial cells *(177,187)*.

4.6.2. Rat Liver Epithelial Stem Cells: Oval Cell Lines

Oval cells have been isolated from diseased liver and established in culture by several laboratories. Morphologically, cultured oval cells are cuboidal and grow in a monolayer *(188,189)*. Some established lines of oval cells are stably diploid or pseudodiploid, are nontumorigenic, and do not proliferate in soft agar *(188–191)*. Ultrastructurally, cultured oval cells exhibit catalase-positive peroxisomes that proliferate in response to treatment with clofibrate *(189,190)*. Cultured oval cells generally express glucose-6-phosphatase activity and lactate dehydrogenase isozymes 2–5, and they are variably positive for albumin and α-fetoprotein *(189–192)*. A few oval cell lines have been characterized for cytokeratin expression and express CK8 and CK18 and variably express or not express CK7 and CK19 *(189,191)*. As with some other liver epithelial cell lines, cultured oval cells tend to be antigenically simple. Cultured LE/6 oval cells do not express antigens for mono-

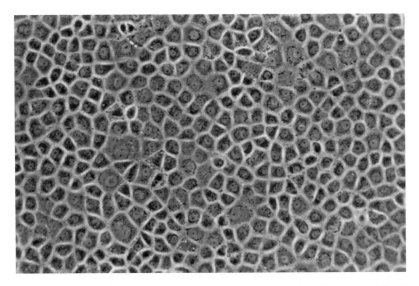

Fig. 2. Morphology of cultured WB-F344 rat liver epithelial stem cells. Low-passage WB-F344 cells viewed by phase contrast microscopy.

clonal antibodies OC.1, OC.2, BD.1, H.1, or H.2 (reviewed in ref. *21*). The presence of peroxisomes and glucose-6-phosphatase activity in cultured oval cells suggests that these cells are part of the hepatocyte lineage.

4.6.3. Rat Liver Epithelial Stem Cells: The WB-F344 Line

Several lines of rat liver epithelial cells have been established from the livers of normal adult rats *(21)*. The WB-F344 rat liver epithelial cell line represents one such propagable rat liver epithelial cell line clonally derived from a single epithelial cell *(20)*. WB-F344 cells are phenotypically similar to other established rat liver epithelial cell lines. A detailed comparison of the phenotypic properties of several rat liver epithelial cell lines was reviewed in ref. *21*. WB-F344 cells are small (9- to 15-μm diameter), polygonal cells that grow in a monolayer (Fig. 2).

Ultrastructurally, WB-F344 cells exhibit a relatively simple cytoplasm with few organelles. Adjacent cells in confluent monolayers are joined by numerous desmosomes *(20)* and nexus junctions containing connexins 26 and 43 *(193–196)* and are dye coupled *(194,196,197)*. Cells are polarized, surfaces directed to the growth medium interface contain microvilluslike projections, and a basement membranelike material containing fibronectin is deposited at the substrate interface *(20)*. WB-F344 cells possess a stable diploid or quasidiploid karyotype *(20)*. They do not proliferate in soft agar culture and are nontumorigenic following transplantation into neonatal syngeneic rats (20). WB-F344 cells share some phenotypic traits with both hepatocytes and biliary epithelial cells, but their overall phenotype differs distinctively from either differentiated cell type (20). Most notably, WB-F344 cells are null for the major antigens that typify and distinguish hepatocytes or biliary epithelial cells (198; A. E. Wennerberg and J. W. Grisham, 1993, unpublished observations).

4.6.4. Progenitor Cells From the Human Liver

Very few studies have appeared that reported the isolation and culture of human liver cells. Nussler and colleagues isolated a cell population from human liver and established it in propagable culture (199). The resulting cell line, AKN-1, has been characterized in culture and shows many characteristics of biliary epithelial cells (199). It is tempting to speculate that AKN-1 cells represent a cultured counterpart to a putative undifferentiated human liver stem cell. However, these cells contain chromosomal abnormalities, display an aneuploid DNA content, and are tumorigenic following transplantation into nude mice (199), indicating that the AKN-1 cell line may not represent propagable normal human liver stem cells.

In a similar study, cells expressing c-Kit and CD34 were isolated and cultured from diseased human liver (170). These cells were localized to portal tracts close to bile ducts in cirrhotic livers (170). In cell culture, these cells expressed markers (such as CK19) that suggested differentiation toward the biliary epithelial cell lineage (170). Of great significance, cells positive for c-Kit and CD34 were also isolated from normal human liver, albeit in smaller numbers, and these cells also acquired biliary epithelial differentiation in vitro (170).

5. EVIDENCE FOR THE MULTIPOTENTIAL DIFFERENTIATION OF ADULT LIVER STEM CELLS IN VIVO

The ultimate proof that a rat liver epithelial cell line represents cultured stemlike cells requires the demonstration that these cells can give rise to hepatocytes or biliary epithelial cells following transplantation into appropriate sites in host animals. Such studies have been carried out by transplanting various cultured liver epithelial cell lines into livers or extrahepatic sites of syngeneic animals and nude mice.

5.1. Transplantation of Neoplastically Transformed Liver Epithelial Cells

Early studies of the differentiation of tumors produced following the subcutaneous or intraperitoneal transplantation of uncloned chemically transformed rat liver epithelial cells demonstrated that most tumors were poorly differentiated, although some tumors expressed morphologic features of hepatocellular carcinomas or biliary adenocarcinomas *(200,201)*. Definitive evidence that rat liver epithelial cells possess a wide differentiation potential came from studies of the tumorigenicity of cloned lines of transformed cells derived from WB-F344 cells *(202)*.

WB-F344 rat liver epithelial cells have been neoplastically transformed in vitro by infrequent passaging of cultures to induce their spontaneous transformation *(203,204)* and following exposure to *N*-methyl-*N'*-nitro-*N*-nitrosoguanidine to transform them chemically *(205)*. The histological types of tumors that result from the subcutaneous or intraperitoneal transplantation of chemically transformed WB-F344 cells and subcloned cell lines include hepatocellular carcinomas, adenocarcinomas (biliary and intestinal types), adenocarcinomas with sarcomatous elements, epidermoid (squamous) carcinomas, hepatoblastomas (containing cartilage and osteoid elements), and poorly differentiated tumors of both epithelial and mesenchymal morphologies *(202)*. Spontaneously transformed WB-F344 cell lines also produce wide variety of tumor cell types, including well-differentiated hepatocellular carcinomas, biliary adenocarcinomas, and hepatoblastomas, as well as mesenchymal tumors, including osteosarcomas *(203,204)*. These results demonstrate that transformed WB-F344 cells are multipotent for the major differentiated epithelial cell types of the rat liver and suggest the possibility that their diploid counterpart may possess a broad differentiation potential.

5.2. Transplantation of Rat Liver Stem Cells Into Livers of Syngeneic Rats

To investigate the differentiation potential of cultured rat liver epithelial cells, WB-F344 cells have been transplanted into livers or extrahepatic sites of syngeneic animals to examine their fate in vivo. Several weeks following the transplantation of WB-F344 cells into the interscapular fat pads of syngeneic Fischer 344 rats, small clusters of cells morphologically resembling hepatocytes were identified *(178)*. However, whether these cells possessed functional attributes of hepatocytes was not determined. Nonetheless, this observation suggested that transplanted WB-F344 cells could acquire characteristics of hepatocytes in vivo.

To demonstrate a precursor–product relationship between transplanted cells and differentiated cell types in the liver, methods had to be established that would allow the definitive identification of the progeny of the transplanted cells among host cells in the adult liver. Three different strategies were utilized to examine the fate of WB-F344 rat liver epithelial cells following transplantation into adult rat livers: (1) introduction of a genetic tag/ marker enzyme into WB-F344 cells *(178,206)*, (2) transplantation of normal WB-F344 cells into the livers of rats deficient for DPPIV enzyme activity *(207)*, and (3) transplantation of normal WB-F344 cells into the livers of Nagase analbuminemic rats *(208)*. In all of these model systems, transplanted WB-F344 cells (or BAG2-WB cells) integrated into hepatic plates and morphologically and functionally differentiated into hepatocytes. The results from these studies combined show that WB-F344 cells are multipotent and differentially responsive to the tissue microenvironment of the transplantation site.

5.2.1. Hepatocytic Differentiation by Transplanted WB-F344 Rat Liver Stem Cells

To facilitate transplantation studies, WB-F344 rat liver epithelial cells were genetically modified by infection with the CRE BAG2 retrovirus, which encodes the *Escherichia coli* β-galactosidase gene and the Tn5 neomycin resistance gene *(209)*. The resulting cells, termed BAG2-WB, were transplanted into the livers of adult Fischer 344 rats; the livers of these rats were examined for the presence of β-galactosidase-positive cells at various times following transplantation (Fig. 3). In these studies, β-galactosidase-positive hepatocytelike cells were detected in the hepatic plates of recipient rats among the host hepatocytes *(206)*. The size and morphologic appearance of these cells is indistinguishable from that of the host hepatocytes *(178,206)*. The β-galactosidase-positive cells were observed at all times examined (Fig. 3), up to more than 1 yr following transplantation. Subsequent studies demonstrated that the β-galactosidase-positive hepatocytelike cells express functional differentiation typical of hepatocytes, including expression of albumin, transferrin, α1-antitrypsin, and tyrosine aminotransferase *(178)*.

Several methods were utilized to demonstrate further that the β-galactosidase-positive cells observed in these livers were derived from the transplanted BAG2-WB cells. In some studies, cells were labeled with the lipophilic fluorescent membrane dye PKH26-GL *(206)*. Examination of liver sections demonstrated the presence of fluorescent cells in the hepatic plates of the host livers, consistent with the observations made using the β-galactosidase marker enzyme *(206)*. In addition, the neomycin resistance gene of the CRE BAG2 retroviral construct could be detected by polymerase chain reaction (PCR) in genomic DNA prepared from livers of rats that received BAG2-WB cell transplants *(206)*.

Together, these studies demonstrate that transplanted WB-F344 rat liver epithelial cells incorporate into the hepatic plates of the host liver, morpho-

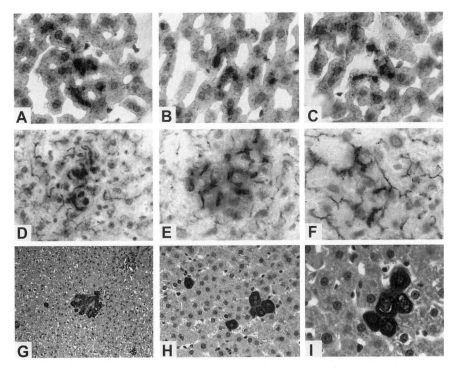

Fig. 3. Hepatocytic differentiation of WB-F344 cells following transplantation into the livers of adult rats. **(A–C)** Liver tissue obtained 128 d following transplantation of WB-F344 cells into syngeneic Fischer 344 rats. β-galactosidase-positive WB-F344-derived hepatocytes are identified by the presence of X-gal reaction product. **(D–F)** Representative liver cryosections from DPPIV-deficient rats demonstrating DPPIV-positive WB-F344-derived hepatocytes 30–60 d after transplantation. Liver sections were histochemically stained for both DPPIV and bile canalicular ATPase. **(G–I)** Albumin immunostaining of representative paraffin sections of liver tissue from Nagase analbuminemic rats at 28 d following transplantation of WB-F344 cells.

logically and functionally differentiate into hepatocytes, and remain a stable component of the hepatic parenchyma over long periods of time.

In other transplantation studies *(207)*, normal WB-F344 cells were transplanted into the livers of German Fischer 344 rats, which are deficient for dipeptidylpeptidase IV enzyme activity *(210,211)*, to examine the fate of these cells in a transplantation model that does not depend on the use of exogenous marker enzymes. Dipeptidylpeptidase IV is a bile canalicular enzyme expressed by mature hepatocytes in normal rats *(212,213)*. The WB-F344 cell line was isolated from an American strain adult Fischer 344 rat *(20)* that expresses normal levels of dipeptidylpeptidase IV activity in hepatocytes. Following transplantation into the rats deficient in dipeptidylpeptidase IV, WB-F344 cells incorporated into hepatic plates and morphologically differentiated into hepatocytelike cells that expressed dipeptidylpeptidase IV enzyme activity *(207)*.

These dipeptidylpeptidase IV-positive hepatocytes (Fig. 3) were easily distinguished from the host hepatocytes using a histochemical staining reaction for dipeptidylpeptidase IV activity *(214)*. The dipeptidylpeptidase IV-positive hepatocytes in hepatic plates were comparable to adjacent host hepatocytes in size and morphology (Fig. 3). Close physical contact between the differentiated progeny of the transplanted cells and host hepatocytes was verified through colocalization of dipeptidylpeptidase IV staining and adenosine triphosphatase (ATPase) staining of hybrid bile canaliculi (Fig. 3). In addition, the localization of dipeptidylpeptidase IV staining to bile canaliculi showed that the surface membranes of differentiating WB-F344 cells acquired the polarization characteristic of fully differentiated hepatocytes *(207)*. These results provide additional evidence that WB-F344 cells morphologically and functionally differentiate into hepatocytes following their transplantation into the liver microenvironment of adult rats.

In a third transplantation model, wild-type WB-F344 cells were transplanted into Nagase analbuminemic rats *(215,216)* to examine the efficacy of liver stemlike cell transplant for phenotypic correction of a genetic liver defect *(208)*. In previous studies, transplanted WB-F344 cells (or BAG2-WB cells) gave rise to differentiated hepatocyte progeny that expressed albumin *(178,207)*. Therefore, in this transplantation model, albumin served dual roles: (1) as a marker for detection of the progeny of transplanted WB-F344 cells and (2) as a metabolic marker for monitoring phenotypic correction of analbuminemia. Albumin-positive hepatocytes were detected in the hepatic plates among albumin-negative host hepatocytes in all rats receiving cell transplants (Fig. 3). In some cases, individual albumin-positive hepatocytes were observed, whereas in other instances, clusters of albumin-positive hepatocytes were detected (Fig. 3). The WB-F344 cells were transplanted into Nagase rats treated with cyclosporin to minimize rejection of the transplanted cells because of strain-specific differences between the Fischer 344 rat cells and the Sprague-Dawley rat hosts *(208)*. However, once engrafted into the livers of these rats, albumin-positive WB-F344 hepatocyte progeny could be detected for up to 4 wk following the cessation of cyclosporin treatment *(208)*.

5.2.2. Differentiation of WB-F344 Rat Liver Stem Cells Into Biliary Epithelial Cells

Until recently, studies aimed at determination of the potential for WB-F344 cells to serve as a progenitor for biliary epithelial cells and to partici-pate in bile duct formation were lacking. However, Hixson and colleagues *(217)* developed techniques for introducing rat liver epithelial cells into the bile ducts. Using these methods, they showed that WB-F344 cells trans-planted into the bile ducts differentiated into biliary epithelial cells and par-ticipated in bile duct formation *(217)*.

5.3. Transplantation of Rat Liver Stem Cells Into Extrahepatic Sites

Several types of stem cells from adult tissues express a capacity for multipotential differentiation. The stem cells respond to inductive signals from the tissue microenvironment in which they engraft and differentiate into cells that express a phenotype characteristic of cells in that tissue micro-environment. To examine directly the possibility that rat liver stem cells possess the ability for multipotential differentiation in vivo, WB-F344 cells were transplanted into extrahepatic sites of nude mice. In response to sig-nals in the various niches of the heart, transplanted WB-F344 cells differen-tiated into cells unique to each of these niches *(218)*. Furthermore, early results indicated that transplanted WB-F344 cells can differentiate into cells of the hematopoietic lineage in vivo. Together, these results suggest that the WB-F344 cells have a broad differentiation potential that, depending on the tissue microenvironment at the transplantation site, can give rise to cells of several nonhepatic lineages. These studies are reviewed briefly.

5.3.1. WB-F344 Cells Differentiate Into Various Kinds of Heart Cells Following Transplantation into the Heart

To investigate the possibility that adult-derived WB-F344 rat liver stem cells can respond to signals in the microenvironment of the heart in vivo and give rise to cells of cardiac lineages, male WB-F344 cells that carry the *E. coli* β-galactosidase reporter gene were transplanted into the hearts of adult female nude mice *(218)*. Six weeks following intracardiac injection, β-galactosidase-positive myocytes were identified, by light *(218)* and electron microscopy, in the myocardium of recipient mice among host myocytes (Fig. 4).

Engrafted cells ranged from small undifferentiated cells to long striated cells measuring up to 110 μm in length *(218)*. By electron microscopy, the WB-F344-derived myocytes were demonstrated to be at various stages of differentiation. The longer cells contained well-organized and differentiated

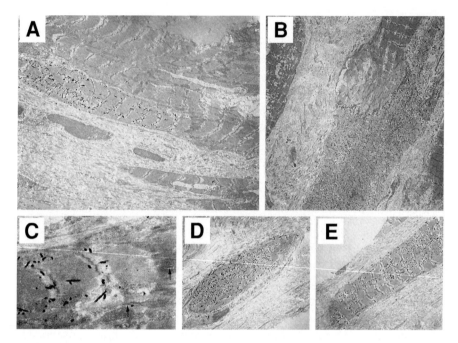

Fig. 4. Transmission electron microscopy of WB-F344-derived cardiac myocytes. β-Galactosidase-positive WB-F344-derived cardiomyocytes are identified by the presence of electron-dense crystalloid X-gal reaction product precipitate in the cytoplasm of well-differentiated (**A, B,** and **E**) and differentiating (**D**) cardiac myocytes. The well-differentiated myocytes contain striations (**A** and **C**) and are coupled to adjacent cells through intercalated disks and gap junctions (**C**).

sarcomeres, with intercalated disks and apparent gap junctional connections to adjacent host myocytes (Fig. 4), consistent with a more differentiated (mature) cell phenotype. The presence of anatomical couplings between stem cell-derived myocytes and host myocytes suggests that the WB-F344-derived myocytes participate in the function of the cardiac syncytium *(218)*. Smaller, less-differentiated cells, some as little as 20 μm in length, demonstrated nascent sarcomeres, suggesting that they were immature cardiac lineage committed, WB-F344-derived myocytes (Fig. 4). These developing cells were isolated in the cardiac connective tissue and did not show any apparent contact with native well-differentiated myocytes.

Taking advantage of the male origin of the WB-F344 cells, β-galactosidase-positive myocytes, engrafted into the hearts of female host mice, were shown to contain a Y chromosome by *in situ* hybridization of tissue sections *(218)*. Individual β-galactosidase-positive myocytes were microdissected from heart using laser capture microdissection and were used to prepare DNA. PCR analysis of these DNA samples demonstrated the presence of rat Y chromosome-specific repetitive DNA sequences *(218)*. Furthermore, the β-galactosidase-positive myocytes expressed cardiac-specific troponin T using a monoclonal antibody that recognizes an epitope on this protein specific for the cardiac isoform and that is conserved across species *(219)*.

It has been suggested that the capacity of many types of stem cells to differentiate into various lineages may be explained by nuclear fusion between these cells and native cells in the tissue microenvironment *(220,221)*. To test this possibility, a mouse L1 repetitive DNA element was identified *(222)*, and a fluorescence *in situ* hybridization (FISH) analysis was performed using this DNA sequence as a probe. Sections of the heart of a recipient mouse that contained a donor WB-F344-derived myocyte expressing β-galactosidase were analyzed. Unlike the nuclei of host cells, the nuclei from the donor WB-F344-derived myocytes did not demonstrate any fluorescence by confocal microscopy using this probe. Together with the presence of immature nascent myocytes isolated in the cardiac connective tissue, this observation indicates that WB-F344 cells may not have fused with adult well-differentiated recipient myocytes, but rather differentiated directly into cardiomyocytes in response to signals from the cardiac microenvironment. Further studies will be needed to exclude the possibility that some transplanted cells fuse with host cells.

In addition to the observation of β-galactosidase-positive cardiomyocytes in recipient mice, transplanted WB-F344 cells gave rise to cells of other cell lineages that participate in the cardiac structure. In one mouse, a cartilaginous mass was found in the left ventricle. This mass was lined with WB-F344-derived endocardial cells that expressed von Willebrand factor and β-galactosidase and contained a rat Y chromosome (Fig. 5). The cells in the cartilaginous mass also contained a Y chromosome (Fig. 5), indicating that they also originated from donor WB-F344 cells. We suspect that, at the time of transplantation, the WB-F344 cells had aggregated into a suspended bolus too large to pass through the aortic valve. Extrapolating from the morphologic gradient concept in development, it is tempting to speculate that local diffusible signals activated different genes of WB-F344 cells located at different points in a concentration gradient *(223,224)*, with cells on the inside and on the outside of the aggregate expressing different phenotypes. Alternatively, it is possible that suspended WB-F344 cells behaved like ES cells, forming an "embryoid body" that differentiated along a default mesenchymal cell lineage. On the other hand, the needle tip used to inject the WB-F344 cells into the left ventricle through the rib cage might have picked up some cartilage cells from a costochondral junction, which formed the nidus of a cartilage microenvironment that dominated the phenotypic differentiation of adjacent and surrounding WB-F344 cells, driving them to differentiate into cartilage cells.

In addition to the cell fates described above, when WB-F344 cells were injected in the pericardial sac, β-galactosidase-positive progeny cells displaying a flat mesothelial-like phenotype lined the epicardium (data not shown). This result suggests that, when these cells are introduced into the tissue niche represented by the pericardial sac, they are directed to engraft into the mesothelial cell lining and acquire a cell phenotype characteristic for cells of that niche of the heart.

Together, the results of intracardiac transplantation of clonal WB-F344 liver stem cells show that they engraft and differentiate in the heart in a niche-specific manner *(218)*. WB-F344 cells acquire a myocytic phenotype in the myocardium, an endocardial phenotype in the endocardium, and a visceral pericardial phenotype in the pericardial space.

5.3.2. WB-F344 Cells Differentiate Into Hematopoietic Cell Lineages Following Transplantation Into Bone Marrow

To determine the capacity for WB-F344 cells to engraft in the bone marrow and differentiate into hematopoietic cells, WB-F344 cells (carrying genes for *E. coli* β-galactosidase and green fluorescent protein) were transplanted via tail vein injection into sublethally irradiated female nude mice. Bone marrow and spleen were harvested from recipient mice 7 to 9 wk later for analysis by clonogenic assays in the presence of the antibiotic G418. These assays demonstrated that WB-F344-derived hematopoietic cells collected from the bone marrow and spleen of irradiated mice receiving cell transplants produced colonies that contained cells with typical characteristics of neutrophils, monocytes, megakaryocytes, macrophages, erythroid, and pre-B cells *(225)*. Verification that these hematopoietic cells were derived from transplanted WB-F344 cells was accomplished by PCR amplification of the rat Y chromosome repetitive sequences. These results suggest that transplanted WB-F344 cells can home into the bone marrow of host animals and differentiate into hematopoietic progenitor cells in response to instructive signals in the microenvironment of the bone marrow.

6. MULTIPOTENTIAL DIFFERENTIATION OF ADULT LIVER STEM CELLS IN CULTURE

Various approaches have been used to examine the differentiation potential of cultured rat liver stem cells. These include modification of culture media to contain differentiation-promoting agents and the use of combina-

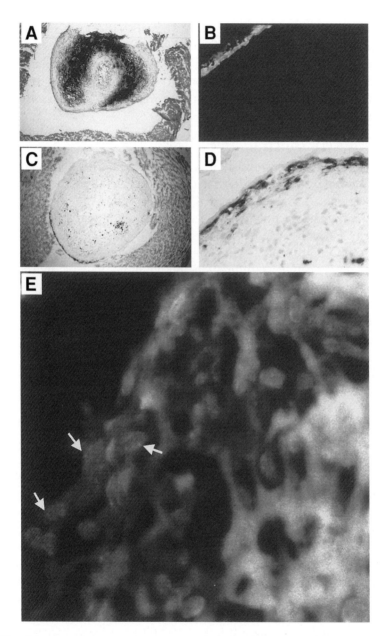

Fig. 5. Endocardial differentiation of WB-F344 cells following transplantation into the heart. A cartilaginous structure in the ventricular chamber of the heart (**A, C**) is lined by flat cells that express β-galactosidase (**C, D**) and von Willebrand factor (**B**). Cells on the surface of the structure, as well as cells in the mass, contain a Y chromosome (*arrows*, **E**).

tions of extracellular matrix materials as culture substrates. These studies have yielded evidence that cultured stemlike epithelial cells can be induced to express characteristics of differentiated liver cell types in culture. In the following sections, we review the results of our studies with WB-F344 rat liver epithelial cells as well as some studies from the literature on the differentiation of oval cells in culture.

6.1. Hepatocytic Differentiation of Oval Cells In Vitro

Oval cell cultures established from rats treated with 3'-methyl-4-dimethylaminoazobenzene exhibit typical epithelial morphology in culture, express various cytokeratins, vimentin, γ-glutamyltranspeptidase, and BDS7 antigen *(226)*. The phenotype of these cells can be modified by culturing them on fibronectin-coated dishes in medium containing various differentiation-promoting agents. Inclusion of sodium butyrate in the growth medium inhibits cellular proliferation and produces dramatic morphological alterations in the cultured oval cells *(226)*. In the presence of sodium butyrate alone or in combination with dexamethasone, cultured oval cells synthesize albumin and express tyrosine aminotransferase activity *(226)*.

Pack and coworkers have also examined the effects of differentiation-promoting chemicals on the phenotypic characteristics of cultured oval cell lines *(191)*. Oval cell lines (OC/CDE) were established from cells isolated from rats maintained on a choline-deficient diet supplemented with ethionine *(191)*. Exposure of OC/CDE cell lines to either sodium butyrate or dimethylsulfoxide resulted in cessation of cell proliferation, increased cell size, expression of albumin (in 35–40% of cells), and enhanced glucose-6-phosphatase, γ-glutamyltranspeptidase, and alkaline phosphatase activities *(191)*. Tyrosine aminotransferase activity was not detected in OC/CDE cell cultures treated with either sodium butyrate or dimethylsulfoxide *(191)*. These studies combined demonstrate that cultured oval cells can be induced to express some characteristics of differentiated hepatocytes.

Lazaro and colleagues *(227)* developed a three-dimensional cell culture system that supports the hepatocytic differentiation of oval cell lines. In this system, LE/2 and LE/6 oval cells were cultured in a collagen I gel matrix, supported by a fibroblast feeder layer. After several weeks in culture, these oval cells acquired a phenotype characterized by typical hepatocyte morphology and ultrastructure, expression of a hepatocytic cytokeratin pattern (CK8+, CK18+, CK19−), and production of albumin *(227)*. In the absence of a fibroblast feeder layer, oval cells cultured in this model system with defined growth factors HGF or keratinocyte growth factor (KGF) produced ductal structures, suggesting differentiation toward the biliary epithelial cell lineage *(227)*. This study showed that cultured oval cells are bipotent, and that their differentiation fate is influenced by soluble factors (growth factors and others) produced by stromal cells.

6.2. Hepatocytic Differentiation by RLE-13 Rat Liver Epithelial Cells in Culture

A number of different simple epithelial cell lines have been established in culture from normal rat livers (reviewed in ref. *15*), including RLE-13, which was established from a normal adult Fischer 344 rat *(228)*. Schroeder and colleagues induced hepatocytic differentiation of RLE-13 cells by treatment with 5-aza deoxycytidine, followed by culture in defined growth medium containing FGF1/2, oncostatin M, HGF, and dexamethasone *(229)*. Culture of the RLE-13 cells under these conditions resulted in significant enlargement of cell size, increased organelle complexity, and decreased proliferation. Concurrent with the morphological alterations, differentiating RLE-13 cells expressed hepatocyte-specific markers, including tyrosine aminotransferase, and liver-enriched transcription factors, including HNF4 *(229)*. These results combine to suggest that RLE-13 cells acquire a hepatocytic phenotype when maintained in culture under defined conditions.

6.3. Hepatocytic Differentiation of WB-F344 Rat Liver Stem Cells in Culture

Exposure to sodium butyrate in culture inhibits proliferation, alters normal cellular morphology (increased cell size and decreased nuclear/cytoplasmic ratio), and dramatically increases cellular protein synthesis in WB-F344 rat liver epithelial cells *(230)*. Ultrastructurally, WB-F344 cells treated with sodium butyrate demonstrate complex cytoplasm with extensive rough endoplasmic reticulum, numerous mitochondria, and large numbers of primary and secondary lysosomes. These cells also express dexamethasone-inducible tyrosine aminotransferase enzyme activity *(230)*, which is a marker of hepatocyte differentiation *(231,232)*. The dexamethasone-inducible tyrosine aminotransferase activity developing in these WB-F344 cells treated with sodium butyrate responds to the modulating effects of insulin and L-tyrosine in a manner that closely resembles that of cultured hepatocytes and hepatoma cell lines *(233,234)*. These studies demonstrated that WB-F344 cells can be induced to express traits of differentiated hepatocytes in vitro.

Using a similar cell culture system, Couchie and colleagues evaluated the ability of WB-F344 cells to differentiate into the biliary epithelial cell lineage *(235)*. WB-F344 cells cultured on dishes coated with laminin-rich

Matrigel in the presence of growth medium containing sodium butyrate demonstrated reduced proliferation and formed cordlike structures between islets of cells *(235)*. WB-F344 cells in the cordlike structures were elongated, and expressed several biliary markers, including BDS7, CK19, and γ-glutamyl transpeptidase *(235)*. These results suggest that WB-F344 cells adopt a biliary epithelial cell phenotype in response to specific factors/signals in cell culture.

6.4. WB-F344 Cells Differentiate Into Cardiac Myocytes in Cell Culture

To examine the capacity for WB-F344 cells to differentiate into myocytes in culture, a cardiac tissue microenvironment was established ex vivo using primary cultures of neonatal rat cardiac myocytes. After 3–4 d, these cultures were seeded with WB-F344 cells (carrying genes for *E. coli* β-galactosidase and green fluorescent protein). The β-galactosidase-positive or fluorescent WB-F344-derived myocytes were identified in these cocultures 6–9 d later *(236)*. WB-F344-derived myocytes were binucleated and demonstrated nascent myofibrils, sarcomeres, and sarcoplasmic reticulum by electron microscopy. The cardiac phenotype of these cells was verified by immunodetection of cardiac-specific proteins. These WB-F344-derived myocytes demonstrated calcium transients and were electrically coupled with the established neonatal-derived myocytes, as evidenced by fluorescence recovery after photobleaching (FRAP) experiments *(236)*. Evidence of fusion between WB-F344 cells and cultured cardiac myocytes was not found. These results suggest that WB-F344 cells can be induced to differentiate into cardiac myocytes in culture, and that a myocyte lineage can be modeled ex vivo, allowing for systematic investigation of the mechanistic basis for this differentiation event.

7. STEM CELL POTENTIAL AND PLASTICITY

The molecular mechanisms that regulate the formation of differentiated progeny from stem cells have not been well characterized, but are thought to be driven by signals from tissue microenvironments *(206,237)*. Of intense current interest to stem cell biologists is the question of stem cell potential and plasticity. Several literature reports suggested that bone marrow-derived stem cells can give rise both to hematopoietic cell types and to a number of other cell types *(121–127)*, including hepatocytes *(129–133)*. It is not clear from the studies reported to date whether this reflects the ability of one bone marrow-derived stem cell to give rise to multiple tissue types, termed *stem cell plasticity*, or whether the bone marrow contains multiple populations of cells with more restricted differentiation potential.

In contrast to bone marrow-derived stem cells, the WB-F344 rat liver stem cell is a clonal cell line *(20)* with demonstrated multipotentiality and plasticity *(178,206–208,218)*. The multipotentiality of this cell line is manifest in its ability to produce both hepatocytes and biliary epithelial cells under the appropriate conditions in vivo or in culture *(178,206–208,230)*. The plasticity of the WB-F344 cell line outside the liver is demonstrated by their ability to give rise to cardiac myocytes, endocardial cells, and pericardial cells in vivo *(218)* and cardiac myocytes in culture *(236)* and to cells of the hematopoietic lineage in vivo *(225)*. Given the clonal nature of this cell line and its derivation from epithelial cells of the adult rat liver, we have suggested that this line represents an adult (cloned) stem cell with multipotential differentiation capacity.

It has been suggested that cells maintained in culture for long periods of time undergo modifications (DNA methylation, mutations) that may affect their differentiation potential, which may be explained by such culture-acquired changes *(19)*. However, the WB-F344 cell line has remained karyotypically and phenotypically stable throughout numerous passages. Furthermore, if cell culture were an impermissible technique to apply to stem cells, then embryonic stem cells would not have been developed or applied to the generation of progeny expressing particular transgenes or lacking particular deleted genes. If stem cells, including embryonic stem cells, are to be used to replace injured tissues, for tissue engineering, or for gene therapy, they must be propagated and manipulated in culture. It is unfortunate that hematopoietic stem cells cannot be propagated in culture. Transplantation of primary cells, such as hematopoietic stem cells, has so far been demonstrated to replace depleted hematopoietic cells efficiently, but engraftment into other tissues, including liver, has not yet been shown to be robust *(136)*. It has been suggested that the differentiation potential or plasticity of some stem cells may be related to fusion of the genetically marked undifferentiated progenitor cells with differentiated cells of interest *(220,221)*. However, recent studies suggesting multipotentiality of adult tissue stem cells have not addressed this issue directly.

In our own studies with WB-F344 cells, we have employed labeling dyes (PHK26), genetic marking (β-galactosidase and neomycin resistance), transplantation into mutant rat strains (DPPIV deficient and Nagase analbuminemic), and male–female transplant mismatch (Y chromosome *in situ* hybridization) to enable identification of progeny of transplanted cells. All of these methods facilitated lineage analysis of transplanted cells, but did not eliminate the possibility that fusion could account for the observed results. However, in our studies of WB-F344 cell transplantation into the cardiac microenvironment, we were able to exploit genetic differences

between the transplanted rat cells and the mouse hosts to show in preliminary studies that WB-F344 cell-derived cardiac myocytes were not genetic
rat–mouse chimeras and therefore were not formed by cell fusion. Additional studies are under way that will address this question in other WB-
F344 cell differentiation models in vivo, ex vivo, and in culture.

8. SHIFTING PARADIGMS IN LIVER STEM CELL BIOLOGY

Our review of the literature highlights numerous developments in liver
stem cell biology that result in several significant shifts in the stem cell paradigm, including the general acceptance of the concept that liver stem cells
exist. This shifting paradigm is the result of an influx of new information
from experimental studies, as well as a new interpretation of historical data.
In addition, liver biologists have accepted the revision of some basic concepts in liver stem cell biology, including the working definitions of stem
and progenitor cell.

Among the conclusions that can be drawn from this review include, (1)
the liver is not a typical stem cell-fed lineage renewal system, (2) multiple
liver cell types can serve as progenitor cells in response to liver injury or
deficit, (3) some stem/progenitor cells from extrahepatic tissues can serve
as progenitor cells for liver under certain pathophysiological conditions, and
(4) some liver stemlike progenitor cells express a broader (multipotential)
differentiation capacity and are able to give rise to cells of other tissue lineages. It is now recognized that the differentiated hepatocytes and biliary
epithelial cells of the adult liver participate in lineage renewal during liver
regeneration, and that these cells represent unipotential liver progenitor cells
(15). In addition, the liver contains a population of undifferentiated, multipotential stem cells that can participate in lineage renewal under some circumstances through the oval cell response *(15,21)*. Furthermore, cells from
extrahepatic tissues have been shown to produce hepatocytes in some model
systems; perhaps the most intriguing extrahepatic source of these cells is the
bone marrow *(18)*.

Several investigators have suggested that the bone marrow-derived liver
stem cell is related to a circulating adult tissue stem cell that possesses broad
(perhaps pluripotent) differentiation potential. In fact, Petersen and colleagues suggested that oval cells originate from bone marrow-derived stem
cells of this type *(128)*. However, from published studies, it is not yet clear
that bone marrow-derived progenitor cells give rise to all oval cells in rat
models of liver injury. The liver stem cell compartment may coexist with
hematopoietic stem cells in the adult organism that are retained from the
period of hepatic embryogenesis when the liver was the primary hema-

topoietic organ. Likewise, the mechanisms through which the bone marrow-derived stem cells may be recruited or the circulating hematopoietic stem cells may home to the injured liver are not known.

Although the location of the stem cell niche in the normal adult liver has not been unequivocally identified and cells possessing stemlike properties have not been identified microscopically in the normal liver, a considerable amount of evidence suggests that the stem cells responsible for oval cell proliferation reside in the periportal regions of the liver lobule, probably in or around the terminal bile ductules in portal tracts (reviewed in ref. *15*).

Despite recent advances in stem cell biology, numerous questions remain to be answered before complete understanding of the complexities of specific stem cell lineage systems is achieved. When these questions are answered, additional modifications of the stem cell paradigm may be required. Several basic questions remain that limit our ability to define the role of liver stem cells in liver injury, repair, and regeneration. How many cell types (hepatic and extrahepatic) represent liver stem cells? What are the ultimate sources of these liver stem cells? Where is the liver stem cell niche, and how does it protect undifferentiated stem cells? What factors or signals determine the extent of involvement of liver stem cells in tissue repair and regeneration? How are liver stem cells related to other stem cells? What is the extent of phenotypic (differentiative) plasticity of liver stem cells? What factors or signals determine liver stem cell lineage commitment? Successful investigation of these and other questions in appropriate model systems will significantly improve the understanding of liver stem cells and their potential use in therapeutic applications *(238)*.

REFERENCES

1. Sell, S. (1990). Is there a liver stem cell? Cancer Res 50, 3811–3815.
2. Thorgeirsson, S.S. (1993). Hepatic stem cells. Am J Pathol 142, 1331–1333.
3. Orkin, S. H. (2001). Hematopoietic stem cells: molecular diversification and developmental interrelationships. In: Marshak, D. R., Gardner, R. L., and Gottlieb, D., eds., Stem Cell Biology. Cold Spring Harbor, NY: Cold Spring Harbor Press, pp. 289–306.
4. Watt, F. M. (2001). Epidermal stem cells. In: Marshak, D. R., Gardner, R. L., and Gottlieb, D., eds., Stem Cell Biology. Cold Spring Harbor, NY: Cold Spring Harbor Press, pp. 439–453.
5. Winton, D. J. (2001). Stem cells in the epithelium of the small intestine and colon. In: Marshak, D. R., Gardner, R. L., and Gottlieb, D., eds., Stem Cell Biology. Cold Spring Harbor, NY: Cold Spring Harbor Press, pp. 515–536.
6. Grompe, M., Laconi, E., and Shafritz, D. A. (1999). Principles of therapeutic liver repopulation. Semin Liver Dis 19, 7–14.
7 Shafritz, D. A. (2000). Rat liver stem cells: Prospects for the future. Hepatology 31, 1399–1400.

8. Faris, R. A., Konkin, T., and Halpert, G. (2001). Liver stem cells: a potential source of hepatocytes for the treatment of human liver disease. Artif Organs 25, 509–512.

9. Feldmann, G. (2001). Liver transplantation of hepatic stem cells: potential use for treating liver diseases. Cell Biol Toxicol 17, 77–85.

10. Grompe, M. (2001). Liver repopulation for the treatment of metabolic diseases. J Inherit Metabol Dis 24, 231–244.

11. Sell, S. (2001). The role of progenitor cells in repair of liver injury and in liver transplantation. Wound Rep Reg 9, 467–482.

12. Robey, P.G. (2000). Stem cells near the century mark. J Clin Invest 105, 1489–1491.

13. Marshak, D. R., Gottlieb, D., and Gardner, R. L. (2001). Introduction: Stem cell biology. In: Marshak, D. R., Gardner, R. L., and Gottlieb, D., eds., Stem Cell Biology. Cold Spring Harbor, NY: Cold Spring Harbor Press, pp. 1–16.

14. Grisham, J. W., and Coleman, W. B. (1996). Neoformation of liver epithelial cells: progenitor cells, stem cells, and phenotypic transitions. Gastroenterology 110, 1311–1313.

15. Coleman, W. B., and Grisham, J. W. (1998). Epithelial stem-like cells of the rodent liver. In: Strain, A. J., and Diehl, A. M., eds., Liver Growth and Repair: From Basic Science to Clinical Practice. London: Chapman and Hall, pp. 50–99.

16. Wulf, G. G., Jackson, K. A., and Goodell, M. A. (2001). Somatic stem cell plasticity: current evidence and emerging concepts. Exp Hematol 29, 1361–1370.

17. D'Amour, K. A., and Gage, F. H. (2002). Are somatic stem cells pluripotent or lineage-restricted? Nature Med 8, 213–214.

18. Strain, A. J. (1999). Changing blood into liver: adding further intrigue to the hepatic stem cell story. Hepatology 30, 1105–1107.

19. Anderson, D. J., Gage, F. H., and Weissman, I. L. (2001). Can stem cells cross lineage boundaries? Nature Med 7, 393–395.

20. Tsao, M.-S., Smith, J. D., Nelson, K. G., and Grisham, J. W. (1984). A diploid epithelial cell line from normal adult rat liver with phenotypic properties of "oval" cells. Exp Cell Res 154, 38–52.

21. Grisham, J. W., and Thorgeirsson, S. S. (1997). Liver stem cells. In: Potten, C. S., ed., Stem Cells. London: Academic, pp. 233–282.

22. MacCallum, W. G. (1902). Regenerative changes in the liver after acute yellow atrophy. Johns Hopkins Hosp Rep 10, 375–379.

23. MacCallum, W. G. (1904). Regenerative changes in cirrhosis of the liver. JAMA 43, 649–654.

24. Muir, R. (1908). On proliferation of the cells of the liver. J Pathol Bacteriol 12, 287–305.

25. Milne, L. (1909). The histology of liver tissue regeneration. J Pathol Bacteriol 13, 127–160.

26. Price, J. M., Harman, J. W., Miller, E. C., and Miller J. M. (1952). Progressive microscopic alterations in the livers of rats fed the hepatic carcinogens 3'-

methyl-4-dimethylaminoazobenzene and 4'-fluoro-4-dimethylaminoazo-benzene. Cancer Res 12, 192–200.

27. Firminger, H. I. (1955). Histopathology of carcinogenesis and tumors of the liver in rats. J Natl Cancer Inst 15, 1427–1441.
28. Wilson, J. W., and Leduc, E. H. (1958). Role of cholangioles in restoration of the liver of the mouse after dietary injury. J Pathol Bacteriol 76, 441–449.
29. Sell, S. (1994). Liver stem cells. Mod Pathol 7, 105–112.
30. Fausto, N. (1990). Hepatocyte differentiation and liver progenitor cells. Curr Opin Cell Biol 2, 1036–1042.
31. Aterman, K. (1992). The stem cells of the liver-A selective review. J Cancer Res Clin Oncol 118, 87–115.
32. Fausto, N. (1994). Liver stem cells. In: Arias, I. M., Boyer, J. L., Fausto, N., Jakoby, W. B., Schachter, D. A., and Shafritz, D. A., eds., The Liver: Biology and Pathobiology. New York: Raven Press, pp. 1501–1518.
33. Golding, M., Sarraf, C., Lalani, E.-N., and Alison, M. R. (1996). Reactive biliary epithelium: the product of a pluripotential stem cell compartment? Human Pathol 27, 872–884.
34. Grisham, J. W. (1994). Migration of hepatocytes along hepatic plates and stem cell-fed hepatocyte lineages. Am J Pathol 144, 849–854.
35. Potten, C. S., and Morris, R. J. (1988). Epithelial stem cells in vivo. J Cell Sci Suppl 10, 45–62.
36. Potten, C. S., and Loeffler, M. (1990). Stem cells: attributes, cycles, spirals, pitfalls and uncertainties. Lessons for and from the crypt. Development 110, 1001–1020.
37. Sun, T. T., Eichner, R., Nelson, W. G., Vidrich, A., and Woodcock-Mitchell, J. (1983). Keratin expression during normal epidermal differentiation. Curr Probl Dermatol 11, 277–291.
38. Cotsarelis, G., Cheng, S., Dong, G., Sun, T., and Lavker, RM. (1989). Existence of slow-cycling limbal epithelial basal cells that can be preferentially stimulated to proliferate: implications on epithelial stem cells. Cell 57, 201–209.
39. Hall, P. A., and Watt, F. M. (1989). Stem cells: the generation and maintenance of cellular diversity. Development 106, 619–633.
40. Marks, P. A., Sheffery, M., and Rifkind, R.A. (1985). Modulation of gene expression during terminal cell differentiation. Prog Clin Biol Res 191, 185–203.
41. Heimfeld, S., and Weissman, I. (1991). Development of mouse hematopoietic lineages. Curr Top Dev Biol 25, 155–175.
42. Grisham, J. W. (1980). Cell types in long-term propagable cultures of rat liver. Ann N Y Acad Sci 349, 128–137.
43. Alison, M., and Sarraf, C. (1998). Hepatic stem cells. J Hepatol 29, 676–682.
44. Grompe, M., and Finegold, M. J. (2001). Liver stem cells. In: Marshak, D. R., Gardner, R. L., and Gottlieb, D., eds., Stem Cell Biology. Cold Spring Harbor, NY: Cold Spring Harbor Press, pp. 455–497.
45. Vessey, C. J., and Hall, P. M. (2001). Hepatic stem cells: a review. Pathology 33, 130–141.

46. Slack, J. M. W. (2000). Stem cells in epithelial tissues. Science 287, 1431–1433.
47. Lois, C., and Alvarez-Buylla, A. (1993). Proliferating subventricular zone cells in the adult mammalian forebrain can differentiate into neurons and glia. Proc Natl Acad Sci U S A 90, 2074–2077.
48. Weiss, S., Reynolds, B. A., Vescovi, A. L., Morshead, C., Craig, C. G., and van der Kooy, D. (1996). Is there a neural stem cell in the mammalian forebrain? Trends Neurosci 19, 387–393.
49. Johansson, C. B., Momma, S., Clarke, D. L., Risling, M., Lendahl, U., and Frisen, J. (1999). Identification of a neural stem cell in the adult mammalian central nervous system. Cell 96, 25–34.
50. Morrison, S. J., White, P. M., Zock, C., and Anderson, D. J. (1999). Prospective identification, isolation by flow cytometry, and in vivo self-renewal of multipotent mammalian neural crest stem cells. Cell 96, 737–749.
51. Grisham, J. W. (1962). A morphologic study of deoxyribonucleic acid synthesis and cell proliferation in regenerating rat liver. Autoradiography with thymidine-H^3. Cancer Res 26, 842–849.
52. Tatematsu, M., Ho, R. H., Kaku, T., Ekem, J. K., and Farber, E. (1984). Studies on the proliferation and fate of oval cells in the liver of rats treated with 2-acetylaminofluorene and partial hepatectomy. Am J Pathol 114, 418–430.
53. Tatematsu, M., Kaku, T., Medline, A., and Farber, E. (1985). Intestinal metaplasia as a common option of oval cells in relation to cholangiofibrosis in liver of rats exposed to 2-acetylaminofluorene. Lab Invest 52, 354–362.
54. Gordon, G. J., Coleman, W. B., Hixson, D. C., and Grisham, J. W. (2000). Liver regeneration in rats with retrorsine-induced hepatocellular injury reveals the existence of a novel liver progenitor cell population. Am J Pathol 156, 607–619.
55. Michalopoulous, G. K. (1990). Liver regeneration: molecular mechanisms of growth control. FASEB J 4, 176–187.
56. Fausto, N., and Webber, E. M. (1994). Liver regeneration. In: Arias, I. M., Boyer, J. L., Fausto, N., Jakoby, W. B., Schachter, D. A., and Shafritz, D. A., eds., The Liver: Biology and Pathobiology. New York: Raven Press, pp. 1059–1084.
57. Klinman, N. R., and Erslev, A. J. (1963). Cellular response to partial hepatectomy. Proc Soc Exp Biol Med 112, 338–340.
58. Dabeva, M. D., Alpini, G., Hurston, E., and Shafritz, D. A. (1993). Models for hepatic progenitor cell activation. Proc Soc Exp Biol Med 204, 242–252.
59. Grisham, J. W. (1969). Cellular proliferation in the liver. Recent Results Cancer Res 17, 28–43.
60. Fabrikant, J. I. (1968). The spatial distribution of parenchymal cell proliferation during regeneration of the liver. Johns Hopkins Med Bull 120, 137–147.
61. Fabrikant, J. I. (1968). The kinetics of cellular proliferation in regenerating liver. J Cell Biol 36, 551–565.
62. Wright, N., and Alison, M. (1984). The Biology of Epithelial Cell Populations. Vol. 2. Oxford, UK: Clarendon Press, pp. 880–980.
63. Marucci, L., Sregliati Baroni, G., Mancini, R., et al. (1993). Cell proliferation

following extrahepatic biliary obstruction. Evaluation by immuno-histochemical methods. J Hepatol 17, 163–169.

64. Polimeno, L., Azzarone, A., Zeng, Q. H., et al. (1995). Cell proliferation and oncogene expression after bile-duct ligation in the rat: evidence of a specific growth effect on bile duct cells. Hepatology 21, 1070–1078.

65. Fabrikant, J. I. (1969). Size of proliferating pools in regenerating liver. Exp Cell Res 55, 277–279.

66. Stöcker, E., and Heine, W.-D. (1971). Regeneration of liver parenchyma under normal and pathological conditions. Beitr Pathol 144, 400–408.

67. Simpson, G. E. C., and Finckh, E. S. (1963). The pattern of regeneration of rat liver after repeated partial hepatectomies. J Pathol Bacteriol 86, 361–370.

68. Sandgren, E. P., Palmiter, R. D., Heckel, J. L., Daugherty, C. C., Brinster, R. L., and Degen, J. L. (1991). Complete hepatic regeneration after somatic deletion of an albumin-plasminogen activator transgene. Cell 66, 245–256.

69. Rhim, J. A., Sandgren, E. P., Degen, J. L., Palmiter, R. D., and Brinster, R. L. (1994). Replacement of diseased mouse liver by hepatic cell transplantation. Science 263, 1149–1152.

70. Overturf, K., Al-Dhalimy, M., Tanguay, R., et al. (1996). Hepatocytes corrected by gene therapy are selected in vivo in a murine model of hereditary tyrosinaemia type I. Nat Genet 12, 266–273.

71. Grompe, M., Al-Dhalimy, M., Finegold, M., et al. (1993). Loss of fumarylacetoacetate hydrolase is responsible for the neonatal hepatic dysfunction phenotype of lethal albino mice. Genes Dev 7, 2298–2307.

72. Overturf, K., Al-Dhalimy, M., Ou, C.-N., Finegold, M., and Grompe, M. (1997). Serial transplantation reveals the stem-cell-like regenerative potential of adult mouse hepatocytes. Am J Pathol 151, 1273–1280.

73. Laconi, E., Oren, R., Mukhopadhyay, D. K., et al. (1998). Long-term, near-total liver replacement by transplantation of isolated hepatocytes in rats treated with retrorsine. Am J Pathol 153, 319–329.

74. Dabeva, M. D., Laconi, E., Oren, R., Petkov, P. M., Hurston, E., and Shafritz, D. A. (1998). Liver regeneration and alpha-fetoprotein messenger RNA expres-sion in the retrorsine model for hepatocyte transplantation. Cancer Res 58, 5825–5834

75. Oren, R., Dabeva, M., Karnezis, A. N., et al. (1999). Role of thyroid hormone in stimulating liver repopulation in the rat by transplanted hepatocytes. Hepatology 30, 903–913

76. Jago, M. V. (1969). The development of hepatic megalocytosis of chronic pyrrolizidine alkaloids poisoning. Am J Pathol 56, 405–422.

77. Gordon, G. J., Coleman, W. B., and Grisham, J. W. (2000). Temporal analysis of hepatocyte differentiation by small hepatocyte-like progenitor cells during liver regeneration in retrorsine-exposed rats. Am J Pathol 157, 771–786.

78. Rogers, A. E. (1978). Dietary effects on chemical carcinogenesis in the livers of rats. In: Newberne, P. M., and Butler, W. H., eds., Rat Hepatic Neoplasia. Cambridge, MA: MIT Press, pp. 243–262.

79. Farber, E., and Cameron, R. (1980). The sequential analysis of cancer development. Adv Cancer Res 31, 125–226.

80. Sell, S., and Leffert, H. L. (1982). An evaluation of the cellular lineages in the pathogenesis of experimental hepatocellular carcinomas. Hepatology 2, 77–86.

81. Lesch, R., Reutter, W., Keppler, D., and Decker, K. (1970). Liver restitution after acute galactosamine hepatitis. Autoradiographic and biochemical studies in rats. Exp Mol Pathol 13, 58–69.

82. Tournier, I., Legres, L., Schoevaert, D., Feldman, G., and Bernuau, D. (1988). Cellular analysis of α-fetoprotein gene activation during carbon tetrachloride and D-galactosamine-induced acute liver injury in rats. Lab Invest 59, 657–665.

83. Dabeva, M. D., and Shafritz, D. A. (1993). Activation, proliferation, and differentiation of progenitor cells into hepatocytes in the D-galactosamine model of liver regeneration. Am J Pathol 143, 1606–1620.

84. Yavorkovsky, L., Lai, E., Ilic, Z., and Sell, S. (1995). Participation of small intraportal stem cells in the restitutive response of the liver to periportal necrosis induced by allyl alcohol. Hepatology 21, 1702–1712.

85. Smith, P. G. J., Tee, L. B. G., and Yeoh, G. C. T. (1996). Appearance of oval cells in the liver of rats after long-term exposure to ethanol. Hepatology 23, 145–154.

86. Inaoka, Y. (1967). Significance of the so-called oval cell proliferation during azo-dye hepatocarcinogenesis. Gann 58, 355–366.

87. Shinozuka, H., Lombardi, B., Sell, S., and Iammarino, R. M. (1978). Early histological and functional alterations of ethionine liver carcinogenesis in rats fed a choline-deficient diet. Cancer Res 38, 1092–1098.

88. Sell, S., Leffert, H. L., Shinozuka, H., Lombardi, B., and Goochman, N. (1981). Rapid development of large numbers of a-fetoprotein-containing "oval" cells in the liver of rats fed N-2-fluorenylacetamide in a choline-devoid diet. Gann 72, 479–487.

89. Sell, S., Osborn, K., and Leffert, H. L. (1981). Autoradiography of "oval cells" appearing rapidly in the livers of rats fed N-2-fluorenylacetamide in a choline devoid diet. Carcinogenesis 2, 7–14.

90. Farber, E. (1956). Similarities in the sequence of early histologic changes induced in the liver of the rat by ethionine, 2-acetylaminofluorene, and 3'-methyl-4-dimethylaminoazobenzene. Cancer Res 16, 142–148.

91. Uryvaeva, I. V., and Factor, V. M. (1988). Total replacement of parenchymal liver cells induced by Dipin and partial hepatectomy. Bull Exp Biol Med 105, 96–98.

92. Factor, V. M., and Radaeva, S. A. (1993). Oval cells—hepatocytes relationships in Dipin-induced hepatocarcinogenesis in mice. Exp Toxicol Pathol 45, 239–244.

93. Factor, V. M., Radaeva, S. A., and Thorgeirsson, S. S. (1994). Origin and fate of oval cells in dipin-induced hepatocarcinogenesis in the mouse. Am J Pathol 145, 409–422.

94. He, X. Y., Smith, G. J., Enno, A., and Nicholson, R. C. (1994). Short-term diethylnitrosamine-induced oval cell responses in three strains of mice. Pathology 26, 154–160.

95. Dunsford, H. A., Sell, S., and Chisari, F. V. (1990). Hepatocarcinogenesis due to chronic liver cell injury in hepatitis B virus transgenic mice. Cancer Res 50, 3400–3407.

96. Bennoun, M., Rissel, M., Engelhardt, N., Guillouzo, A., Briand, P., and Weber-Benarous, A. (1993). Oval cell proliferation in early stages of hepatocarcino-genesis in Simian virus 40 large T antigen transgenic mice. Am J Pathol 143, 1326–1336.

97. Richards, W. G., Yoder, B. K., Isfort, R. J., et al. (1996). Oval cell proliferation associated with the murine insertional mutation TgN737Rpw. Am J Pathol 149, 1919–1930.

98. Fujio, K., Evarts, R. P., Hu, Z., Marsden, E. R., and Thorgeirsson, S. S. (1994). Expression of stem cell factor and its receptor, c-kit, during liver regeneration from putative stem cells in adult rat. Lab Invest 70, 511–516.

99. Yaswen, P., Hayner, N. T., and Fausto, N. (1984). Isolation of oval cells by centrifugal elutriation and comparison with other cell types purified from normal and preneoplastic livers. Cancer Res 44, 324–331.

100. Fausto, N., Lemire, J. M., and Shiojiri, N. (1992). Oval cells in liver carcinogenesis: cell lineages in hepatic development and the identification of faculative stem cells in normal liver. In: Sirica, A. E., ed., The Role of Cell Types in Carcinogenesis. Boca Raton, FL: CRC Press, pp. 89–108.

101. Grisham, J. W., and Hartroft, W. (1961). Morphologic identification by electron microscopy of "oval" cells in experimental hepatic degeneration. Lab Invest 2, 317–332.

102. Lenzi, R., Liu, M. H., Tarsetti, F., et al. (1992). Histogenesis of bile duct-like cells proliferating during ethionine hepatocarcinogenesis. Evidence for a biliary epithelial nature of oval cells. Lab Invest 66, 390–402.

103. Sarraf, C., Lalani, E., Golding, M., Anilkumar, T. V., Poulsom, R., and Alison, M. (1994). Cell behavior in the acetylaminofluorene-treated regenerating rat liver. Light and electron microscopic observations. Am J Pathol 145, 1114–1126.

104. Dunsford, H. A., Maset, R., Salman, J., and Sell, S. (1985). Connection of ductlike structures induced by a chemical hepatocarcinogen to portal bile ducts in the rat liver detected by injection of bile ducts with a pigmented barium gelatin medium. Am J Pathol 118, 218–224.

105. Makino, Y., Yamamoto, K., and Tsuji, T. (1988). Three-dimensional arrangement of ductular structures formed by oval cells during hepatocarcinogenesis. Acta Med Okayama 42, 143–150.

106. Alpini, G., Aragona, E., Dabeva, M., Salvi, R., Shafritz, D. A., and Tavoloni, N. (1992). Distribution of albumin and alpha-fetoprotein mRNAs in normal, hyperplastic, and preneoplastic rat liver. Am J Pathol 141, 623–632.

107. Betto, H., Kaneda, K., Yamamoto T., Kojima, A., and Sakurai, M. (1996). Development of intralobular bile ductules after spontaneous hepatitis in Long-Evans mutant rats. Lab Invest 75, 43–53.

108. Evarts, R. P., Nagy, P., Marsden, E., and Thorgeirsson, S. S. (1987). A precursor-product relationship exists between oval cells and hepatocytes in rat liver. Carcinogenesis 8, 1737–1740.

109. Evarts, R. P., Nagy, P., Nakatsukasa, H., Marsden, E., and Thorgeirsson, S. S. (1989). In vivo differentiation of rat liver oval cells into hepatocytes. Cancer Res 49, 1541–1547.

110. Lemire, J. M., Shiojiri, N., and Fausto, N. (1991). Oval cell proliferation and the origin of small hepatocytes in liver injury induced by D-galactosamine. Am J Pathol 139, 535–552.

111. Kinosita, R. (1937). Studies on the cancerogenic chemical substances. Trans Soc Pathol Jpn 27, 665–727.

112. Onoé, T., Dempo, K., Kaneko, A., and Watabe, H. (1973). Significance of α-fetoprotein appearance in the early stage of azo-dye carcinogenesis. Gann Monogr Cancer Res 14, 233–247.

113. Golding, M., Sarraf, C. E., Lalani, E.-N., et al. (1995). Oval cell differentiation into hepatocytes in the acetylaminofluorene-treated regenerating rat liver. Hepatology 22, 1243–1253.

114. Alison, M. R., Golding, M., Sarraf, C. E., Edwards, R. J., and Lalani, E.-N. (1996). Liver damage in the rat induces hepatocyte stem cells from biliary epithelial cells. Gastroenterology 110, 1182–1190.

115. Spangrude, G. J., Heimfield, S., and Weissman, I. L. (1988). Purification and characterization of mouse hematopoietic stem cells. Science 241, 58–62.

116. Baum, C. M., Weissman, I. L., Tsukamoto, A. S., Buckle, A. M., and Peault, B. (1992). Isolation of a candidate human hematopoietic stem-cell population. Proc Natl Acad Sci U S A 89, 2804–2808.

117. Prockop, D. J. (1998). Marrow stromal cells as stem cells for nonhematopoietic tissues. Matrix Biol 16, 519–528.

118. Pereira, R. F., Halford, K. W., O'Hara, M. D., et al. (1997). Cultured adherant cells from marrow can serve as long-lasting precursor cells for bone, cartilage, and lung in irradiated mice. Science 276, 71–74.

119. Reyes, M., and Verfaillie, C. M. (2001). Characterization of multipotent adult progenitor cells, a subpopulation of mesenchgymal stem cells. Ann N Y Acad Sci 938, 231–235.

120. Reyes, M., Lund, T., Lenvik, T., Aguiar, D., Koodie, L., and Verfaillie, C. M. (2001). Purification and ex vivo expansion of postnatal human marrow meso-dermal progenitor cells. Blood 98, 2615–2625.

121. Orlic, D., Kajstura, J., Chimenti, S., et al. (2001). Bone marrow cells regenerate infarcted myocardium. Nature 410, 701–705.

122. Ferrari, G., Cusella-De Angelis, G., Coletta, M., et al. (1998). Muscle regeneration by bone marrow-derived myogenic progenitors. Human Immunol 59, 137–148.

123. Gussoni, E., Soneoka, Y., Strickland, C. D., et al. (1999). Dystrophin expression in the mdx mouse restored by stem cell transplantation. Nature 40, 390–394.

124. Kopen, G. C., Prockop, D. J., and Phinney, D. G. (1999). Marrow stromal cells migrate throughout forebrain and cerebellum, and they differentiate into astocytes after injection into neonatal mouse brains. Proc Natl Acad Sci U S A 96, 10,711–10,716.

125. Brazelton, T. R., Rossi, F. M., Keshet, G. I., and Blau, H. M. (2000). From bone marrow to brain: expression of neuronal phenotypes in adult mice. Science 290, 1775–1779.

126. Mezey, E., Chandross, K. J., Harta, G., Maki, R. A., McKercher, S. R. (2000). Turning blood into brain: cells bearing neuronal antigens generated in vivo from bone marrow. Science 290, 1959–1962.

127. Krause, D. S., Theise, N. D., Collector, M. I., et al. (2001). Multi-organ, multi-lineage engraftment by a single bone marrow-derived stem cell. Cell 105, 369–377.

128. Petersen, B. E., Bowen, W. C., Patrene, K. D., et al. (1999). Bone marrow as a potential source of hepatic oval cells. Science 284, 1168–1170.

129. Theise, N. D., Badve, S., Saxena, R., et al. (2000). Derivation of hepatocytes from bone marrow cells in mice after radiation-induced myeloablation. Hepatology 31, 235–240.

130. Theise, N. D., Nimmakayalu, M., Gardner, R., et al. (2000). Liver from bone marrow in humans. Hepatology 32, 11–16.

131. Alison, M. R., Poulsom, R., Jeffery, R., et al. (2000). Hepatocytes from non-hepatic adult stem cells. Nature 406, 257.

132. Lagasse, E., Connors, H., Al-Dhalimy, M., et al. (2000). Purified hematopoietic stem cells can differentiate into hepatocytes in vivo. Nature Med 6, 1229–1234.

133. Wang, X., Montini, E., Al-Dhalimy, M., Lagasse, E., Finegold, M., and Grompe, M. (2002). Kinetics of liver repopulation after bone marrow transplantation. Am J Pathol 161, 565–574.

134. Mallet, V. O., Mitchell, C., Mezey, E., et al. (2002). Bone marrow transplantation in mice leads to a minor population of hepatocytes that can be selectively amplified in vivo. Hepatology 35, 799–804.

135. Gao, Z., McAlister, V. C., and Williams, G. M. (2001). Repopulation of liver endothelium by bone marrow-derived cells. Lancet 357, 932–933.

136. Wagers, A. J., Sherwood, R. I., Christensen, J. L., and Weissman, I. L. (2002). Little evidence for developmental plasticity of adult hematopoietic stem cells. Science 297, 2256–2259.

137. Schwartz, R. E., Reyes, M., Koodie, L., et al. (2002). Multipotent adult progenitor cells from bone marrow differentiate into functional hepatocyte-like cells. J Clin Invest 109, 1291–1302.

138. Scarpelli, D. G., and Rao, M. S. (1981). Differentiation of regenerating pancreatic cells into hepatocyte-like cells. Proc Natl Acad Sci U S A 78, 2577–2581.

139. Rao, M. S., Dwivedi, R., Subbarao, V., et al. (1988). Almost total conversion of pancreas to liver in the adult rat: a reliable model to study transdifferentiation. Biochem Biophys Res Commun 156, 131–136.

140. Rao, M. S., Dwivedi, R. S., Yeldandi, A. V., et al. (1989). Role of periductal and ductular epithelial cells of the adult rat pancreas in pancreatic hepatocyte lineage. A change in the differentiation commitment. Am J Pathol 134, 1069–1086.

141. Makino, T., Usuda, N., Rao, S., Reddy, J. K., and Scarpelli, D. G. (1990). Transdifferentiation of ductular cells into hepatocytes in regenerating hamster pancreas. Lab Invest 62, 552–561.

142. Bisgaard, H. C., and Thorgeirsson, S. S. (1991). Evidence for a common cell of origin for primitive epithelial cells isolated from rat liver and pancreas. J Cell Physiol 147, 333–343.

143. Dabeva, M. D., Hwang, S.-G., Vasa, S. R. G., et al. (1997). Differentiation of pancreatic epithelial progenitor cells into hepatocytes following transplantation into rat liver. Proc Natl Acad Sci U S A 94, 7356–7361.

144. Wang, X., Al-Dhalimy, M., Lagasse, E., Finegold, M., and Grompe, M. (2001). Liver repopulation and correction of metabolic disease by transplanted adult mouse pancreatic cells. Am J Pathol 158, 571–579.

145. Chen, J.-R., Tsao, M.-S., and Duguid, W. P. (1995). Hepatocytic differentiation of cultured rat pancreatic ductal epithelial cells after in vivo implantation. Am J Pathol 147, 707–717.

146. Shepherd, J. G., Chen, J., Tsao, M.-S., and Duguid, W. P. (1993). Neoplastic transformation of propagable cultured rat pancreatic duct epithelial cells by azaserine and streptozotocin. Carcinogenesis 14, 1027–1033.

147. Gage, F. H. (2000). Mammalian neural stem cells. Science 287, 1433–1438.

148. Gage, F. H., Coates, P. W., Palmer, T. D., et al. (1995). Survival and differentiation of adult neuronal progenitor cells transplanted to the adult brain. Proc Natl Acad Sci U S A 92, 11,879–11,883.

149. Gaiano, N., and Fishell, G. (1998). Transplantation as a tool to study progenitors within the vertebrate nervous system. J Neurobiol 36, 152–161.

150. Fricker, R. A., Carpenter, M. K., Winkler, C., Greco, C., Gates, M. A., and Bjorklund, A. (1999). Site-specific migration and neuronal differentiation of human neural progenitor cells after transplantation in the adult rat brain. J Neurosci 19, 5990–6005.

151. Bjornson, C. R., Rietze, R. L., Reynolds, B. A., Magli, M. C., and Vescovi, A. L. (1999). Turning brain into blood: a hematopoietic fate adopted by adult neural stem cells in vivo. Science 283, 534–537.

152. Clarke, D. L., Johansson, C. B., Wilbertz, J., et al. (2000). Generalized potential of adult neural stem cells. Science 288, 1660–1663.

153. Sell, S. (1998). Comparison of liver progenitor cells in human atypical ductular reactions with those seen in experimental models of liver injury. Hepatology 27, 317–331.

154. Lai, Y. H., Thung, S. N, Gerber, M. A., Chen, M. L., and Schaffner, F. (1989). Expression of cytokeratins in normal and diseased livers and in primary liver carcinomas. Arch Pathol Lab Med 113, 134–138.

155. Koukoulis, G., Rayner, A., Tan, K. C., Williams, R., and Portmann, B. (1992). Immunolocalization of regenerating cells after submassive liver necrosis using PCNA staining. J Pathol 166, 359–368.

156. De Vos, R., and Desmet, V. (1992). Ultrastructural characteristics of novel epithelial cell types identified in human pathological liver specimens with chronic ductular reaction. Am J Pathol 140, 1441–1450.

157. Hsai, C. C., Evarts, R. P., Nakatsukasa, H., Marsden, E. R., and Thorgeirsson, S. S. (1992). Occurrence of oval-type cells in hepatitis B virus-associated human hepatocarcinogenesis. Hepatology 16, 1327–1333.

158. Haque, S., Haruna, Y., Saito, K., et al. (1996). Identification of bipotential progenitor cells in human liver regeneration. Lab Invest 75, 699–705.

159. Demetris, A. J., Seaberg, E. C., Wennerberg, A., Ionellie, J., and Michalopoulos, G. (1996). Ductular reactions after submassive necrosis in humans. Am J Pathol 149, 439–448.

160. Roskams, T., De Vos, R., and Desmet, V. (1996). Undifferentiated progenitor cells in focal nodular hyperplasia of the liver. Histopathology 28, 291–299.

161. Roskams, T., De Vos, R., van Eyken, P., Myazaki, H., Van Damme, B., and Desmet, V. (1998). Hepatic OV-6 expression in human liver disease and rat experiments: evidence for hepatic progenitor cells in man. J Hepatol 29, 455–463.

162. Crosby, H. A., Hubscher, S., Fabris, L., et al. (1998). Immunolocalization of putative human liver progenitor cells in livers from patients with end-stage primary biliary cirrhosis and sclerosing cholangitis using the monoclonal antibody OV-6. Am J Pathol 152, 771–779.

163. Crosby, H. A., Hubscher, S. G., Joplin, R. E., Kelly, D. A., and Strain, A. J. (1998). Immunolocalization of OV-6, a putative progenitor cell marker in human fetal and diseased pediatric liver. Hepatology 28, 980–985.

164. Ruck, P., Xiao, J.-C., Pietsch, T., von Schweinitz, D., and Kaiserling, E. (1997). Hepatic stem-like cells in hepatoblastoma: expression of cytokeratin 7, albumin and oval cell associated antigens detected by OV-1 and OV-6. Histopathology 31, 324–329.

165. Baumann, U., Crosby, H. A., Ramani, P., Kelly, D. A., and Strain, A. J. (1999). Expression of the stem cell factor c-kit in normal and diseased pediatric liver: identification of a human hepatic progenitor cell? Hepatology 30, 112–117.

166. Xiao, J.-C., Ruck, P., and Kaiserling, E. (1999). Small epithelial cells in extrahepatic biliary atresia: electron microscopic and immunoelectron microscopic findings suggest a close relationship to liver progenitor cells. Histopathology 35, 454–460.

167. Libbrecht, L., Desmet, V., van Damme, B., and Roskams, T. (2000). Deep intralobular extension of human hepatic "progenitor cells" correlates with parenchymal inflammation in chronic viral hepatitis: can "progenitor cells" migrate? J Pathol 192, 373–378.

168. Libbrecht, L., Desmet, V., van Damme, B., and Roskams, T. (2000). The immu-nohistochemical phenotype of dysplastic foci in human liver: correlation with putative progenitor cells. J Hepatol 33, 76–84.

169. Van den Heuvel, M. C., Slooff, M. J. H., Visser, L., et al. (2001). Expression of anti-OV6 antibody and anti-N-CAM antibody along the biliary line of normal and diseased human livers. Hepatology 33, 1387–1393.

170. Crosby, H. A., Kelly, D. A., and Strain, A. J. (2001). Human hepatic stem-like cells isolated using c-kit or CD34 can differentiate into biliary epithelium. Gastroenterology 120, 534–544.

171. Korbling, M., Katz, R. L., Khanna, A., et al. (2002). Hepatocytes and epithelial cells of donor origin in recipients of peripheral-blood stem cells. N Engl J Med 346, 738–746.

172. Fogt, F., Beyser, K. H., Poremba, C., Zimmerman, R. L., Khettry, U., and Ruschoff, J. (2002). Recipient-derived hepatocytes in liver transplants: a rare event in sex-mismatched transplants. Hepatology 36, 173–176.

173. Grisham, J. W. (1983). Cell types in rat liver cultures: their identification and isolation. Mol Cell Biochem 53/54, 23–33.

174. Alpini, G., Phillips, J. O., Vroman, B., and LaRusso, N. F. (1994). Recent advances in the isolation of liver cells. Hepatology 20, 494–514.

175. Joplin, R. (1994). Isolation and culture of biliary epithelial cells. Gut 35, 875–878.

176. Strain, A. J. (1994). Isolated hepatocytes: use in experimental and clinical hepatology. Gut 35, 433–436.

177. Grisham, J. W., Thal, S. B., and Nagel, A. (1975). Cellular derivation of continuously cultured epithelial cells from normal rat liver. In: Gershenson, L. E., and Thompson, E. B., eds. Gene Expression and Carcinogenesis in Cultured Liver. New York: Academic Press, pp. 1–23.

178. Grisham, J. W., Coleman, W. B., and Smith, G. J. (1993). Isolation, culture, and transplantation of rat hepatocytic precursor (stem-like) cells. Proc Soc Exp Biol Med 204, 270–279.

179. Alexander, R. W., and Grisham, J. W. (1970). Explant culture of rat liver. I. Method, morphology and cytogenesis. Lab Invest 22, 50–62.

180. Berry, M. N., and Friend, D. S. (1966). High-yield preparation of isolated rat liver parenchymal cells: a biochemical and fine structural study. J Cell Biol 43, 506–520.

181. Seglen, P. O. (1976). Preparation of isolated rat liver cells. Methods Cell Biol 13, 29–83.

182. Williams, G. M. (1976). Primary and long-term culture of adult rat liver epithelial cells. Methods Cell Biol 14, 357–364.

183. Mills, D. M., and Zucker-Franklin, D. (1969). Electron microscopic study of isolated Kupffer cells. Am J Pathol 54, 147–166.

184. Munthe-Kaas, A. C., Berg, T., Seglen, P. O., and Seljelid, R. (1975). Mass isolation and culture of rat Kupffer cells. J Exp Med 141, 1–10.

185. Knook, D. L., Blansjaar, N., and Slyster, E. C. (1977). Isolation and characterization of Kupffer and endothelial cells from the rat liver. Exp Cell Res 109, 317–329.

186. Emeis, J. J., and Planque, B. (1976). Heterogeneity of cells isolated from rat liver by pronase digestion: ultrastructure, cytochemistry, and cell culture. J Reticuloendo Soc 20, 11–29.

187. Furukawa, K., Shimada, T., England, P., Mochizuki, Y., and Williams, G. M. (1987). Enrichment and characterization of clonogenic epithelial cells from adult rat liver and initiation of epithelial cell strains. In Vitro Cell Dev Biol 23, 339–348.

188. Fausto, N., Thompson, H. L., and Braun, L. (1987). Purification and culture of oval cells from rat liver. In: Pretlow, T. G., II, and Pretlow, T. R., eds., Cell

Separation Methods and Selected Applications. Vol. 4. Orlando, FL: Academic Press, pp. 45–77.

189. Radaeva, S., and Steinberg, P. (1995). Phenotype and differentiation patterns of the oval cell lines OC/CDE 6 and OC/CDE 22 derived from the livers of carcinogen-treated rats. Cancer Res 55, 1028–1038.

190. Plenat, F., Braun, L., and Fausto, N. (1988). Demonstration of glucose-6-phosphatase and paroxisomal catalase activity by ultrastructural cytochemistry in oval cells from livers of carcinogen-treated rats. Am J Pathol 130, 91–102.

191. Pack, R., Heck, R., Dienes, H. P., Oesch, F., and Steinberg, P. (1993). Isolation, biochemical characterization, long-term culture, and phenotype modulation of oval cells from carcinogen-fed rats. Exp Cell Res 204, 198–209.

192. Hayner, N. T., Braun, L., Yaswen, P., Brooks, M., and Fausto, N. (1984). Isozyme profiles of oval cells, parenchymal cells, and biliary cells isolated by centrifugal elutriation from normal and preneoplastic livers. Cancer Res 44, 332–338.

193. Stutenkamper, R., Geisse, S., Schwarz, H. J., et al. (1992). The hepatocyte-specific phenotype of murine liver cells correlates with high expression of connexin 32 and connexin 26 but very low expression of connexin 43. Exp Cell Res 201, 43–54.

194. Spray, D. C., Chanson, M., Moreno, A. P., Dermietzel, R., and Meda, P. (1991). Distinctive gap junction channel types connect WB cells, a clonal cell line derived from rat liver. Am J Physiol 260, C513–C527.

195. Ren, D., deFeijter, A. W., Paul, D. L., and Ruch, R. J. (1994). Enhancement of liver cell gap junction protein expression by glucocorticoids. Carcinogenesis 15, 1801–1813.

196. Neveu, M. J., Sattler, C. A., Sattler, G. L., et al. (1994). Differences in the expression of connexin genes in rat hepatomas in vivo and in vitro. Mol Carcinogenesis 11, 145–154.

197. El-Fouly, M. H., Trosko, J. E., and Chang, C. C. (1987). Scrape-loading and dye transfer: a rapid and simple technique to study gap junctional intercellular communication. Exp Cell Res 168, 422–430.

198. Marceau, N., Blouin, M.-J., Noël, M., Török, N., and Loranger, A. (1992). The role of bipotential progenitor cells in liver ontogenesis and neoplasia. In: Sirica, A. E., ed., The Role of Cell Types in Hepatocarcinogenesis. Boca Raton, FL: CRC Press, pp. 121–149.

199. Nussler, A. K., Vergani, G., Gollin, S. M., et al. (1999). Isolation and characterization of a human hepatic epithelial-like cell line (AKN-1) from a normal liver. In Vitro Cell Dev Biol 35, 190–197.

200. Montesano, R., Saint Vincent, L., and Tomatis, L. (1973). Malignant transformation in vitro of rat liver cells by dimethylnitrosamine and *N*-methyl-*N*'-nitro-*N*-nitrosoguanidine. Br J Cancer 28, 215–220.

201. Williams, G. M., Elliott, J. M., and Weisburger, J. H. (1973). Carcinoma after malignant conversion in vitro of epithelial-like cells from rat liver following exposure to chemical carcinogens. Cancer Res 33, 606–612.

202. Tsao, M.-S., and Grisham, J. W. (1987). Hepatocarcinomas, cholangio-

carcinomas, and hepatoblastomas produced by chemically transformed cultured rat liver epithelial cells—A light- and electron-microscopic analysis. Am J Pathol 127, 168–181.

203. Lee, L. W., Tsao, M.-S., Grisham, J. W., and Smith, G. J. (1989). Emergence of neoplastic transformants spontaneously or after exposure to N-methyl-N'-nitro-N-nitrosoguanidine in populations of rat liver epithelial cells cultured under selective and nonselective conditions. Am J Pathol 135, 63–71.

204. Hooth, M. J., Coleman, W. B., Presnell, S. C., Borchert, K. M., Grisham, J. W., and Smith, G. J. (1998). Spontaneous neoplastic transformation of WB-F344 rat liver epithelial cells. Am J Pathol 153, 1913–1921.

205. Tsao, M.-S., Grisham, J. W., Chou, B. B., and Smith, J. D. (1985). Clonal isolation of populations of g-glutamyl transpeptidase-positive and -negative cells from rat liver epithelial cells chemically transformed in vitro. Cancer Res 45, 5134–5138.

206. Coleman, W. B., Wennerberg, A. E., Smith, G. J., and Grisham, J. W. (1993). Regulation of the differentiation of diploid and some aneuploid rat liver epithelial (stemlike) cells by the hepatic microenvironment. Am J Pathol 142, 1373–1382.

207. Coleman, W. B., McCullough, K. D., Esch, G. L., et al. (1997). Evaluation of the differentiation potential of WB-F344 rat liver epithelial stem-like cells in vivo: differentiation to hepatocytes following transplantation into dipeptidyl peptidase IV-deficient rat liver. Am J Pathol 151, 353–359.

208. Coleman, W. B., Butz, G. M., Howell, J. A., and Grisham, J. W. (2002). Transplantation and differentiation of rat liver epithelial stem-like cells. In: Gupta, S., Jansen, P. L. M., Klempnauer, J., and Manns, M. P., eds., Hepatocyte Transplantation. Proceedings of the International Falk Workshop, Dordrecht, The Netherlands: Kluwer Academic, pp. 38–50.

209. Price, J., Turner, D., and Cepko, C. (1987). Lineage analysis in the vertebrate nervous system by retrovirus-mediated gene transfer. Proc Natl Acad Sci U S A 84, 156–160.

210. Watanbe, Y., Kojima, T., and Fujimoto, Y. (1987). Deficiency of membrane-bound dipeptidyl aminopeptidase IV in a certain rat strain. Experientia 43, 400–401.

211. Thompson, N. L., Hixson, D. C., Callanan, H., et al. (1991). A Fischer rat substrain deficient in dipeptidyl peptidase IV activity makes normal steady-state RNA levels and an altered protein. Biochem J 273, 497–502.

212. Hong, W., and Doyle, D. (1987). cDNA cloning for a bile canaliculus domain-specific membrane glycoprotein of rat hepatocytes. Proc Natl Acad Sci U S A 84, 7962–7966.

213. Fukui, Y., Yamamoto, A., Kyoden, T., Kato, K., and Tashiro, Y. (1990). Quantitative immunogold localization of dipeptidylpeptidase IV (DPPIV) in rat liver cells. Cell Struct Funct 15, 117–125.

214. Lodja, Z. (1979). Studies on dipeptidyl(amino)peptidase IV (glycyl-proline napthylamidase). II. Blood vessels. Histochemistry 59, 153–166.

215. Nagase, S., Shimamura, K., and Shumiya, S. (1979). Albumin-deficient rat mutant. Science 205, 590–591.

216. Ogawa, K., Ohta, T., Inagaki, M., and Nagase, S. (1993). Identification of F344 rat hepatocytes transplanted within the liver of congenic analbuminemic rats by the polymerase chain reaction. Transplantation 56, 9–15.
217. Chapman, L., Coleman, W. B., and Hixson, D. (1999). Integration, survival and differentiation of liver epithelial cells in hepatic and pancreatic ducts. FASEB J 13, A160.
218. Malouf, N. N., Coleman, W. B., Grisham, J. W., et al. (2001). Adult-derived liver stem cells become heart cells in the heart in vivo. Am J Pathol 158, 1929–1935.
219. Malouf, N. N., McMahon, D., Oakeley, A. E., and Anderson, P. A. W. (1992). A cardiac troponin T epitope conserved across phyla. J Biol Chem 267, 9269–9274.
220. Terada, N., Hamazaki, T., Oka, M., et al. (2002). Bone marrow cells adopt the phenotype of other cells by spontaneous cell fusion. Nature 416, 542–545.
221. Ying, Q.-L., Nichols, J., Evans, E. P., and Smith, A. G. (2002). Changing potency by spontaneous fusion. Nature 416, 545–548.
222. Loeb, D. D., Padgett, R. W., Hardies, S. C., et al. (1986). The sequence of a large L1Md element reveals a tandemly repeated 5' end and several features found in retrotransposons. Mol Cell Biol 6, 168–182.
223. Gurdon, J. B., Mitchell, A., and Mahony, D. (1995). Direct and continuous assessment by cells of their position in a morphogen gradient. Nature 376, 520–521.
224. Dyson, S., and Gurdon, J. B. (1998). The interpretation of position in a morphogen gradient as revealed by occupancy of activin receptors. Cell 93, 557–568.
225. Kirby, S., Bentley, S., Frye, J., et al. (2002). Hematopoietic cell transdifferentiation of cloned adult liver-derived stem cells. Mol Biol Cell 13, 121A.
226. Germain, L., Noel, M., Gourdeau, H., and Marceau, N. (1988). Promotion of growth and differentiation of rat ductular oval cells in primary culture. Cancer Res 48, 368–378.
227. Lazaro, C. A., Rhim, J. A., Yamada, Y., and Fausto, N. (1998). Generation of hepatocytes from oval cell precursors in culture. Cancer Res 58, 5514–5522.
228. McMahon, J. B., Richards, W. L., delCampo, A. A., Song, M. K., and Thorgeirsson, S. S. (1986). Differential effects of transforming growth factor-b on proliferation of normal and malignant rat liver epithelial cells in culture. Cancer Res 46, 4665–4671.
229. Schroeder, I. S., Hironaka, K., Munoz, A. S., et al. (2002). In vitro differentiation of adult liver stem cells into the hepatocytic lineage. Proc Am Assoc Cancer Res 43, 462A.
230. Coleman, W. B., Smith, G. J., and Grisham, J. W. (1994). Development of dexamethasone-inducible tyrosine aminotransferase activity in WB-F344 rat liver epithelial stemlike cells cultured in the presence of sodium butyrate. J Cell Physiol 161, 463–469.
231. Hargrove, J. L., and Mackin, R. B. (1984). Organ specificity of glucocorticoid-sensitive tyrosine aminotransferase: separation from aspartate aminotransferase isozymes. J Biol Chem 259, 386–393.

232. Hargrove, J. L., and Granner, D. K. (1985). Biosynthesis and intracellular processing of tyrosine aminotransferase. In: Christen, P., and Metzler, P. E., eds., Transaminases. New York: John Wiley and Sons, pp. 511–532.

233. Ernst, M. J., Chen, C.-C., and Feigelson, P. (1977). Induction of tyrosine aminotransferase synthesis in isolated liver cell suspensions: absolute dependence of induction on glucocorticoids and glucagon or cyclic AMP. J Biol Chem 252, 6783–6791.

234. Ho, K. K. W., Cake, M. H., Yeoh, G. C. T., and Oliver, I. T. (1981). Insulin antagonism of glucocorticoid induction of tyrosine aminotransferase in cultured foetal hepatocytes. Eur J Biochem 118, 137–142.

235. Couchie, D., Holic, N., Chobert, M.-N., Corlu, A., and Laperche, Y. (2002). In vitro differentiation of WB-F344 rat liver epithelial cells into the biliary lineage. Differentiation 69, 209–215.

236. Cascio, W. E., Brackhan, J. A., Muller-Borer, B. J., et al. (2002). Adult-derived stem cells from the liver respond to signals from the microenvironment and transdifferentiate into cardiac myocytes in culture. Circulation 106(Suppl. II), II–220.

237. McCullough, K. D., Coleman, W. B., Ricketts, S. L., Wilson, J. W., Smith, G. J., and Grisham, J. W. (1998). Plasticity of the neoplastic phenotype in vivo is regulated by epigenetic factors. Proc Natl Acad Sci U S A 95, 15,333–15,338.

238. Weissman, I. L. (2000). Translating stem and progenitor cell biology to the clinic: barriers and opportunities. Science 287, 1442–1446.

8

Endothelial Progenitor Cells

Takayuki Asahara and Jeffrey M. Isner

1. STEM AND PROGENITOR CELLS FOR TISSUE REGENERATION

Tissue regeneration for organ recovery in the adult has two physiological mechanisms. One is the replacement of differentiated cells by newly generated populations derived from residual cycling stem cells. Hematopoietic cells are a typical example for this kind of regeneration. Whole hematopoietic lineage cells are derived from a few self-renewing stem cells by regulated differentiation under the influence of appropriate cytokines or growth factors. The second mechanism is the self-repair of differentiated functioning cells, preserving their proliferative activity. Hepatocytes, endothelial cells (ECs), smooth muscle cells (SMCs), keratinocytes, and fibroblasts are considered to possess this ability. Following physiological stimulation or injury, factors secreted from surrounding tissues stimulate cell replication and replacement.

Although most cells in adult organs are composed of differentiated cells, which express a variety of specific phenotypic genes adapted to each organ's environment, quiescent stem or progenitor cells are maintained locally or in the systemic circulation and are activated by environmental stimuli for physiological and pathological tissue regeneration. These reserved quiescent stem or progenitor cells are mobilized in response to environmental stimuli when an emergent regenerative process is required; neighboring differentiated cells are relied on during a minor event. In the past decade, the stem or progenitor cells have been defined from various tissues, including bone marrow, peripheral blood, brain, liver, and reproductive organs, in both adult animals and humans.

Among these stem/progenitor cells, the endothelial progenitor cell (EPC) has been identified recently and investigated to elucidate its biology for therapeutic applications. As recent reports demonstrated that endothelial lineage cells play a critical role in the early stage of liver or pancreatic differ-

From: *Adult Stem Cells*
Edited by: K. Turksen © Humana Press Inc., Totowa, NJ

entiation, prior to the blood vessel *(1,2)*, the significance of vascular development in organogenesis has become a crucial issue in regenerative medicine. In this review, the profiles of embryonic and postnatal EPCs and therapeutic application designs are introduced.

2. EMBRYONIC ENDOTHELIAL PROGENITOR CELLS

Embryonic stem cell research has opened a novel door for vascular biology, as for any medical field, to elucidate the history of vascular development. Embryonic endothelial progenitor cells, so-called angioblasts, for blood vessel development arise from migrating mesodermal cells. EPCs have the capacity to proliferate, migrate, and differentiate into endothelial lineage cells, but have not yet acquired characteristic mature endothelial markers. Available evidence suggests that hematopoietic stem cells (HSCs) and EPCs *(3,4)* are derived from a common precursor (hemangioblast) *(5–7)*. Growth and fusion of multiple blood islands in the yolk sac of the embryo ultimately give rise to the yolk sac capillary network *(8)*; after the onset of blood circulation, this network differentiates into an arteriovenous vascular system *(4)*. The integral relationship between the elements that circulate in the vascular system (the blood cells) and the cells principally responsible for the vessels themselves (the ECs) is implied by the composition of the embryonic blood islands. The cells destined to generate hematopoietic cells are situated in the center of the blood island and are termed HSCs. EPCs are located at the periphery of the blood islands.

The key molecular players determining the fate of the hemangioblast are not fully elucidated. However, several factors have been identified that may play a role in this early event. Studies in quail/chick chimeras have shown that the fibroblast growth factor 2 (FGF-2) mediates the induction of EPCs from the mesoderm *(9)*. These embryonic EPCs express flk-1, the receptor 2 for vascular endothelial growth factor (VEGFR-2), and respond to a pleiotropic angiogenic factor, vascular endothelial growth factor (VEGF), for proliferation and migration. Deletion of the flk-1 gene in mice results in embryonic lethality because of a lack of both hematopoietic and endothelial lineage development, supporting the critical importance of flk-1 at that developmental stage, although not defining the steps regulating differentiation into endothelial vs hematopoietic cells.

Mesodermal cells expressing flk-1 have also been defined as an embryonic common vascular progenitor that differentiates into endothelial and SMCs *(10)*. The vascular progenitors differentiated to ECs in response to VEGF, whereas they developed into SMCs in response to PDGF-BB. It remains to be determined whether embryonic EPCs or vascular progenitor

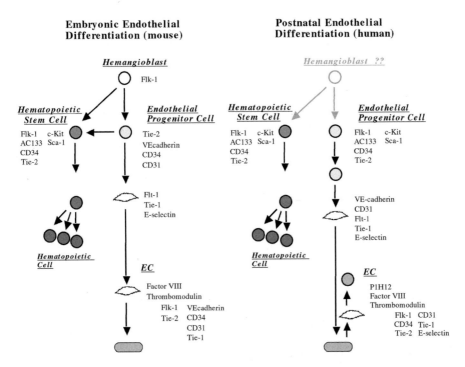

Fig. 1. Embryonic vs postnatal differentiation profiles of the endothelial lineage cell.

cells persist with an equivalent capability during adult life and whether these cells contribute to postnatal vessel growth (*see* below).

3. ADULT ENDOTHELIAL PROGENITOR CELLS

3.1. Identification of Adult Endothelial Progenitor Cells

The identification of putative HSCs in peripheral blood and bone marrow and the demonstration of sustained hematopoietic reconstitution with these HSC transplants have constituted inferential evidence for HSCs in adult tissues (*11–14*). The related descendents (EPCs) have been isolated along with HSCs in hematopoietic organs. Flk-1 and CD34 antigens were used to detect putative EPCs from the mononuclear cell (MNC) fraction of peripheral blood (*15*). This is supported by the former findings that embryonic HSCs and EPCs share certain antigenic determinants, including Flk-1, Tie-2, c-Kit, Sca-1, CD133, and CD34. These progenitor cells have consequently been considered derived from a common precursor, putatively termed a hemangioblast (*5–7*) (Fig. 1).

In vitro, these cells differentiated into endothelial lineage cells and, in animal models of ischemia, heterologous, homologous, and autologous EPCs were shown to incorporate into sites of active neovascularization. This finding was followed by diverse identifications of EPCs using equivalent or different methodologies *(16–20)*. EPCs were subsequently shown to express VE-cadherin, a junctional molecule, and AC133, an orphan receptor specifically expressed on EPCs, but with expression that is lost once they differentiate into more mature ECs *(19)*. Their high proliferation rate distinguishes circulating marrow-derived EPCs in the adult from mature ECs shed from the vessel wall *(18)*. Thus far, a bipotential common vascular progenitor, giving rise to both ECs and SMCs, has not been documented in the adult.

These findings have raised important questions regarding fundamental concepts of blood vessel growth and development in adult subjects. Does the differentiation of EPCs *in situ* (vasculogenesis) play an important role in adult neovascularization, and would impairments in this process lead to clinical diseases? There is now a strong body of evidence suggesting that vasculogenesis does in fact make a significant contribution to postnatal neovascularization. Recent studies with animal bone marrow transplantation models in which bone marrow (donor)-derived EPCs could be distinguished have shown that the contribution of EPCs to neovessel formation may range from 5 to 25% in response to granulation tissue formation *(21)* or growth factor-induced neovascularization *(22)*.

3.2. Diverse Identifications of Human EPCs and Their Precursors

Since the initial report of EPCs *(15)*, a number of groups have set out to define this cell population better. Because EPCs and HSCs share many surface markers and no simple definition of EPCs exists, various methods of EPC isolation have been reported *(15–20,23–33)* (Table 1). The term EPC may therefore encompass a group of cells that exist in a variety of stages, ranging from hemangioblasts to fully differentiated ECs. Although the true differentiation lineage of EPCs and their putative precursors remains to be determined, there is overwhelming evidence in vivo that a population of EPCs exists in humans.

Lin et al. *(18)* cultivated peripheral MNCs from patients receiving gender-mismatched bone marrow transplantation and studied their growth in vitro. This study identified a population of bone marrow (donor)-derived ECs with high proliferative potential (late outgrowth); these bone marrow cells likely represent EPCs. Gunsilius et al. *(17)* investigated a chronic myelogenous leukemia model and disclosed that bone marrow-derived EPCs

Table 1
Methods of Endothelial Progenitor Cell Isolation

Reference				
18	PB	MNCs	On type I collagen, 1 mo (outgrowth)	acLDL, vWF, CD144, KDR, CD36
19	G-CSF-PB, CB	Nonadherent CD34+ cells	bFGF, VEGF, heparin on collagen, 2 wk	vWF, CD144
26	PB	MNCs	VEGF, bFGF, IGF-1, EGF, ascorbate on FN, 7–10 d	acLDL, UEA-1, KDR, CD144, CD31
17	BM, Hydroxyurea/ G-CSF-PB	MNCs	BBE, VEGF, SCGF on FN, 10 d	CD31, factor VIII, vWF, UEA-1, acLDL
16	G-CSF-PB	CD133+ cells	Hydrocortisone, SCGF, VEGF on FN, 2 wk	CD31, CD144, KDR, Tie-2, UEA-1, vWF, Weibel-Palade body
24	PB	CD14+ cells	VEGF, bFGF, EGF, IGF-1, hydrocortisone, heparin, ascorbate on FN, 1 wk	vWF, CD144, CD105, acLDL, CD36, flt-1, KDR, Weibel-Palade body
28	CB	MNCs	BBE, heparin on FN, 7 d	CD31, CD144, KDR, eNOS, vWF, acLDL
32	PB	CD34+ cells	BBE on FN, 2 wk	Tie-2, acLDL
33	PB	Negative cells for CD3, CD7, CD19, CD34, CD45RA, CD56, and IgE	VEGF, bFGF, IGF-1 on FN, 2–4 wk	vWF, CD144, eNOS
27	CB	MNCs	PHA, 7 d	KDR, CD31, CD144
25	PB	CD34- cells	BBE on FN, 10–12 d	Tie-2, eNOS, CD144
30	BM	CD133+ cells	VEGF, bFGF, IGF-1 on FN, 3 wk, then UEA-1 selection	vWF, CD105, KDR, CD31, CD144
31	BM	CD45- and glycophorin A- cells	Insulin, transferrin, selenium, linoleic acid, dexamethasone, ascorbate, VEGF on FN cells	vWF, P1H12, Tie-2, CD144, KDR

Abbr: acLDL, acetylated low-density lipoprotein; BBE, bovine brain extract; BM, bone marrow; CB, cord blood; CM, conditioned media; eNOS, endothelial nitric oxide synthase; FN, fibronectin; G-CSF-PB, granulocyte colony-stimulating factor-mobilized PB; IgE, immunoglobulin E; MNCs, mononuclear cells; PB, peripheral blood; SCGF, stem cell growth factor.

contribute to postnatal neovascularization in humans. Multipotent adult progenitor cells (MAPCs) were isolated from bone marrow MNCs *(34)*, differentiated into EPCs, and were proposed as the origin of EPCs *(31)*. These studies therefore provided evidence to support the presence of bone marrow-derived EPCs that take part in neovascularization.

3.3. Kinetics of Endothelial Progenitor Cells in the Adult Body

Given the result of common antigenicity, bone marrow has been considered the origin of EPCs as HSCs in adult. The bone marrow transplantation experiments have demonstrated the incorporation of bone marrow-derived EPCs into foci of physiological and pathological neovascularization *(35)*. Wild-type mice were lethally irradiated and transplanted with bone marrow harvested from transgenic mice, in which constitutive LacZ expression is regulated by an EC-specific promoter, Flk-1 or Tie-2. The tissues in growing tumor, healing wound, ischemic skeletal, and cardiac muscles and cornea micropocket surgery have shown localization of Flk-1 or Tie-2 expressing endothelial lineage cells derived from bone marrow in blood vessels and stroma around vasculatures. The similar incorporation was observed in physiological neovascularization in uterus endometrial formation following induced ovulation and estrogen administration *(35)*.

Investigators have shown that wound trauma causes mobilization of hematopoietic cells, including pluripotent stem or progenitor cells in spleen, bone marrow, and peripheral blood. Consistent with EPC/HSC common ancestry, recent data from our laboratory have shown that mobilization of bone marrow-derived EPCs constitutes a natural response to tissue ischemia *(36)*. The murine bone marrow transplantation model presented direct evidence of enhanced bone marrow-derived EPC incorporation into foci of corneal neovascularization following the development of hind limb ischemia. Light microscopic examination of corneas excised 6 d after micropocket injury and concurrent surgery to establish hind limb ischemia demonstrated a statistically significant increase in cells expressing β-galactosidase in the corneas of mice with an ischemic limb vs those without one *(36)*. The finding indicates that circulating EPCs are mobilized endogenously in response to tissue ischemia, following which they may be incorporated into neovascular foci to promote tissue repair. This was confirmed by clinical findings of EPC mobilization in patients of coronary artery bypass grafting, burns *(37)*, and acute myocardial infarction *(38)*.

Having demonstrated the potential for endogenous mobilization of bone marrow-derived EPCs, we considered that iatrogenic expansion and mobilization of this putative EC precursor population might represent an effec-

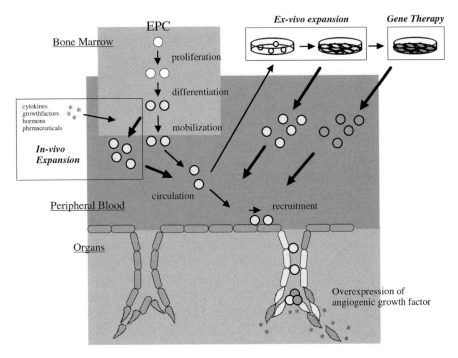

Fig. 2. Endothelial progenitor cell kinetics in the postnatal body.

tive means to augment the resident population of ECs competent to respond to administered angiogenic cytokines. Such a program might thereby address the issue of endothelial dysfunction or depletion, which may compromise strategies of therapeutic neovascularization in older, diabetic, or hypercholesterolemic animals and patients. Granulocyte macrophage colony-stimulating factor (GM-CSF), which stimulates hematopoietic progenitor cells, myeloid lineage cells, and nonhematopoietic cells, including bone marrow stromal cells and ECs, has been shown to exert a potent stimulatory effect on EPC kinetics *(36)*. Such cytokine-induced EPC mobilization could enhance neovascularization of severely ischemic tissues as well as *de novo* corneal vascularization *(36)* (Fig. 2).

The mechanisms by which these EPCs are mobilized to the peripheral circulation are in their early definitions. Among other growth factors, VEGF is the most critical factor for vasculogenesis and angiogenesis *(39–41)*. Recent data indicate that VEGF is an important factor for the kinetics of EPC as well. Our studies, performed first in mice *(42)* and subsequently in patients undergoing VEGF gene transfer for critical limb ischemia *(43)* and myocardial ischemia *(44)*, established that a previously unappreciated

mechanism by which VEGF contributes to neovascularization is via mobilization of bone marrow–derived EPCs. Similar EPC kinetics modulation has
been observed in response to other hematopoietic stimulators, such as granulocyte colony-stimulating factor (G-CSF) *(16)*, angiopoietin 1 *(45)* and
stroma-derived factor 1 (SDF-1) *(19)*.

This therapeutic strategy of EPC mobilization has been implicated not
only by natural hematopoietic or angiogenic stimulants, but also by recombinant pharmaceuticals. Statins, the 3-hydroxy-3-methylglutaryl coenzyme
A (HMG-CoA) reductase inhibitors, inhibit the activity of HMG-CoA
reductase, which catalyzes the synthesis of mevalonate, a rate-limiting step
in cholesterol biosynthesis. The statins rapidly activate Akt signaling in ECs,
and this stimulates EC bioactivity in vitro and enhances angiogenesis in vivo
(46). Our group and Dimmeler et al. demonstrated a novel function for
HMG-CoA reductase inhibitors that contributes to postnatal neovascularization by augmented mobilization of bone marrow-derived EPCs through
stimulation of the Akt signaling pathway *(47–49)*. Regarding its pharmacological safety and the effectiveness on hypercholesterolemia, one of the risk
factors for atherogenesis, the statin might be a potent medication against
atherosclerotic vascular diseases.

4. ENDOTHELIAL PROGENITOR CELLS FOR THERAPEUTIC VASCULOGENESIS

4.1. Endothelial Progenitor Cell Transplantation

The regenerative potential of stem cells has been under intense investigation. In vitro, stem and progenitor cells possess the capability of self-renewal
and differentiation into organ-specific cell types. In vivo, transplantation of
these cells may reconstitute organ systems, as shown in animal models of
diseases *(15,26,50–53)*. In contrast, differentiated cells do not exhibit such
characteristics. Human EPCs have been isolated from the peripheral blood
of adult individuals, expanded in vitro, and committed into an endothelial
lineage in culture *(15)*. The transplantation of these human EPCs has facilitated the successful salvage of limb vasculature and perfusion in athymic
nude mice with severe hind limb ischemia; differentiated ECs (human
microvascular ECs) failed to accomplish limb-saving neovascularization
(26) (Fig. 2).

These experimental findings call into question certain fundamental concepts regarding blood vessel growth and development in adult organisms.
Postnatal neovascularization has been considered synonymous with proliferation and migration of preexisting, fully differentiated ECs resident in
parent vessels, that is, angiogenesis *(54)*. The finding that circulating EPCs

may home to sites of neovascularization and differentiate into ECs *in situ* is consistent with "vasculogenesis" *(4)*, a critical paradigm for the establishment of the primordial vascular network in the embryo. Although the proportional contributions of angiogenesis and vasculogenesis to postnatal neovascularization remain to be clarified, our findings and the reports from other investigators *(20,55)* suggest that growth and development of new blood vessels in the adult are not restricted to angiogenesis, but encompass both embryonic mechanisms. As a corollary, augmented or retarded neovascularization—whether endogenous or iatrogenic—likely includes enhancement or impairment of vasculogenesis.

We therefore considered a novel strategy of EPC transplantation to provide a source of robust ECs that might supplement fully differentiated ECs thought to migrate and proliferate from preexisting blood vessels according to the classic paradigm of angiogenesis developed by Folkman *(56)*. Our studies indicated that ex vivo cell therapy, consisting of culture-expanded EPC transplantation, successfully promotes neovascularization of ischemic tissues, even when administered as "sole therapy" (i.e., in the absence of angiogenic growth factors).

Such a "supply side" version of therapeutic neovascularization in which the substrate (ECs as EPCs) rather than ligand comprises the therapeutic agent was first demonstrated, using donor cells from human volunteers, in the hind limb ischemia model of immunodeficient mice *(26)*. These findings provided novel evidence that exogenously administered EPCs augment naturally impaired neovascularization in an animal model of experimentally induced critical limb ischemia. Not only did heterologous cell transplantation improve neovascularization and blood flow recovery, but also important biological consequences—notably limb necrosis and autoamputation—were reduced by 50% in comparison with mice receiving differentiated ECs or control mice receiving media in which harvested cells were expanded ex vivo prior to transplantation.

A similar strategy applied in a model of myocardial ischemia in the nude rat demonstrated that transplanted human EPCs incorporated into rat myocardial neovascularization differentiated into mature ECs in ischemic myocardium, enhanced neovascularization, preserved left ventricular (LV) function, and inhibited myocardial fibrosis *(57)*.

Shatteman et al. locally injected freshly isolated human CD34[+] MNCs into diabetic nude mice with hind limb ischemia and showed an increase in the restoration of limb flow *(32)*. Kocher et al. attempted intravenous infusion of freshly isolated (not cultured) human CD34[+] MNCs (EPC-enriched fraction) into nude rats with myocardial ischemia *(58)*. This strategy resulted in preservation of LV function associated with inhibition of cardiomyocyte apoptosis. These experimental findings using immunodeficient animals suggest that both cultured and freshly isolated human EPCs have therapeutic potential in peripheral and coronary artery diseases.

4.2. The Limitation of Primary EPC Transplantation

Despite promising potential for regenerative applications, the fundamental scarcity of EPC populations in the hematopoietic system, combined with possible functional impairment of EPCs associated with a variety of phenotypes, such as aging, diabetes, hypercholesterolemia, and homocysteinemia (*see* Section 4.4.), constitute important limitations of primary EPC transplantation. Ex vivo expansion of EPCs cultured from the peripheral blood of healthy human volunteers typically yields approx 5.0×10^6 cells per 100 mL of blood. Our animal studies *(26)* suggested that heterologous transplantation required $0.5–2.0 \times 10^4$ human EPCs/gram weight (of the recipient mouse) to achieve satisfactory reperfusion of the ischemic hind limb. Rough extrapolation of this experience to humans suggests that a volume of as much as 12 L of peripheral blood may be necessary to harvest the EPCs required to treat critical limb ischemia. Even with the integration of certain technical improvements, the adjustment of species compatibility by autologous transplantation, and adjunctive strategies (e.g., cytokine supplements) to promote EPC mobilization *(36,42)*, the primary scarcity of a viable and functional EPC population constitutes a potential limitation of therapeutic vasculogenesis based on the use of ex vivo expansion alone.

4.3. Impact of Clinical Phenotype on EPCs

Preliminary clinical findings in patients with critical limb ischemia indicated that the response to phVEGF gene transfer was most robust and expeditious in young patients with premature atherosclerosis involving the lower extremities, so-called Buerger's disease *(59)*. This clinical observation was supported by experiments performed in live animal models, specifically young (6 to 8 mo) vs old (4 to 5 yr) rabbits and young (8 wk) vs old (2 yr) mice. In both cases, native neovascularization of the ischemic hind limb was markedly retarded in old vs young animals. Retardation of neovascularization in old animals appeared in part to result from reduced expression of VEGF in tissue sections harvested from the ischemic limb *(60)*.

Endogenous cytokine expression, however, is not the only factor contributing to impaired neovascularization. Older, diabetic, and hypercholesterolemic animals—like human subjects *(61–68)*—also exhibit evidence of age-related endothelial dysfunction. Although endothelial dysfunction does not necessarily preclude a favorable response to cytokine replacement therapy, indices of limb perfusion fail to reach ultimate levels recorded in wild-type animals, reflecting limitations imposed by a less-responsive EC substrate *(60,69–71)*.

It is then conceivable that unfavorable clinical situations (such as aging) might be associated with dysfunctional EPCs, defective vasculogenesis, and thus impaired neovascularization. Indeed, preliminary results from our laboratory indicated that replacement of native bone marrow (including its compartment of progenitor cells) of young mice with bone marrow transplanted from old animals led to marked reduction in neovascularization following corneal micropocket injury in comparison to young mice transplanted with young bone marrow *(72)*. These studies thus established evidence of an age-dependent impairment in vasculogenesis (as well as angiogenesis) and the origin of progenitor cells as a critical parameter influencing neovascularization. Moreover, analysis of clinical data from older patients at our institution disclosed a significant reduction in the baseline levels and the population of EPCs mobilized in response to VEGF165 gene transfer *(73)*; specifically, the number of EPCs in the systemic circulation of young patients with critical limb ischemia was five times more than the number circulating in old individuals. Impaired EPC mobilization or activity in response to VEGF may thus contribute to the age-dependent defect in postnatal neovascularization.

4.4. Gene Therapy of Endothelial Progenitor Cell

Given these findings and the limited quantity of EPCs available even under healthy, physiological conditions, a strategy must be considered that addresses this shortfall and mitigates the possibility of dysfunctional EPCs for therapeutic vasculogenesis in ischemic disorders complicated by aging, diabetes, hypercholesterolemia, or hyperhomocysteinemia. Genetic modification of EPCs to overexpress angiogenic growth factors enhance the signaling activity of the angiogenic response, and rejuvenating the bioactivity or extending the lifespan of EPCs constitutes one potential strategy that might address these limitations of EPC transplantation, thereby optimizing therapeutic neovascularization (Fig. 2).

Our recent findings provide the first evidence that exogenously administered gene-modified EPCs augment naturally impaired neovascularization in an animal model of experimentally induced limb ischemia *(74)*. Transplantation of heterologous EPCs transduced with adenovirus encoding VEGF not only improved neovascularization and blood flow recovery, but also had meaningful biological consequences: Limb necrosis and autoamputation were reduced by 63.7% in comparison with controls. The dose of EPCs used in the in vivo experiments was subtherapeutic; that is, this dose of EPCs was 30 times less than that required in previous experiments to improve the rate of limb salvage above that seen in untreated controls. Adenoviral VEGF EPC–gene transfer, however, accomplished a therapeutic effect, as evidenced by a functional outcome despite a subtherapeutic dose of EPCs. Thus, VEGF EPC–gene transfer constitutes one option to address the limited number of EPCs that can be isolated from peripheral blood prior to ex vivo expansion and subsequent autologous readministration.

5. ENDOTHELIAL PROGENITOR CELLS IN OTHER FIELDS

EPCs have been applied to the field of tissue engineering to improve biocompatibility of vascular grafts. Artificial grafts first seeded with autologous CD34+ cells from canine bone marrow and then implanted into the aortas had increased surface endothelialization and vascularization compared with controls (75). Similarly, when cultured autologous bovine EPCs were seeded onto carotid interposition grafts, the EPC-seeded grafts achieved physiological motility and remained active for 130 d vs 15 d in nonseeded grafts (76).

EPCs have also been investigated in the cerebrovascular field. Embolization of the middle cerebral artery in Tie2/lacZ/bone marrow transplantation mice disclosed that the formation of new blood vessels in the adult brain after stroke involves vasculogenesis/EPCs (77). Similar data were reported using gender-mismatched wild-type mice transplanted with bone marrow from green fluorescent protein (GFP) transgenic mice (78). However, whether autologous EPC transplantation would augment cerebral revascularization has yet to be examined.

To date, the role of EPCs in tumor angiogenesis has been demonstrated by several groups. Davidoff et al. showed that bone marrow-derived EPCs contribute to tumor neovasculature, and that bone marrow cells transduced with an antiangiogenic gene can restrict tumor growth in mice (79). Lyden et al. recently used angiogenic defective, tumor-resistant Id-mutant mice and showed the restoration of tumor angiogenesis with bone marrow (donor)-derived EPCs throughout the neovessels following the transplantation of wild-type bone marrow into these mice (80).

These data demonstrate that EPCs are not only important, but also critical to tumor neovascularization. Although it is not known whether local administration of exogenous EPCs may augment tumor neovascularization, this issue should be carefully considered for the clinical application of EPC cell therapy to treat cardiovascular diseases.

REFERENCES

1. Lammert, E., Cleaver, O., and Melton, D. (2001). Induction of pancreatic differentiation by signals from blood vessels. Science 294, 564–567.
2. Matsumoto, K., Yoshitomi, H., Rossant, J., and Zaret, K. S. (2001). Liver organogenesis promoted by endothelial cells prior to vascular function. Science 294, 559–563.
3. Pardanaud, L., Altman, C., Kitos, P., and Dieterien-Lievre, F. (1989). Relationship between vasculogenesis, angiogenesis and haemopoiesis during avian ontogeny. Development 105, 473–485.
4. Risau, W., Sariola, H., Zerwes, H.-G., et al. (1988). Vasculogenesis and angiogenesis in embryonic stem cell-derived embryoid bodies. Development 102, 471–478.
5. Flamme, I., and Risau, W. (1992). Induction of vasculogenesis and hematopoiesis in vitro. Development 116, 435–439.
6. His, W. (1900). Leoithoblast und angioblast der wirbelthiere. Abhandl K S Ges Wiss Math Phys 22, 171–328.
7. Weiss, M., and Orkin, S. H. (1996). In vitro differentiation of murine embryonic stem cells: new approaches to old problems. J Clin Invest 97, 591–595.
8. Risau, W., and Flamme, I. (1995). Vasculogenesis. Annu Rev Cell Dev Biol 11, 73–91.
9. Poole, T. J., Finkelstein, E. B., and Cox, C. M. (2001). The role of FGF and VEGF in angioblast induction and migration during vascular development. Dev Dyn 220, 1–17.
10. Yamashita, J., Itoh, H., Hirashima, M., et al. (2000). Flk1-positive cells derived from embryonic stem cells serve as vascular progenitors. Nature 408, 92–96.
11. Brugger, W., Heimfeld, S., Berenson, R. J., Mertelsmann, R., and Kanz, L. (1995). Reconstitution of hematopoiesis after high-dose chemotherapy by autologous progenitor cells generated ex vivo. N Engl J Med 333, 283–287.
12. Kessinger, A., and Armitage, J. O. (1991). The evolving role of autologous peripheral stem cell transplantation following high-dose therapy for malignancies. Blood 77, 211–213.
13. Sheridan, W. P., Begley, C. G., and Juttener, C. (1992). Effect of peripheral-blood progenitor cells mobilised by filgrastim (G-CSF) on platelet recovery after high-dose chemotherapy. Lancet 339, 640–644.
14. Shpall, E. J., Jones, R. B., and Bearman, S. I. (1994). Transplantation of enriched CD34-positive autologous marrow into breast cancer patients following high-dose chemotherapy. J Clin Oncol 12, 28–36.
15. Asahara, T., Murohara, T., Sullivan, A., et al. (1997). Isolation of putative progenitor endothelial cells for angiogenesis. Science 275, 964–967.
16. Gehling, U. M., Ergun, S., Schumacher, U., et al. (2000). In vivo differentia-

tion of endothelial cells from AC133-positive progenitor cells. Blood 95, 3106–3112.

17. Gunsilius, E., Duba, H. C., Petzer, A. L., et al. (2000). Evidence from a leukaemia model for maintenance of vascular endothelium by bone-marrow–derived endothelial cells. Lancet 355, 1688–1691.

18. Lin, Y., Weisdorf, D. J., Solovey, A., and Hebbel, R. P. (2000). Origins of circulating endothelial cells and endothelial outgrowth from blood. J Clin Invest 105, 71–77.

19. Peichev, M., Naiyer, A. J., Pereira, D., et al. (2000). Expression of VEGFR-2 and AC133 by circulating human CD34⁺ cells identifies a population of functional endothelial precursors. Blood 95, 952–958.

20. Shi, Q., Rafii, S., Wu, M. H.-D., et al. (1998). Evidence for circulating bone marrow-derived endothelial cells. Blood 92, 362–367.

21. Crosby, J. R., Kaminski, W. E., Schatteman, G., et al. (2000). Endothelial cells of hematopoietic origin make a significant contribution to adult blood vessel formation. Circ Res 87, 728–730.

22. Murayama, T., Tepper, O. M., Silver, M., et al. (2002). Determination of bone marrow-derived endothelial progenitor cell significance in angiogenic growth factor-induced neovascularization in vivo. Exp Hematol 30, 967–972.

23. Boyer, M., Townsend, L. E., Vogel, L. M., et al. (2000). Isolation of endothelial cells and their progenitor cells from human peripheral blood. J Vasc Surg 31, 181–189.

24. Fernandez Pujol, B., Lucibello, F. C., Gehling, U. M., et al. (2000). Endothelial-like cells derived from human CD14 positive monocytes. Differentiation 65, 287–300.

25. Harraz, M., Jiao, C., Hanlon, H. D., Hartley, R. S., and Schatteman, G. C. (2001). CD34⁻ blood-derived human endothelial cell progenitors. Stem Cells 19, 304–312.

26. Kalka, C., Masuda, H., Takahashi, T., et al. (2000). Transplantation of ex vivo expanded endothelial progenitor cells for therapeutic neovascularization. Proc Natl Acad Sci U S A 97, 3422–3427.

27. Kang, H. J., Kim, S. C., Kim, Y. J., et al. (2001). Short-term phytohaemagglutinin-activated mononuclear cells induce endothelial progenitor cells from cord blood CD34⁺ cells. Br J Haematol 113, 962–969.

28. Murohara, T., Ikeda, H., Duan, J., et al. (2000). Transplanted cord blood–derived endothelial precursor cells augment postnatal neovascularization. J Clin Invest 105, 1527–1536.

29. Nieda, M., Nicol, A., Denning-Kendall, P., Sweetenham, J., Bradley, B., and Hows, J. (1997). Endothelial cell precursors are normal components of human umbilical cord blood. Br J Haematol 98, 775–777.

30. Quirici, N., Soligo, D., Caneva, L., Servida, F., Bossolasco, P., and Deliliers, G. L. (2001). Differentiation and expansion of endothelial cells from human bone marrow CD133(+) cells. Br J Haematol 115, 186–194.

31. Reyes, M., Dudek, A., Jahagirdar, B., Koodie, L., Marker, P. H., and Verfaillie, C. M. (2002). Origin of endothelial progenitors in human postnatal bone marrow. J Clin Invest 109, 337–346.

32. Schatteman, G. C., Hanlon, H. D., Jiao, C., Dodds, S. G., and Christy, B. A. (2000). Blood-derived angioblasts accelerate blood-flow restoration in diabetic mice. J Clin Invest 106, 571–578.
33. Schmeisser, A., Garlichs, C. D., Zhang, H., et al. (2001). Monocytes coexpress endothelial and macrophagocytic lineage markers and form cord-like structures in Matrigel under angiogenic conditions. Cardiovasc Res 49, 671–680.
34. Jiang, Y., Jahagirdar, B. N., Reinhardt, R. L., et al. (2002). Pluripotency of mesenchymal stem cells derived from adult marrow. Nature 418, 41–49.
35. Asahara, T., Masuda, H., Takahashi, T., et al. (1999). Bone marrow origin of endothelial progenitor cells responsible for postnatal vasculogenesis in physiological and pathological neovascularization. Circ Res 85, 221–228.
36. Takahashi, T., Kalka, C., Masuda, H., et al. (1999). Ischemia- and cytokine-induced mobilization of bone marrow–derived endothelial progenitor cells for neovascularization. Nat Med 5, 434–438.
37. Gill, M., Dias, S., Hattori, K., et al. (2001). Vascular trauma induces rapid but transient mobilization of VEGFR(+)AC133(+) endothelial precursor cells. Circ Res 88, 167–174.
38. Shintani, S., Murohara, T., Ikeda, H., et al. (2001). Augmentation of postnatal neovascularization with autologous bone marrow transplantation. Circulation 103, 897–903.
39. Carmeliet, P., Ferreira, V., Breier, G., et al. (1996). Abnormal blood vessel development and lethality in embryos lacking a single VEGF allele. Nature 380, 435–439.
40. Ferrara, N., Carver-Moore, K., Chen, H., et al. (1996). Heterozygous embryonic lethality induced by targeted inactivation of the VEGF gene. Nature 380, 439–442.
41. Shalaby, F., Rossant, J., Yamaguchi, T. P., et al. (1995). Failure of blood-island formation and vasculogenesis in Flk-1 deficient mice. Nature 376, 62–66.
42. Asahara, T., Takahashi, T., Masuda, H., et al. (1999). VEGF contributes to postnatal neovascularization by mobilizing bone marrow-derived endothelial progenitor cells. EMBO J 18, 3964–3972.
43. Kalka, C., Masuda, H., Takahashi, T., et al. (2000). Vascular endothelial growth factor 165 gene transfer augments circulating endothelial progenitor cells in human subjects. Circ Res 86, 1198–1202.
44. Kalka, C., Tehrani, H., Laudenberg, B., et al. (2000). Mobilization of endothelial progenitor cells following gene therapy with $VEGF_{165}$ in patients with inoperable coronary disease. Ann Thorac Surg 70, 829–834.
45. Hattori, K., Dias, S., Heissig, B., et al. (2001). Vascular endothelial growth factor and angiopoietin-1 stimulate postnatal hematopoiesis by recruitment of vasculogenic and hematopoietic stem cells. J Exp Med 193, 1005–1014.
46. Kureishi, Y., Luo, Z., and Shiojima, I. (2000). The HMG-CoA reductase inhibitor simvastatin activates the protein kinase Akt and promotes angiogenesis in normocholesterolemic animals. Nature Medicine 6, 1004–1010.
47. Llevadot, J., Murasawa, S., Kureishi, Y., et al. (2001). HMG-CoA reductase inhibitor mobilizes bone-marrow derived endothelial progenitor cells. J Clin Invest 108, 399–405.

48. Dimmeler, S., Aicher, A., Vasa, M., et al. (2001). HMG-CoA-reductase inhibitors (statins) increase endothelial progenitor cells via the P13 kinase/Akt pathway. J Clin Invest 108, 391–397.

49. Vasa, M., Breitschopf, K., Zeiher, A. M., and Dimmeler, S. (2000). Nitric oxide activates telomerase and delays endothelial cell senescence. Circ Res 87, 540–542.

50. Evans, J. T., Kelly, P. F., O'Neill, E., and Garcia, J. V. (1999). Human cord blood CD34⁺CD38⁻ cell transduction via lentivirus-based gene transfer vectors. Hum Gene Ther 10, 1479–1489.

51. Flax, J. D., Aurora, S., Yang, C., et al. (1998). Engraftable human neural stem cells respond to developmental cues, replace neurons, and express foreign genes. Nature Biotechnol 16, 1033–1039.

52. Lindvall, O., Brundin, P., Widner, H., et al. (1990). Grafts of fetal dopamine neurons survive and improve motor function in Parkinson's disease. Science 247, 574–577.

53. Anklesaria, P., Kase, K., Glowacki, J., et al. (1987). Engraftment of a clonal bone marrow stromal cell line in vivo stimulates hematopoietic recovery from total body irradiation. Proc Natl Acad Sci U S A 84, 7681–7685.

54. Folkman, J. (1971). Tumor angiogenesis: therapeutic implications. N Engl J Med 285, 1182–1186.

55. Hatzopoulos, A. K., Folkman, J., Vasile, E., Eiselen, G. K., and Rosenberg, R. D. (1998). Isolation and characterization of endothelial progenitor cells from mouse embyros. Development 125, 1457–1468.

56. Folkman, J. (1993). Tumor angiogenesis. In: Holland, J. F., Frei, E., III, Bast, R. C., Jr., Kute, D. W., Morton, D. L., and Weichselbaum, R. R., eds., Cancer Medicine. 3rd ed. Philadelphia: Lea and Febiger, pp. 153–170.

57. Kawamoto, A., Gwon, H.-C., Iwaguro, H., et al. (2001). Therapeutic potential of ex vivo expanded endothelial progenitor cells for myocardial ischemia. Circulation 103, 634–637.

58. Kocher, A. A., Schuster, M. D., Szabolcs, M. J., et al. (2001). Neovascularization of ischemic myocardium by human bone marrow-derived angioblasts prevents cardiomyocyte apoptosis, reduces remodeling and improves cardiac function. Nat Med 7, 430–436.

59. Isner, J. M., Baumgartner, I., Rauh, G., et al. (1998). Treatment of thromboangiitis obliterans (Buerger's disease) by intramuscular gene transfer of vascular endothelial growth factor: preliminary clinical results. J Vasc Surg 28, 964–975.

60. Rivard, A., Fabre, J.-E., Silver, M., et al. (1999). Age-dependent impairment of angiogenesis. Circulation 99, 111–120.

61. Chauhan, A., More, R. S., Mullins, P. A., Taylor, G., Petch, M. C., and Schofield, P. M. (1996). Aging-associated endothelial dysfunction in humans is reversed by L-arginine. J Am Coll Cardiol 28, 1796–1804.

62. Cosentino, F., and Luscher, T. F. (1998). Endothelial dysfunction in diabetes mellitus. J Cardiovasc Pharm 32, S54–S61.

63. Drexler, H., Zeiher, A. M., Meinzer, K., and Just, H. (1991). Correction of

endothelial dysfunction in coronary microcirculation of hypercholesterolemic patients by L-arginine. Lancet 338, 1546–1550.

64. Gerhard, M., Roddy, M.-A., Creager, S. J., and Creager, M. A. (1996). Aging progressively impairs endothelium-dependent vasodilation in forearm resistance vessles of humans. Hypertension 27, 849–853.

65. Johnstone, M. D., Creager, S. J., Scales, K. M., Cusco, J. A., Lee, B. K., and Creager, M. A. (1993). Impaired endothelium-dependent vasodilation in patients with insulin-dependent diabetes mellitus. Circulation 88, 2510–2516.

66. Luscher, T. F., and Tshuci, M. R. (1993). Endothelial dysfunction in coronary artery disease. Annu Rev Med 44, 395–418.

67. Taddei, S., Virdis, A., Mattei, P., et al. (1995). Aging and endothelial function in normotensive subjects and patients with essential hypertension. Circulation 91, 1981–1987.

68. Tschudi, M. R., Barton, M., Bersinger, N. A., et al. (1996). Effect of age on kinetics of nitric oxide release in rat aorta and pulmonary artery. J Clin Invest 98, 899–905.

69. Couffinhal, T., Silver, M., Kearney, M., et al. (1999). Impaired collateral vessel development associated with reduced expression of vascular endothelial growth factor in ApoE^{-1-} mice. Circulation 99, 3188–3198.

70. Rivard, A., Silver, M., Chen, D., et al. (1999). Rescue of diabetes related impairment of angiogenesis by intramuscular gene therapy with adeno-VEGF. Am J Pathol 154, 355–364.

71. Van Belle, E., Rivard, A., Chen, D., et al. (1997). Hypercholesterolemia attenuates angiogenesis but does not preclude augmentation by angiogenic cytokines. Circulation 96, 2667–2674.

72. Rivard, A., Asahara, T., Takahashi, T., Chen, D., and Isner, J. M. (1998). Contribution of endothelial progenitor cells to neovascularization (vasculogenesis) is impaired with aging. Circulation 98, 1–39.

73. Kalka, C., Masuda, H., Gordon, R., Silver, M., and Asahara, T. (1999). Age dependent response in mobilization of endothelial progenitor cells (EPC) to VEGF gene therapy in human subjects. Circulation 100, 1–40.

74. Iwaguro, H., Yamaguchi, J., Kalka, C., et al. (2002). Endothelial progenitor cell vascular endothelial growth factor gene transfer for vascular regeneration. Circulation 105, 732–738.

75. Bhattacharya, V., McSweeney, P. A., Shi, Q., et al. (2000). Enhanced endothelialization and microvessel formation in polyester graft seeded with CD34$^+$ bone marrow cells. Blood 95, 581–585.

76. Kaushal, S., Amiel, G. E., Guleserian, K. J., et al. (2001). Functional small-diameter neovessels created using endothelial progenitor cells expanded ex vivo. Nat Med 7, 1035–1040.

77. Zhang, Z. G., Zhang, L., Jiang, Q., and Chopp, M. (2002). Bone marrow–derived endothelial progenitor cells participate in cerebral neovascularization after focal cerebral ischemia in the adult mouse. Circ Res 90, 284–288.

78. Hess, D. C., Hill, W. D., Martin-Studdard, A., Carroll, J., Brailer, J., and Carothers, J. (2002). Bone marrow as a source of endothelial cells and NeuN-expressing cells after stroke. Stroke 33, 1362–1368.

79. Davidoff, A. M., Ng, C. Y., Brown, P., et al. (2001). Bone marrow–derived cells contribute to tumor neovasculature and, when modified to express an angiogenesis inhibitor, can restrict tumor growth in mice. Clin Cancer Res 7, 2870–2879.

80. Lyden, D., Hattori, K., Dias, S., et al. (2001). Impaired recruitment of bone-marrow-derived endothelial and hematopoietic precursor cells blocks tumor angiogenesis and growth. Nat Med 7, 1194–1201.

9

Prostate Epithelial Stem Cells

Anne T. Collins and David E. Neal

1. INTRODUCTION

Prostate cancer is increasingly prevalent in our aging Western society. Despite recent advances in the detection of early prostate cancer, there remains little effective therapy for patients with locally advanced or metastatic disease. The majority of patients with advanced disease respond initially to androgen ablation therapy; however, most go on to develop androgen-independent tumors that are inevitably fatal. To seek a deeper understanding of the cellular origins of cancer, the normal pathways of cellular differentiation and tissue renewal must be understood.

Central to this understanding is an appreciation of the nature of stem cells that must be present in each renewable tissue compartment. These are cells that self-renew with each division and commit others of their progeny to differentiate into one or more of the mature cell types that define each tissue. Classically, stem cells have been studied in tissues that undergo rapid cell turnover, such as bone marrow, skin, and the gastrointestinal tract *(1–3)*. Now, however, it has been recognized that stem cells are also present in tissues that normally undergo very limited regeneration or turnover, such as the prostate. The ultrastructural studies of Mao and Angrist *(4)*, Dermer *(5)*, and Heatfield et al. *(6)* and the experimental evidence from androgen cycling studies of Brandes *(7)* and Isaacs and Coffey *(8)* argue for the existence of stem cells in prostate epithelium.

This review focuses on more recent data, on expression patterns of genes in the proliferation compartment, and on recent attempts to characterize and isolate the stem and transit-amplifying cells of the prostate epithelium. We also discuss emerging evidence on the plasticity of adult stem cells and the concept that stem cell biology could provide new insights into the biology of prostate cancer.

From: *Adult Stem Cells*
Edited by: K. Turksen © Humana Press Inc., Totowa, NJ

2. GENERAL FEATURES OF STEM CELLS

Although still a contentious issue, the prevailing view is that stem cells have the capacity for unlimited or prolonged self-renewal and the ability to produce at least one type of highly differentiated descendent *(9–12)*. It is also generally accepted that, between the stem cell and its terminally differentiated progeny, there is an intermediate population of committed progenitors with limited proliferative capacity and restricted differentiation potential, sometimes defined as *transit-amplifying cells*. The primary function of this transit population is to increase the number of differentiated cells produced by each stem cell division. Thus, a stem cell, which has high self-renewal capacity, may actually divide relatively infrequently.

Stem cells are thought to undergo asymmetric division to yield one stem cell daughter and one differentiated daughter (progenitor). This may be true in certain situations, but the population of stem cells can still be self-maintaining when some cell divisions yield two stem cell daughters and others yield two differentiating daughters *(10)*. Symmetric divisions allow the size of the stem cell pool to be regulated by factors that control the probability of self-renewing vs differentiative divisions *(3)*.

Another characteristic attributed to stem cells is that they divide slowly or rarely. This is thought to be true for stem cells in the skin *(13)* and bone marrow *(14)*. Other kinds of stem cells, however, divide more rapidly. Stem cells in the mammalian intestinal crypt have been estimated to divide every 12 h *(3)*.

3. THE ARCHITECTURE OF THE PROSTATE EPITHELIUM AND ITS PATTERNING DURING DEVELOPMENT

The prostate is a complex tubulo-alveolar gland composed of an epithelial parenchyma embedded in a connective tissue matrix. The epithelial cells are arranged in glands composed of ducts that branch out from the urethra and terminate into acini. From the 20th wk of gestation until puberty, the immature prostatic acini and ducts are lined with multiple layers of immature cells with round nuclei and scant cytoplasm. In the immature epithelium, cytokeratins of simple and stratified epithelium are expressed (primary cytokeratins; numbers 8, 18, and 19 and the large molecular weight forms, numbers 4 to 7, 10, 11, 14, 15) *(15)*. During puberty, this immature, multilayered epithelium differentiates into a two-layer epithelium consisting of peripheral flattened-to-cuboidal basal cells and inner secretory cylindrical epithelium. This androgen-induced differentiation process is associated with an alteration in the expression of several cytokeratins. Thus, the high molecular weight cytokeratins are expressed in the basal cells, whereas

cytokeratins 8 and 18 selectively occur in the secretory epithelium *(15–17)*. However, there are subpopulations of intermediate phenotypes (coexpressing simple and high molecular weight cytokeratins) present in the basal and luminal layers in the mature prostate (*see* Section 5.2.).

4. CELLULAR COMPARTMENTS IN MATURE PROSTATIC EPITHELIA

Three main cell types are discernible in normal, mature prostatic epithelium: basal, secretory luminal, and neuroendocrine. The luminal or glandular cells constitute the exocrine compartment of the prostate, secreting prostate-specific antigen (PSA) and prostatic acid phosphatase (PAP) into the glandular lumina. They are terminally differentiated and represent the major cell type in normal and hyperplastic epithelium. They express high levels of the androgen receptor (AR) *(18,19)* and are dependent on androgens for their survival *(20)*.

In contrast, basal cells are relatively undifferentiated and without secretory activity. As the name suggests, basal cells rest on the basement membrane; morphologically, they range from small, flattened to cuboidal cells. They express low or undetectable levels of AR *(21)* and are independent of androgens for their survival *(20)*. Basal cells focally express the estrogen receptor and proliferate under estrogen therapy *(22)*.

Significant populations of neuroendocrine cells also reside among the more abundant secretory epithelium in the normal prostate gland. These cells are found in the epithelium of the acini and in ducts of all parts of the gland. The major type of neuroendocrine cell contains serotonin and thyroid-stimulating hormone. Neuroendocrine cells are terminally differentiated, postmitotic cell types that are androgen insensitive *(23)*.

5. EVIDENCE FOR STEM CELLS IN PROSTATE EPITHELIA

5.1. Androgen Cycling Studies

The existence of stem cells in the prostate is probably best illustrated by animal studies investigating the effect of androgen on the prostate. The majority of prostatic epithelial cells in the adult gland are androgen dependent for their survival *(20)*. Thus, castration of male rats leads to rapid involution of the gland, with loss of up to 90% of the total epithelial cells *(20)*. The remaining epithelial cells do not require androgen for survival, yet some of these androgen-independent cells are sensitive to androgen because subsequent administration of exogenous androgen results in induction of proliferation and regeneration of the prostate to its original size and function *(24,25)*. By cyclically inducing prostate involution and regeneration, it is

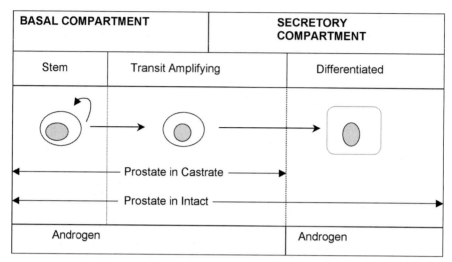

Fig. 1. Stem cell model for the organization of the prostate epithelium. Stem cells in the basal cell layer give rise to a population of transit-amplifying cells, which in turn differentiate into the functional, secretory luminal cells. The survival of the secretory luminal cells is dependent on androgens.

possible to induce more than 60 population doublings in the rat ventral prostate *(26)*.

These results led Isaacs and Coffey *(8)* to propose a stem cell model for prostate epithelia (Fig. 1) by which androgen-independent stem cells give rise to a population of androgen-independent amplifying cells. Although these cells are androgen independent, they can respond to androgens and amplify the number of androgen-dependent, secretory luminal cells. This point is emphasized by the fact that it is possible to castrate adult male rats and allow an extended period (i.e., >3 yr) before replacing androgen and still fully restore the gland *(8)*.

However, this model for the prostate epithelium has not been universally accepted. For example, cell kinetic and morphological investigation in the prostate gland of the rat, in which involution is induced by castration, suggest that basal and luminal secretory cells are self-replicating cell types *(27,28)*. Both groups observed that, in the presence of castrate levels of androgen, the basal cells persist, as does a population of cuboidal glandular cells. When androgen levels were restored, both populations proliferated simultaneously, but the glandular cells proliferated at a higher rate compared to the basal population; that is, the glandular and the basal compartments were responsible for restoration of the gland.

Ki67 antigen, which is expressed in late G1, S, G2, and M phases of the cell cycle *(29)*, is expressed specifically in the basal cell compartment in the normal prostate *(30)*. However, Van der Kwast et al. *(31)* observed that, under complete androgen blockade, luminal cells also express Ki67. Based on the evidence, it was inferred that the two populations comprise independent and separate lineages. However, this observation does not preclude derivation of luminal cells from basal cells because the glandular cells that persist postcastration are analogous to the androgen-independent amplifying cells hypothesized in the stem cell model of Isaacs and Coffey *(8)*.

5.2. Phenotypic Relationship Between Prostatic Epithelial Cell Types

Of relevance to the determination of lineage(s) is the finding that cells morphologically and phenotypically intermediate between basal and luminal cells have been identified in the normal prostatic epithelium. For example, electron microscopic studies of the human prostate have identified foci of cells with morphological features typical of basal and secretory cells *(4,5)*. Brandes *(7)* noted similar transitional forms of basal cells with similarities to luminal cells following experimental castration and androgen administration.

Patterns of cytokeratin expression serve as valuable markers for epithelial phenotypes. Analysis of cytokeratins in normal, hyperplastic, and malignant prostate has identified cell phenotypes intermediate between basal and luminal cells *(32–36)*. PSA-expressing cells have also been identified in the basal layer *(37)*. Bonkhoff and coworkers also reported simultaneous expression of neuroendocrine markers with PSA and neuroendocrine cells expressing basal-specific cytokeratins *(37)*.

In primary cultures of human prostatic epithelia, the majority of cells in the initial outgrowth have a phenotype intermediate between basal and luminal ($CK5^+/CK14^+/CK18^+$). After this initial period of proliferation, the cell layer becomes confluent, and organization of cells is observed concurrent with multilayering and morphological differentiation *(35,38,39)*. Glandular buds appear on the uppermost layer and are typified by the presence of numerous secretory vacuoles and increased expression of cytokeratins 18 and 19, the androgen receptor, and PSA *(35,38)*. In organ cultures and three-dimensional Matrigel cultures derived from human and rat ventral prostate, the initial epithelial buds coexpress cytokeratins 5 and 14 as well as the secretory cytokeratins 8 and 18. Under androgenic stimulation, the ducts canalize, and basal and luminal cells become morphologically and phenotypically distinct *(40,41)*.

Although the above observations demonstrate that basal and luminal cells are linked in a hierarchical pathway, they do not actually show that they are derived from a common stem cell. Despite this, some investigators have hypothesized the lineage of prostate epithelium based on such studies *(35,36)*. To resolve the issue of lineage, it will be necessary to track the progeny and differentiation of a marked or isolated stem cell—either as a clonal regeneration assay (regenerating a culture from a single cell) or using a transfected marker.

5.3. Clonogenic Rapidly Adherent Basal Cells and Formation of Fully Differentiated Glands In Vivo

Stem cells of different tissues show certain similarities in biological behavior. For instance, stem cells are usually on the basement membrane situated in a protected region, or niche, among supporting cells. Thus, certain tissue stem cells adhere to basement membrane proteins more than other basal cells (mediated through differential expression of specific integrins); this can be used to locate and isolate these cells *(42,43)*. This is true for prostate epithelial stem cells *(44)*. Although there is no single definition for a stem cell, there is general agreement that such a cell would exhibit clonogenicity and, more important, the ability to regenerate the different cell types that constitute the tissue in which it exists. Thus, transplanted cells should be capable of self-renewal and produce progeny that differentiate into a fully functional epithelium.

Basal cells directly isolated from prostatic tissue, on the basis of rapid adherence to type I collagen, are clonogenic *(39,44)*, whereas basal cells that do not adhere rapidly to type I collagen do not form actively growing colonies. The cells that found the actively growing colonies express basal-specific markers (CK5$^+$/CK14$^+$) and not markers of differentiation, namely, CK19, CK18, PSA, and AR *(44)*. Moreover, this selected population is distinct from other basal cells by their potential to generate prostatelike glands in vivo with morphologic and immunohistochemical evidence of prostate-specific differentiation, properties consistent with a stem cell origin *(44)*.

6. MAINTENANCE OF PROSTATIC STEM CELLS AND REGULATION OF HOMEOSTASIS IN PROSTATE EPITHELIUM

Continued tissue function during the lifetime of a multicellular organism depends on the ability to turn over its cell population. This is accomplished by a delicate balance between apoptosis, through cell removal, and replacement via proliferation and differentiation. The current view is that every cell in a multicellular organism is programmed to die unless it receives external survival signals *(45)*. In the prostate epithelium, the highest rate of cell death is observed in the luminal cell layer, whereas the basal cell compartment undergoes a much lower rate of apoptosis *(46)*. Several molecular mechanisms appear to protect stem cells from apoptosis, differentiation, and senescence, thereby maintaining the genomic integrity of these cells.

6.1. Bcl-2 Expression in the Basal Cell Compartment

Expression of Bcl-2, an antiapoptotic protein *(47)* is restricted to the basal cell compartment in normal prostate *(48,49)*. In typical self-renewing epithelial (gastrointestinal tract, skin) cells, the expression of Bcl-2 is restricted to stem cells and proliferative zones *(50)*. Downregulation of Bcl-2 in secretory luminal cells, in the prostate, correlates with differentiation, reduced proliferative capacity, and reduction in remaining lifespan. This suggests that, after terminal differentiation, secretory luminal cells undergo programmed cell death induced by androgen or the androgen receptor *(51)* and are replaced by generative stem cells.

A corollary of Bcl-2 expression in the stem cell population is that protection against programmed cell death increases the susceptibility of epithelial tissues to undergo carcinogenesis. For example, altered susceptibility to apoptosis through differential expression of Bcl-2 is seen in different regions of the gut. The maximum level of apoptosis in the small intestine occurs in the stem cell region, yet the stem cells of the colon express Bcl-2. The implication of this is that stem cells that express Bcl-2 have a greater chance of survival in a background of genomic damage and may progress to cancer. Interestingly, Bcl-2, which is restricted to the androgen-independent basal cells of the prostate, is associated with the development of androgen-independent prostate carcinoma *(52)*.

6.2. Telomerase Expression in the Basal Cell Compartment

Telomeres are noncoding repetitive deoxyribonucleic acid (DNA) sequences localized at the ends of chromosomes and are required for chromosomal stability and for faithful replication of DNA. Telomerase is the enzyme responsible for the synthesis and maintenance of telomere repeats, and its expression correlates with self-renewal potential in many cell types *(53)*. Telomerase is normally expressed in embryonic, germline cells and certain adult stem cells *(54)* and is readily detectable in the majority of tumor samples *(55)*. In most somatic cells, however, telomerase expression is repressed, and telomeres shorten after each cell division. This progressive shortening, related to the inability of the lagging strand to replicate the 5' end of a linear DNA molecule, has been proposed as a major control mechanism that determines cellular senescence *(56,57)*. In the normal adult prostate, telomerase is expressed throughout the basal cell compartment *(58)*.

Other tissues capable of extensive renewal have been shown to express at least low levels of telomerase, including human epidermis *(59)*, intestinal epithelium *(60)*, and hematopoietic cells *(61)*. In each of these tissues, telomerase activity is restricted to topographic locations that harbor stem cells. Unlike tumor cells, however, stem cells are mortal and show decreasing telomere length with increasing age *(62)*.

Thus, the function of telomerase in the stem cell or progenitor cell compartment may be to slow the rate of telomere shortening and thus permit more doublings before the cells becomes senescent. For example, telomerase expression is reinduced in several tissues as part of a normal physiological response to hormonal stimulus *(63,64)* or in response to injury *(65)*. Therefore, in these tissues, telomerase activation may be a marker for cellular survival rather than immortality *(66)*. It would seem, in the prostate; for example, that telomerase expression is more ubiquitously expressed than would be expected for a stem cell population. Nevertheless, it may be that the level of telomerase expression may be the factor that distinguishes stem cells from other proliferating cells in these tissues.

6.3. Control of Stem Cell Survival and Expression of p27^{kip1}

The persistence of stem cell populations throughout adulthood likely depends on the survival of quiescent cells and on the ability of cycling cells to self-renew. Immunohistochemical and radiolabeling experiments have shown that the bulk of the proliferating pool in the normal human prostate is restricted to the basal compartment *(30,67,68)*, while the terminally differentiated luminal cells and neuroendocrine cells are considered postmitotic *(23)*. In the basal compartment, however, there is evidence that a subgroup of basal cells are quiescent or slow cycling as they express the cyclin-dependent kinase (cdk) inhibitor p27^{kip1} *(69)*.

The cdk inhibitor p27^{kip1} preferentially inactivates the cyclin E/cdk2 complex, thereby preventing cell cycle progression from G1 to S phase. In contrast to p21, p27^{kip1} expression is highest in quiescent cells *(70)*, and studies have suggested an antiapoptotic role for p27^{kip1} *(71,72)*. In the prostate of castrated rats, p27^{kip1} (normally restricted to a subgroup of basal cells) is upregulated in response to androgen deprivation. Waltregny et al. *(72)* speculated that upregulation of p27^{kip1} may protect basal cells from androgen deprivation-induced apoptosis.

6.4. External Controls of Stem Cell Self-Fate: The Stem Cell Niche

It is now becoming increasingly clear that paracrine factors, cell–matrix interactions, and intercellular interactions from neighboring cells also control stem cell function. The concept that the microenvironment or niche might control stem cell function was first realized in the hematopoietic system *(73)*, in which the survival of stem cells depends on factors secreted by other cell types. A wide range of secreted factors regulates stem cell proliferation and fate, including tumor growth factor-β (TGF-β) *(74)*, fibroblast growth factors (FGFs), epidermal growth factors (EGFs) *(75)*, and Wnts *(76)*. It is not known whether these families are involved in regulating stem cell fate in the prostate, but they show remarkable functional conservation between species and between tissues that self-renew. Moreover, they have been implicated in the genesis of a number of malignancies, including prostate cancer.

The histological structure of classical, self-renewing epithelia is clearly composed of structural units in which the stem cells, transit cells, and postmitotic differentiated cells are located in distinct regions *(11)*. For example, in the intestinal crypt, the stem cells are present near the crypt base, the transit-amplifying cells occupy approximately two-thirds of the height of the crypt, and the differentiated cells line the upper part of the crypt and villi *(3)*.

Similarly, in the epidermis, stem cells reside at the tips of the deep rete ridges in the palm and can be located on the basal layer by their high surface expression of integrins $\alpha_2\beta_1$ and $\alpha_3\beta_1$ *(42)*. Adhesion to the extracellular matrix is mediated by integrins, which hold cells in the right place in a tissue, and loss or alteration of integrin expression ensures departure from the stem cell niche through differentiation *(77)* or apoptosis *(78)*.

A similar histological arrangement is proposed for the prostate; by this arrangement, the stem cells can be distinguished from transit-amplifying cells on the basal layer by their high surface expression of integrin $\alpha_2\beta_1$ *(44)*. This population of integrin-bright cells are not confined to tips of acini, for example, but are randomly located throughout each acinus, with no more than one bright cell found together *(44)*. This is in contrast to the epidermis, where the integrin-bright cells are arranged in clusters *(42)*.

The Notch family of receptors and ligands plays a critical role in cell fate determination in embryonic *(79)* and adult tissue *(80)* and is an excellent example of local signaling that requires cell–cell contact. Notch signaling is best known for its role in lateral inhibition *(81)* and boundary definition *(82)*, in which it acts to maintain a pool of stem cells. However, Notch can work in the opposite way and promote differentiation of the receiving cell *(83)*. Thus, ligand expression in clusters of stem cells may be responsible for blocking Notch activation, thereby preventing these cells from differentiating. It is not known whether Notch is involved in the fate of prostate stem cells; however, a study determined that Notch1 is expressed in prostate epithelial cells during normal mouse development and is elevated in malignant prostate cancer cells of primary and metastatic tumors *(84)*.

7. OTHER PUTATIVE STEM/PROGENITOR CELL MARKERS OF THE PROSTATE

Over the last decade, several putative stem cell markers have been identified in the prostate, including prostate stem cell antigen (PSCA), *p63*, and pp32. Although their cellular location and function are still relatively unclear, their involvement in tissue development and cancer suggests a role in prostate stem cell biology.

7.1. Prostate Stem Cell Antigen

PSCA is a prostate-specific cell surface antigen with homology to stem cell antigen 2 (Sca-2), a marker for early hematopoietic development *(85)*. PSCA was originally identified by its high expression in the LAPC-4 prostatic cancer xenograft model *(86)*. RNA *in situ* hybridization localized PSCA to a subset of basal cells in normal prostate sections *(86)*. However, further work has demonstrated that it marks cells coexpressing basal and secretory cytokeratins *(87)*.

7.2. Homolog p63

The *p53* homolog *p63* encodes for different isotypes able either to transactivate *p53* reporter genes or TO act as *p53*-dominant negatives *(88)*. Homolog *p63* is expressed in the basal cells of many epithelial organs *(89)*, including the prostate, and its germline inactivation in the mouse results in agenesis of the prostate *(90)*. Expression of *p63* is restricted to the basal compartment, but it is expressed throughout the basal layer, suggesting that it is a marker for transit-amplifying cells as well as stem cells. Because of the defect in prostate development in *p63*(–/–) mice and because the expression is restricted to the basal compartment, *p63* may be essential either for maintaining a stem cell population or for differentiation to a secretory phenotype. Interestingly, *p63* is not expressed in prostatic intraepithelial neoplasia (PIN; a premalignant lesion of prostate cancer) or invasive prostatic cancer *(90)*, suggesting that it may have a role as a tumor suppresser.

7.3. pp32 Nuclear Phosphoprotein

The pp32 nuclear phosphoprotein initially was discovered as a constitutively expressed protein in neoplastic B cells *(91)* and subsequently in anatomically defined stem cell compartments, such as intestinal crypts. In the rat prostatic epithelium, pp32 is expressed exclusively in small acini at the periphery of the gland *(25)*, supporting morphological evidence that the stem cells are situated at distal buds in rat prostate *(92)*. In the human prostate, pp32 is expressed throughout the basal layer and is heterogeneously expressed in PIN and at high levels in poorly differentiated cancer *(93)*.

8. STEM CELL PLASTICITY

Until recently, most researchers believed that somatic stem cells were restricted in their potential to an individual organ system. However, reports have founded an emerging concept that adult stem cells may exhibit extraordinary plasticity in their differentiation repertoire, including the production of cell types outside their organ system. This potential has been proposed for stem cells found in the bone marrow, skeletal muscle, skin, adipose tissue, and central nervous system *(94)*. In one of the most definitive studies so far, Lagasse and coworkers *(95)* rescued mice from an otherwise lethal liver defect by transplantation of as few as 50 purified hematopoietic stem cells.

There are two possible mechanisms by which tissue stem cells may give rise to differentiated cells characteristic of another organ. Stem cells obtained from different sites might share properties and represent multipotent cells set aside during development. Alternatively, tissue-restricted stem cells may have the ability to become reprogrammed.

8.1. Inductive Effect of Mesenchyme on Prostate Morphology and Cytodifferentiation

Prostate development occurs via mesenchymal–epithelial interactions in which urogenital sinus mesenchyme induces epithelial morphogenesis *(96)*. Many studies using tissue recombinants have established that the mesenchyme is the target and mediator of androgenic effects on the developing epithelium *(97)*. However, the expression of androgen-dependent secretory proteins is dependent on intraepithelial androgen receptor *(98)*, as is the survival of the secretory luminal cells.

Tissue recombination experiments have been used to define further the characteristics of mesenchymal–epithelial interactions. For example, when mesenchyme is recombined with epithelium (adult or neonatal) from another organ, an instructive induction may occur in which the developmental fate of the epithelium is changed. In instructive inductions, the epithelium commonly takes on the characteristics of the epithelium normally associated with the inducing mesenchyme.

However, instructive inductions of the urogenital tract depend on many factors, including the degree to which the epithelium and mesenchyme share a developmental history. For example, mesenchyme that normally interacts with epithelium from one germ layer can alter the fate of other epithelia from the same germ layer, but is not permissive with epithelia from another germ layer For instance, epithelium from the bladder, urethra, and vagina (derived from the endodermal urogenital sinus) can be instructively induced by mesenchyme from the prostatic anlagen, another region of the urogenital sinus, to form prostate if the tissue recombinant is grown in the intact male host *(99)*.

In contrast, seminal vesicle mesenchyme, normally associated with a mesodermal epithelium, cannot induce bladder or urethral epithelium to develop into seminal vesicle. Interestingly, this interaction results in the development of prostate tissue, but only in the intact host *(100)*.

These results indicate that there are restrictions in the developmental potential of urogenital epithelium because a truly instructive induction would have predicted seminal vesicle development from such an interaction. Nevertheless, these results demonstrate that adult stem cells from the urogenital tract, which have always been thought of as irreversibly determined, can be reprogrammed in the presence of mesenchyme. As proposed above, either adult stem cells retain a developmental plasticity equivalent to their embryonic counterparts or they are reprogrammed to express the induced phenotype.

The role of stromal (adult connective tissue)–epithelial interactions has received considerably less attention than that of mesenchymal–epithelial interactions, probably because of the fact that stromal interactions in adult prostate have more to do with homeostasis than development. Nonetheless, there is evidence that stroma of nonurogenital origin can induce prostate regeneration *(101,102)*. These results imply that, in the adult prostate, epithelial lineage is inherent and can only be re-reprogrammed in the presence of an instructive, embryonic connective tissue.

As with all the reports demonstrating enhanced plasticity from some somatic stem cells, uncertainty remains as to the possible mechanism. It is clear that more work is required to determine which microenvironments can induce stem cells to develop along different pathways.

8.2. Are Somatic Stem Cells Pluripotent or Lineage Restricted?

Despite the overwhelming evidence that some adult stem cells have the ability to generate cells of all the germ layers, there is still controversy regarding the rigor of the experimental methods used to demonstrate lineage heterogeneity *(103)*. So far, most observations have been obtained with populations of cells and, as such, have not demonstrated transdifferentiation at a clonal level. One exception to this was the study by Bjornson and coworkers *(104)* in which clonally derived neural stem cells were observed, after extensive passaging in culture, to generate hematopoietic cells in mice after lethal irradiation.

To investigate whether this was a consistent or unusual feature of neural stem cells, Morshead et al. *(105)* attempted to reproduce these findings. Despite numerous attempts, their experiments failed to show a single event of hematopoietic reconstitution. Morshead and coworkers suggested that the switch from a neural to blood fate was a rare event and was because of genetic or epigenetic alterations in the neurosphere cultures.

Other possible mechanisms for hematopoietic competence of neural stem cells include: (1) the existence of multiple varieties of multipotent neural stem cells, only one of which has the capacity for hematopoietic fates; (2) the epigenetic and genetic alterations that are known to occur in cells in vitro may cause alterations in the differentiation potential of the cells; and (3) apparent lineage heterologous behavior may be the result of host–donor cell fusion *(106)*.

To resolve the issue of stem cell plasticity, D'Amour and Gage *(106)* suggested that future studies must strictly adhere to rigorous experimental methods, such as (1) determining the absolute identity of a stem cell without prior culturing periods; (2) demonstrating functional differentiation, not just morphology or gene expression; and (3) sustaining the contribution to the target tissue over time.

9. STEM CELLS AND CANCER: THE ORIGINS OF PROSTATE CARCINOMAS

Pierce and coworkers *(107,108)* first proposed the concept that tumors might arise from the transformation of normal stem cells. In this model, carcinogenesis results in a block in the differentiation process, allowing accumulation of cycling cells of varying phenotypes. Thus, the tumor retains the basic phenotypic potential of the tissue from which it was derived, but expresses this potential to varying degrees compared to normal tissue.

Although this model has been generally accepted for teratocarinomas *(107,108)* and hematopoietic tumors *(109,110)*, dedifferentiation is still widely accepted as the cellular mechanism for development of carcinomas. This is particularly true for prostate cancer, for which it has long been postulated that these cancers arise from transformed secretory cells *(16,111,112)* or progenitors with an intermediate phenotype *(8,32,87)*. Dedifferentiation has been generally accepted as the model of choice because the majority of malignant prostate cells have a luminal phenotype (PSA[+]/CK18[+]), and the minority of cells coexpress basal and luminal markers *(32)*. Moreover, basal cells are "absent" in high-grade PIN *(113)*.

9.1. Stem Cells As Targets of Mutation

Although the number of mutations necessary to induce neoplastic change is debatable, it is generally accepted that cancer development occupies a significant time span in humans. Stem cells are the only cells maintained throughout the lifetime of an individual. Mainly for this reason, stem cells have been considered carcinogen target cells.

Many pathways that regulate stem cell maintenance are associated with carcinogenesis. For example, *bcl-2* and telomerase (normally restricted to the basal cells in normal prostate) are overexpressed in prostate cancer *(52,58)*. Many other pathways, such as Notch, Sonic hedgehog, and Wnt pathways that regulate stem cell self-renewal, are also associated with oncogenesis (reviewed in ref. *114*). The demonstration that genes associated with stem cell maintenance can be detected in premalignant lesions and at high levels in advanced cancers suggests that stem cell genes are not reexpressed in tumors, but rather the stem cells are targeted in oncogenesis. Furthermore, the observation that the majority of prostate cancer cells express markers specific for the secretory cell compartment simply reflects the ability of the cancer stem cell to give rise to progeny characteristic of the tissue from which they were derived.

9.2. Evidence for Cancer Stem Cells

It has been shown for leukemia and some solid cancers that only a small subset of cells is clonogenic in culture and in vivo *(115–117)*. This led to the hypothesis that only a few cancer cells are actually tumorigenic, and that this subgroup could be considered as cancer stem cells *(117)*. To prove this possibility, subsets of uncultured cancer cells enriched for the ability to form new tumors would have to be purified. This was accomplished by Bonnet and Dick *(118)*, who showed that a small subgroup of human acute myeloid leukemic (AML) cells expressed the stem cell phenotype CD34[+], and only these cells were able to transfer AML to SCID mice.

9.3. Stem Cells and Androgen-Independent Prostate Cancer

If a small population of cancer stem cells drives tumor growth and metastasis, this may explain the failure of existing therapies to eradicate solid tumors. Treatment for prostate cancer, for example, has been developed largely against the bulk population of tumor cells, which are largely androgen dependent. While the majority of patients with advanced disease respond initially to androgen ablation therapy, most develop androgen-independent tumors that are inevitably fatal.

One possibility for the eventual failure of androgen ablation therapy is that androgen-resistant tumor growth arises as a result of alterations in the androgen receptor or its signaling pathway. This hypothesis is based on the observation that most hormone-refractory tumors continue to express androgen receptor and androgen-regulated genes, such as PSA *(119)*. Thus, the tumor cells may have adapted by increasing their sensitivity to androgen through increased androgen receptor expression *(120)*. Androgen receptor amplifications *(121)* and mutations *(122)* have been reported in patients with hormone-refractory cancer, and androgen receptor pathways can be activated by cross talk with other signaling pathways *(123)*.

The other possibility is that androgen ablation fails to kill the tumor stem cell, resulting in expansion of androgen-independent progenitor cells, driven by the stem cells (Fig. 2). Their overproduction in the tumor as a consequence of androgen ablation imparts an undifferentiated appearance to the tumor.

10. CONCLUSIONS AND FUTURE DIRECTIONS

The understanding of prostate stem cells is likely to progress rapidly in the next few years as these elusive cells are defined and isolated, and their properties are identified. Progress will require further characterization of molecular markers and the establishment of in vitro culture systems to maintain the cells. Stem cells can be transfected, and the effect of the transfected gene on the regenerating tissue can be monitored. Ultimately, this technology could be used to manipulate stem cells for clinical benefit; altering stem cell kinetics, stem cell number, and sensitivity to apoptosis.

It is increasingly obvious that, in defining the identity of stem cells at the molecular level, not only the ability of these cells to express different classes of molecules should be considered, but also the relative levels of expression. Integrin $\alpha_2\beta_1$ is a case in point. The properties of stem cells are also likely to be influenced by their physical position in the context of the three-dimensional structure of the gland. Thus, cell–cell and cell–matrix

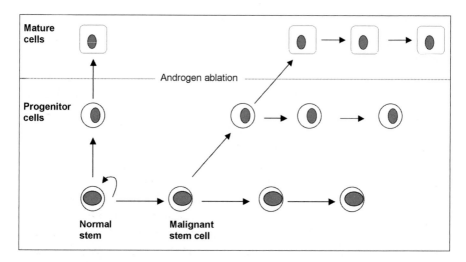

Fig. 2. Stem cell model of normal tissue renewal and prostate cancer. The malignant stem cell arises from transformation of the normal stem cell. The tumor is made up of cells that are arrested in their maturation and that do not die. The histology of the tumor reflects the stage of differentiation arrest. Androgen ablation results in overproduction of malignant stem cells and their progenitors, imparting the undifferentiated, not dedifferentiated, appearance to the tumor.

interactions become important in determining and maintaining stem cell populations.

It was through the investigation of aberrant cell growth that many of the genes involved in stem cell function were first identified. The antiapoptotic gene *bcl-2* and the enzyme telomerase are expressed in high-grade PIN and are restricted to the basal compartment in normal prostate *(48,58)*. Similarly, altered expression of growth factors and their receptors, such as EGFs and FGFs, have been implicated in the etiology of prostate cancer. It remains to be seen which members of these families play a part in stem cell identity and function in the prostate.

The cell fate decisions made by stem cells are likely to be controlled by the *Notch* pathway. *Notch* is expressed during development of the mouse and is elevated in prostate cancer *(84)*. The phenotype of a conditional ablation of the *Notch* gene in the prostate will provide answers as to the role of *Notch* in prostate epithelial stem cell fate and cancer.

If the growth of solid cancers was driven by rare cancer stem cells, it would have profound implications for cancer therapy. Therapies that are specifically directed against cancer stem cells might result in more durable responses and even cures for metastatic disease.

It also remains to be seen if the stem cells of the prostate epithelium, like the neural stem cells *(104)*, are capable of giving rise to differentiated cell lineages in heterologous tissue.

REFERENCES

1. Dexter, T. M., Allen, T. C., and Lajtha, L. G. (1977). Conditions controlling the proliferation of haemopoietic stem cells in vitro. J Cell Physiol 91, 335–344.
2. Hall, P. A., and Watt, F. M. (1989). Stem cells: the generation and maintenance of cellular diversity. Development 106, 619–633.
3. Potten, C. S., and Loeffler, M. (1990). Stem cells–attributes, cycles, spirals, pitfalls and uncertainties–lessons for and from the crypt. Development 110, 1001–1020.
4. Mao, P., and Angrist, A. (1966). The fine structure of the basal cell of human prostate. Lab Invest 15, 1768–1782.
5. Dermer, G. (1978). Basal cell proliferation in benign prostatic hyperplasia. Cancer 41, 1857–1862.
6. Heatfield, B., Sanefuji, H., and Trump, B. (1979). Studies on carcinogenesis of human prostate. IV. Comparison of normal and neoplastic prostate during long-term explant culture. Scanning Microsc 3, 645–656.
7. Brandes, D. (1966). The fine structure and histochemistry of prostatic glands in relation to sex hormones. Int Rev Cytol 20, 207–276.
8. Isaacs, J. T., and Coffey, D. S. (1989). Etiology and disease process of benign prostatic hyperplasia. Prostate Suppl 2, 33–50.
9. Lajtha, L. G. (1979). Stem cell concepts. Differentiation 14, 23–34.
10. Morrison, S. J., Shah, N. M., and Anderson, D. J. (1997). Regulatory mechanisms in stem cell biology. Cell 88, 287–298.
11. Slack, J. M. W. (2000). Stem cells in epithelial tissues. Science 287, 1431–1432.
12. Watt, F. M., and Hogan, B. L. (2000). Out of Eden: stem cells and their niches. Science 287, 1427–1430.
13. Lavker, R. M., Miller, S., Wilson, C., et al. (1993). Hair follicle stem cells: their location, role in hair cycle, and involvement in skin tumor formation. J Invest Dermatol Suppl 101, 16S–26S.
14. Morrison, S. J., and Weissman, I. L. (1994). The long-term repopulating subset of hematopoetic stem cells is deterministic and isolatable by phenotype. Immu-nity 1, 661–673.
15. Wernert, N., Seitz, G., and Achtstätter, T. (1987). Immunohistochemical investigation of different cytokeratins and vimentin in the prostate from the fetal period up to adulthood and in prostate carcinoma. Pathol Res Pract 182, 617–626.
16. Okada, H., Tsubura, A., Okamura, A., et al. (1992). Keratin profiles in normal/hyperplastic prostates and prostate carcinoma. Virchows Arch A 421, 157–161.
17. Achtstätter, T., Moll, R., Moor, B., and Franke, W. W. (1985). Cytokeratin polypeptide patterns of different epithelia of the human male urogenital tract:

immunofluorescence and gel electrophoretic studies. J Histochem Cytochem 33, 415–426.

18. Sar, M., Lubahn, D. B., French, F. S., and Wilson, E. M. (1990). Immuno-histochemical localisation of the androgen receptor in rat and human tissues. Endo-crinology 127, 3180–3186.

19. Leav, I., McNeal, J. E., Kwan, P. W., Komminoth, P., and Merk, F. B. (1996). Androgen receptor expression in prostatic dysplasia (prostatic intraepithelial neoplasia). In the human prostate: an immunohistochamical and *in situ* hybridization study. Prostate 29, 137–145.

20. Kyprianou, N., and Isaacs, J. T. (1988). Activation of programmed cell death in the rat ventral prostate after castration. Endocrinology 122, 552–562.

21. Bonkoff, H., and Remberger, K. (1993). Widespread distribution of nuclear androgen receptors in the basal cell layer of the normal and hyperplastic human prostate. Virchows Arch Pathol Anat Histopathol 422, 35–38.

22. Aumüller, G. (1983). Morphologic and endocrine aspects of prostatic function. Prostate 4, 195–214.

23. Bonkoff, H., Stein, U., and Remberger, K. (1995). Endocrine-paracrine cell types in the prostate and prostatic adenocarcinoma are postmitotic cells. Hum Pathol 26, 167–170.

24. English, H. F., Kyprianou, N., and Isaacs, J. T. (1989). Relationship between DNA fragmentation and apoptosis in the programmed cell-death in the rat prostate following castration. Prostate 15, 233–250.

25. Walensky, L. D., Coffey, D. S., Chen, T.-H., Wu, T.-C., and Pasternack, G. R. (1993). A novel Mr 32,000 nuclear phosphopreotein is selectively expressed in cells competent for self-renewal. Cancer Res 53, 4720–4726.

26. Isaacs, J. T. (1987). Control of cell proliferation and cell death in the normal and neoplastic prostate: a stem cell model. In: Rodgers, C. H., Coffey, D. S., Cunha, G., Grayhack, J. T., Hinman, F., Jr., and Horton, R., eds., Benign Prostatic Hyperplasia. Washington, DC: US Department of Health and Human Services, pp. 85–94. NIH Publication 87-2881.

27. Evans, G. S., and Chandler, J. A. (1987). Cell proliferation studies in the rat prostate: II. The effects of castration and androgen-induced regeneration upon basal and secretory cell proliferation. Prostate 11, 339–351.

28. English, H. F., Santen, R. J., and Isaacs, J. T. (1987). Response of glandular vs basal rat ventral prostate epithelial cells to androgen withdrawal and replacement. Prostate 11, 229–242.

29. Scholzen, T., and Gerdes, J. (2000). The Ki-67 protein: from the known and the unknown. J Cell Physiol 182, 311–322.

30. Bonkhoff, H., Stein, U., and Remberger, K. (1994). The proliferative function of basal cells in the normal and hyperplastic human prostate. Prostate 24, 114–118.

31. Van der Kwast, T. H., Tètu, B., Suburu, E. R., Gomez, J., Lemay, M., and Labrie, F. (1998). Cycling activity of benign prostatic epithelial cells during long-term androgen blockage: evidence for self-renewal of luminal cells. J Pathol 186, 406–409.

32. Verhagen, A. P. M., Ramaekers, F. C. S., Aalders, T. W., Schaafsma, H. E., Debruyne, F. M. J., and Schalken, J. A. (1992). Colocalization of basal and luminal cell-type cytokeratins in human prostate cancer. Cancer Res 52, 6182–6187.
33. Sherwood, E. R., Berg, L. A., Mitchell, N. J., McNeal, J. E., Kozlowski, J. M., and Lee, C. (1990). Differential cytokeratin expression in normal hyperplastic and malignant epithelial cells from human prostate. J Urol 143, 167–171.
34. Xue, Y., Verhofstad, A., Lange, W., et al. (1997). Prostatic neuroendocrine cells have a unique keratin expression pattern and do not express Bcl-2: cell kinetic features of neuroendocrine cells in the human prostate. Am J Pathol 151, 1759–1765.
35. Van Leenders, G., Dijkman, H., Hulsbergen-van de Kaa, C., Ruiter, D., and Schalken, J. (2000). Demonstration of intermediate cells during human prostate epithelial differentiation *in situ* and in vitro using triple-staining confocal scanning microscopy. Lab Invest 80, 1251–1258.
36. Hudson, D. L., Guy, A. T., Fry, P., O'Hare, M. J., Watt, F. M., and Masters J. R. W. (2001). Epithelial cell differentiation pathways in the human prostate: identification of intermediate phenotypes by keratin expression. J Histochem Cytochem 49, 271–278.
37. Bonkhoff, H., Stein, U., and Remberger, K. (1994). Multidirectional differentiation in the normal, hyperplastic, and neoplastic human prostate. Simultaneous demonstration of cell specific epithelial markers. Hum Pathol 25, 42–46.
38. Robinson, E. J., Neal, D. E., and Collins, A. T. (1998). Basal cells are progenitors of luminal cells in primary cultures of differentiating human prostatic epithelium. Prostate 37, 149–160.
39. Hudson, D. L., O'Hare, M. J., Watt, F. M., and Masters, J. R. W. (2000). Proliferative heterogeneity in the human prostate: evidence for epithelial stem cells. Lab Invest 80, 1243–1250.
40. Lipschutz, J. H., Foster, B. A., and Cunha, G. R. (1997). Differentiation of rat neonatal ventral prostates grown in a serum-free organ culture system. Prostate 32, 35–42.
41. Lang, S. H., Stark, M., Collins, A., Paul, A. B., Stower, M. J., and Maitland, N. J. (2001). Experimental prostate epithelial morphogenesis in response to stromal and three-dimensional Matrigel culture. Cell Growth Differ 12, 631–640.
42. Jones, P. H., Harper, S., and Watt, F. M. (1995). Stem cell patterning and fate in human epidermis. Cell 60, 83–93.
43. Shinohara, T., Avarbock, M. R., and Brinster, R. L. (1999). b_1- and a_6-integrin are surface markers on mouse spermatogonial stem cells. Proc Natl Acad Sci U S A 96, 5504–5509.
44. Collins, A. T., Habib, F. K., Maitland, N. J., and Neal D. E. (2001). Identification and isolation of human prostate epithelial stem cells based on a_2b_1 integrin expression. J Cell Sci 114, 3865–3872.
45. Raff, M. C. (1992). Social controls on cell survival and cell death. Nature 356, 397–400.
46. Krajewska, M., Wang, H.-G., Krajewski, S., et al. (1997). Immunohistochemi-

cal analysis of in vivo patterns of expression of CCP32 (Caspase-3), a cell death protease. Cancer Res 57, 1605–1613.

47. Hockenberry, D., Nunez, G., Milliman, C., Schreiber, R., and Korsmeyer, S. J. (1990). Bcl-2 is an inner mitochondrial membrane protein that blocks programmed cell death. Nature 348, 334–336.

48. Columbel, M., Symmans, G., Gil, S., et al. (1993). Detection of the apoptosis-suppressing oncoprotein bcl-2 in hormone-refractory human prostate cancer. Am J Clin Pathol 80, 850–854.

49. Bonkhoff, H., and Remberger, K. (1996). Differentiation pathways and histogenetic aspects of normal and abnormal prostatic growth—a stem cell model. Prostate 28, 98–106.

50. Hockenberry, D., Zutter, M., Hickey, W., Nahm, M., and Korsmeyer, S. J. (1991). Bcl-2 protein is topographically restricted to tissues characterised by apoptotic cell death. Proc Natl Acad Sci U S A 88, 6961–6965.

51. Lin, H. K., Yeh, S. Y., Kang, H. Y., and Chang, C. S. (2001). Akt suppresses androgen-induced apoptosis by phosphorylating and inhibiting the androgen receptor. Proc Natl Acad Sci U S A 98, 7200–7205.

52. McDonnell, T. J., Troncoso, P., Brisbay, S. M., et al. (1992). Expression of the protooncogene bcl-2 in the prostate and its association with emergence of androgen-independent prostatic cancer. Cancer Res 52, 6940–6944.

53. Morrison, S. J., Prowse, K. R., Ho, P., and Weissman, I. L. (1996). Telomerase activity of hematopoietic cells is associated with self-renewal potential. Immunity 5, 207–216.

54. Campisi, J. (1997). The biology of replicative senescence. Eur J Cancer 5, 703–709.

55. Kim, N. W., Piatyszek, M. A., Prowse, K. R., et al. (1994). Specific association of human telomerase activity with immortal cells and cancer. Science 266, 2011–2015.

56. Levy, M. Z., Allsopp, R. C., Futcher, A. B., Greider, C. W., and Harley, C. B. (1992). Telomere end-replication problem and cell-aging. J Mol Biol 22, 951–960.

57. Harley, C. B. (1991). Telomere loss: mitotic clock or genetic time bomb? Mutat Res 256, 271–282.

58. Paradis, V., Dargére, D., Laurendeau, I., et al. (1999). Expression of the RNA component of human telomerase (hTR) in prostatic cancer, prostatic intraepithelial neoplasia, and normal prostate tissue. J Pathol 189, 213–218.

59. Taylor, R. S., Ramirez, R. D., Ogoshi, M., Chaffins, M., Piatyszek, M. A., and Shay, J. W. (1996). Detection of telomerase activity in malignant and nonmalignant skin conditions. J Invest Dermatol 106, 759–765.

60. Hiyama, E., Hiyama, K., Tatsumoto, N., Shay, J. W., and Yokoyama, T. (1996). Telomerase activity in human intestine. Int J Oncol 9, 453–458.

61. Norrback, K. F., and Roos, G. (1997). Telomeres and telomerase in normal and malignant haematopoietic cells. Eur J Cancer 33, 774–780.

62. Vaziri, H., Dragowska, W., Allsopp, R. C., Thomas, T. E., Harley, C. B., and Lansdorp, P. M. (1994). Evidence for a mitotic clock in human hematopoietic cell: loss of telomeric DNA with age. Proc Natl Acad Sci U S A 91, 9857–9860.

63. Tanaka, M., Kyo, S., Takakura, M., et al. (1998). Expression of telomerase activity in human endometrium is localised to epithelial glandular cells and regulated in a menstrual phase-dependent manner correlated with cell proliferation. Am J Pathol 153, 1985–1991.
64. Meeker, A. K., Sommerfeld, H. J., and Coffey, D. S. (1996). Telomerase is activated in the prostate and seminal vesicles of the castrated rat. Endocrinology 137, 5743–5746.
65. Driscoll, B., Buckley, S., Chi Bui, K., Anderson, K. D., and Warburton, D. (2000). Telomerase in alveolar epithelial development and repair. Am J Physiol Lung Cell Mol Physiol 279, L1191–L1198.
66. Belair, C. D., Yeager, T. R., Lopez, P. M., and Reznikoff, C. A. (1997). Telomerase activity: a biomarker of cell proliferation not malignant transformation. Proc Natl Acad Sci U S A 94, 13,677–13,682.
67. McNeal, J. E., Haillot, O., and Yemoto, C. (1995). Cell proliferation in dysplasia of the prostate: analysis by PCNA immunostaining. Prostate 27, 255–268.
68. Merchant, D. J., Clarke, S. M., Ives, K., and Harris, S. (1983). Primary explant culture: an in vitro model of the human prostate. Prostate 4, 523–542.
69. De Marzo, A. M., Meeker, A. K., Epstein, J. I., and Coffey, D. S. (1998). Prostate stem cell compartments. Expression of the cell cycle inhibitor p27kip1 in normal, hyperplastic, and neoplastic cells. Am J Pathol 153, 911–919.
70. Rivard, N., L'Alleimain, G., Bartek, J., and Pouyssegur, J. (1996). Abrogation of p27 by cDNA antisense suppresses quiescence (G0 state) in fibroblasts. J Biol Chem 271, 18,337–18,341.
71. Hiromura, K., Fero, M. L., Roberts, J. M., and Shankland, S. J. (1999). Modulation of apoptosis by the cyclin-dependent kinase inhibitor p27. J Clin Invest 103, 597–604.
72. Waltregny, D., Leav, I., Signoretti, S., et al. (2001). Androgen-driven prostate epithelial cell proliferation and differentiation in vivo involve the regulation of p27. Mol Endocrinol 15, 765–782.
73. Schofield, R. (1978). The relationship between the spleen colony-forming cell and the haematopoietic stem cell. Blood 4, 7–25.
74. Shah, N. M., Groves, D. J., and Anderson, D. J. (1996). Alternative neural crest fates are instructively promoted by TGF beta superfamily members. Cell 85, 331–343.
75. Reynolds, B. A., and Weiss, S. (1996). Clonal and population analyses demonstrate that an EGF-responsive mammalian embryonic CNS precursor is a stem cell. Dev Biol 175, 1–13.
76. Cadigan, K. M., and Nusse, R. (1997). Wnt signalling: a common theme in animal development. Genes Dev 11, 3286–3305.
77. Zhu, A. J., Haase, I., and Watt, F. M. (1999). Signaling via a1 integrins and mitogen-activated protein kinase determines human epidermal stem cell fate in vitro. Proc Natl Acad Sci U S A 96, 6728–6733.
78. Frisch, S. M., and Francis, H. (1994). Disruption of epithelial cell-matrix interactions induces apoptosis. J Cell Biol 124, 619–626.
79. Artavanis-Tsakonas, S., Rand, M. D., and Lake, R. J. (1999). Notch signalling: cell fate control and signal integration in development. Science 284, 770–776.

80. Milner, L. A., Kopan, R., Martin, D. I., and Bernstein, I. D. (1994). A human homolog of the *Drosophila* developmental gene, *Notch*, is expressed in CD34[+] hematopoietic precursors. Blood 83, 2057–2062.

81. Lewis, J. (1998). Notch signalling and the control of cell fate choices in vertebrates. Semin Cell Dev Biol 9, 583–589.

82. Irvine, K. D. (1999). Fringe, Notch, and making developmental boundaries. Curr Opin Genet Dev 9, 434–441.

83. Lowell, S., Jones, P., LeRoux, I., Dunne, J., and Watt, F. M. (2000). Stimulation of human epidermal differentiation by Delta-Notch signalling at the boundaries of stem-cell clusters. Curr Biol 10, 491–500.

84. Shou, J., Ross, S., Koeppen, H., de Suavage, F. J., and Gao, W.-Q. (2001). Dynamics of Notch expression during murine prostate development and tumorigenesis. Cancer Res 61, 7291–7297.

85. Antica, M., Wu, L., and Scolley, R. (1997). Stem cell antigen 2 expression in adult and developing mice. Immunol Lett 55, 47–51.

86. Reiter, R. E., Gu, Z., Watabe, T., et al. (1998). Prostate stem cell antigen: a cell surface marker overexpressed in prostate cancer. Proc Natl Acad Sci U S A 95, 1735–1740.

87. Bui, M., and Reiter, R. E. (1999). Stem cell genes in androgen-independent prostate cancer. Cancer Metastasis Rev 17, 391–399.

88. Yang, A., Kaghad, M., Wang, Y., et al. (1998). *p63*, a *p53* homolog at 3q27–29, encodes multiple products with transactivating, death-inducing, and dominant-negative activities. Mol Cell 2, 305–316.

89. Parsa, R., Yang, A., McKeon, F., and Green, H. (1999). Association of *p63* with proliferative potential in normal and neoplastic human keratinocytes. J Invest Dermatol 113, 714–718.

90. Signoretti, S., Waltregny, D., Dilks, J., et al. (2000). *p63* is a prostate basal cell marker and is required for prostate development. Am J Pathol 157, 1769–1775.

91. Malek, S. N., Katumuluwa, A. I., and Pasternack, G. R. (1990). Identification and preliminary characterisation of two related proliferation-associated nuclear phosphoproteins. J Biol Chem 265, 13,400–13,409.

92. Hayashi, N., Sugimura, Y., Kawamura, J., Donjacour, A. A., and Cunha, G. R. (1991). Morphological and functional heterogeneity in the rat prostatic gland. Biol Reprod 45, 308–321.

93. Kadkol, S. S., Brody, J. R., Epstein, J. I., Kuhajda, F. P., and Pasternack, G. R. (1998). Novel nuclear phosphoprotein pp32 is highly expressed in intermediate- and high-grade prostate cancer. Prostate 34, 231–237.

94. Wulf, G. G., Jackson, K. A., and Goodell, M. A. (2001). Somatic stem cell plasticity. Current evidence and emerging concepts. Exp Hematol 29, 1361–1370.

95. Lagasse, E., Connors, H., Al-Dhalimy, M., et al. (2000). Purified hematopoietic stem cells can differentiate to hepatocytes in vivo. Nature Med 6, 1229–1234.

96. Cunha, G. R. (1976). Epithelial-stromal interactions is development of the urogenital tract. Int Rev Cytol 47, 137–194.

97. Cunha, G. R., Donjacour, A. A., Cooke, P. S., et al. (1987). The endocrinology and developmental biology of the prostate. Endocr Rev 8, 338–363.
98. Cunha, G. R., and Young, P. (1991). Inability of Tfm (testicular feminization) epithelial cells to express androgen-dependent seminal vesicle secretory proteins in chimeric tissue recombinants. Endocr J 128, 3293–3298.
99. Donjacour, A. A., and Cunha, G. R. (1993). Assessment of prostatic protein secretion in tissue recombinants made of urogenital sinus mesenchyme and urothelium from normal or androgen-insensitive mice. Endocrinology 132, 2342–2350.
100. Donjacour, A. A., and Cunha, G. R. (1995). Induction of prostatic morphology and secretion in urothelium by seminal vesicle mesenchyme. Development 121, 2199–2207.
101. Hayward, S. W, Del Buono, R., Deshpande, N., and Hall, P. A. (1992). A functional model of adult human prostate epithelium. J Cell Sci 102, 361–372.
102. Gao, J., Arnold, J. T., and Isaacs, J. T. (2001). Conversion from a paracrine to an autocrine mechanism of androgen-stimulated growth during malignant transformation of prostate epithelial cells. Cancer Res 61, 5038–5044.
103. Anderson, D. J., Gage, F. H., and Weissman, I. L. (2001). Can stem cells cross lineage boundaries? Nature Med 7, 393–395.
104. Bjornson, C. R., Rietze, R. L., Reynolds, B. A., Magli, M. C., and Vescovi, A. L. (1999). Turning brain into blood: a hematopoietic fate adopted by adult neural stem cells in vivo. Science 283, 534–537.
105. Morshead, C. M., Benveniste, P., Iscove, N. N., and ven der Kooy, D. (2002). Hematopoietic competence is a rare property of neural stem cells that may depend on genetic and epigenetic alterations. Nature Med 8, 268–273.
106. D'Amour, K. A., and Gage, F. H. (2002). Are somatic stem cells pluripotent or lineage-restricted. Nature Med 8, 213–214.
107. Pierce, G. B., and Johnson, L. D. (1971). Differentiation and cancer. In vitro 7, 140–145.
108. Pierce, G. B., Shikes, R., and Fink, L. M. (1978). Cancer: A Problem of Developmental Biology. Englewood Cliffs, NJ: Prentice Hall, 1–242.
109. Kersey, J. H. (1981). Lymphoid progenitor cells and acute lymphoblastic leukemia: studies with monoclonal antibodies. J Clin Immunol 1, 210–217.
110. Aisenberg, A. C. (1983). Cell lineage is lymphoproliferative disease. Am J Med 74, 679–685.
111. Nagle, R. B., Ahmann, F. R., McDaniel, K. M., Paquin, M. L., Clark, V. A., and Celniker, A. (1987). Cytokeratin characterisation of human prostatic carcinoma and its derived cell lines. Cancer Res 47, 281–286.
112. De Marzo, A. M., Nelson, W. G., Meeker, A. K., and Coffey, D. S. (1998). Stem cell features of benign and malignant prostate epithelial cells. J Urol 160, 2381–2392.
113. Montironi, R., Bostwick, D., Bonkhoff, H., et al. (1996). Workgroup 1: Origins of prostate cancer. Cancer 78, 362–365.
114. Taipale, J., and Beachy, P. A. (2001). The Hedgehog and Wnt signalling pathways in cancer. Nature 411, 349–354.

115. Wodinsky, I., Swiniarski, J., and Kensler, C. J. (1967). Spleen colony studies of leukemia L1210. I. Growth kinetics of lymphocytic L1210 cells in vivo as determined by spleen colony assay. Cancer Chemother Rep 51, 415–421.
116. Fidler, I. J., and Kripke, M. L. (1977). Metastasis results from preexisting variant cells within a malignant tumor. Science 197, 893–895.
117. Hamburger, A. W., and Salmon, S. E. (1977). Primary bioassay of human tumor stem cells. Science 197, 461–463.
118. Bonnet, D., and Dick, J. E. (1997). Human acute myeloid leukemia is organized as a hierarchy that originates from a primitive hematopoietic cell. Nature Med 3, 730–737.
119. Koivisto, P., Kolmer, M., Visakorpi, T., and Kallioniemi, O. P. (1998). Androgen receptor gene and hormonal therapy failure of prostate cancer. Am J Pathol 152, 1–9.
120. Kokontis, J. M., Hay, N., and Liao, S. (1998). Progression of LNCaP prostate tumor cells during androgen deprivation: hormone-independent growth, repression of proliferation by androgen, and role for p27Kip1 in androgen-induced cell cycle arrest. Mol Endocrinol 12, 941–953.
121. Visakorpi, T., Hyytinen, E., Doivisto, P., et al. (1995). In vivo amplification of the androgen receptor gene and progression of human prostate cancer. Nature Gen 9, 401–406.
122. Tilley, W. D., Buchanan, G., Hickey, T. E., and Bentel, J. M. (1996). Mutations in the androgen receptor gene are associated with progression of human prostate cancer to androgen independence. Clin Cancer Res 2, 277–285.
123. Hobisch, A., Eder, I. E., Putz, T., et al. (1998). Interleukin-6 regulates prostate-specific protein expression in prostate carcinoma cells by activation of the androgen receptor. Cancer Res 58, 4640–4545.

10

Mammary Epithelial Stem Cells

Elizabeth Anderson and Robert B. Clarke

1. INTRODUCTION

Mammary gland development and function would not be possible without tissue-specific stem cells. The cycle of pregnancy-associated proliferation, differentiation, apoptosis, and remodeling, which may occur many times during the mammalian reproductive lifespan, can only be explained by the presence of a long-lived population of stem cells with a near-unlimited capacity to generate functional cells. The aim of this chapter, therefore, is to review the evidence for the presence of stem cells in the mammary gland, to describe the progress in isolating and characterizing these stem cells, to discuss the role of stem cells in mammary gland carcinogenesis, and to speculate whether mammary stem cells share properties with stem cells in other adult tissues.

2. STRUCTURE AND HISTOLOGY OF NORMAL AND MALIGNANT HUMAN BREAST

2.1. Normal Breast Tissue

The human breast, in common with the mammary glands of other species, contains both epithelial and mesenchymal components. The adult human breast contains a number of "treelike" glandular structures derived by dichotomous branching of each of several ducts arising from the nipple. The major functional units of these glands are the lobular structures, situated at the end of the terminal ductules, which comprise several smaller blind-ended ductules often referred to as *terminal ductal lobulo-alveolar units* (TDLUs). The TDLUs are lined by a continuous layer of luminal epithelial cells, which in turn are enmeshed by myoepithelial cells that contact the basement membrane. The whole structure is then surrounded by delimiting fibroblasts and embedded in a specialized intralobular stroma *(1)*.

The luminal epithelial and myoepithelial populations are the two major epithelial cell types in the mammary gland, but there is also a third, far less

From: *Adult Stem Cells*
Edited by: K. Turksen © Humana Press Inc., Totowa, NJ

Table 1
Cell Surface and Cytoskeletal Markers Used to Characterize Epithelial Cell Populations in the Human Mammary Gland

Antigen	Luminal epithelium	Intermediate cells	Myoepithelium	Reference
ESA/Ber-EP4/Ep-CAM	+	+	+/–	*6, 7*
CALLA	–	–	+	*3*
MUC-1 (sialomucin)	+	–	–	*2*
CK8 and CK18	+	–	–	*5*
CK14	–	–	+	*5*
CK19	+	+	–	*5*
α-Smooth muscle actin	–	–	+	*4*
α6-Integrin	+/–	+	+	*32*
ERα	+	+	–	*10, 40*
Proliferation	+	?	–	*8*

Abbr: Ep-CAM, epithelial cell adhesion molecule; ESA, epithelial-specific antigen; CALLA, common acute lymphoblastic leukemia antigen; CK, cytokeratin; ERα, estrogen receptor α

common, "intermediate" population. The characteristics of each of the cell types as described in breast tissue from nonpregnant, nonlactating women are detailed in Table 1. Fortunately, for studies on isolated breast epithelial cells, there are antigens expressed exclusively on one cell type or another.

For example, only luminal epithelial cells express the sialomucin MUC1, which is present on their apical membranes, whereas myoepithelial cells express the common acute lymphoblastic leukemia antigen (CALLA) and smooth muscle actin *(2–4)*. Each of the cell types has a particular cytokeratin profile: Luminal epithelial cells express cytokeratins 8 and 18 (CK8 and CK18, respectively) whereas myoepithelial cells express CK14 *(5)*. All cell types express panepithelial markers such as that recognized by the Ber-EP4 antibody *(6,7)*. It is interesting to note that the majority of proliferating cells are found in the luminal population, whereas cell division or expression of antigens associated with proliferation is exceedingly rare in the myoepithelial cell type *(8)*. In addition, receptors for estradiol and progesterone (ERα and PR, respectively), key regulators of postnatal mammary gland development, are found exclusively in the luminal compartment of the human breast *(9,10)*.

2.2. Malignant Breast Tissue

Careful histological studies indicated that the majority of human breast tumors arise in the TDLU, and that they have a luminal phenotype in terms of appearance and specific protein expression *(2,5,11)*. Therefore, interest in mammary stem cells is not just academic; one implication of the "multihit theory" of carcinogenesis is that cancer is a stem cell disease. As breast cancer is the most common malignancy in Western women, understanding mammary stem cells, their mechanisms of self-renewal, and the lineages arising from them may provide new insights into the carcinogenic process. In addition, improved understanding may also provide an explanation for why the luminal phenotype is predominant in tumors and may eventually lead to new targets for breast cancer prevention.

3. EMBRYONIC MAMMARY GLAND DEVELOPMENT AND ESTABLISHMENT OF A STEM CELL POPULATION

3.1. Embryonic Development of the Mammary Gland

Most studies on mammary stem cells have used mouse models and, to a lesser extent, human material. The development, morphology, and histology of the mammary glands of the two species are surprisingly similar, although there are important differences in the details; these are compared and contrasted if appropriate. The mammary gland is an unusual organ in that it is not fully formed at birth. Instead, rudimentary structures are established during embryogenesis that develop fully only during pregnancy and lactation in adulthood.

During embryonic development in mice, five pairs of lenticular structures called *mammary buds* form from localized thickenings or *milk streaks* that form on both sides of the midventral line in the epidermis. These mammary buds start to develop between embryonic d 10 and 11 (E10 and E11, respectively), and their early enlargement appears to occur by addition of cells migrating from the adjoining epidermis. At later embryonic stages, enlargement results from proliferation of the cells in the mammary bud and invasion into the underlying mammary fat pad precursor until, at birth, there is a rudimentary ductal tree comprising up to 15 canalized branches arising from a single duct attached to each nipple and that terminate in small clublike ends that regress shortly after birth *(12,13)*.

Development of the mammary mesenchyme occurs alongside that of the epithelial or parenchymal component; it is becoming clear that epithelial–mesenchymal interactions are essential for mammary gland development. For example, it would appear that all the ectodermally derived cells in the milk ridge have the capacity to become mammary epithelial cells, but the signals that direct further development in a specific number and position of locations appear to come from the mesenchyme. Some of these signals have been elucidated, but it is still not clear how the mesenchyme "knows" where to induce mammary epithelial development *(14,15)*. It is clear that, by day

E12.5, the cells in the mammary bud are already committed to a mammary fate because they are capable of reconstituting a complete, functional gland when transplanted into the cleared mammary fat pad of a syngeneic host *(14)*.

In humans, just two ectodermal buds toward the cranial end of the milk ridge give rise to clusters of mammary epithelial cells, which, between wk 7 and 8 of gestation, begin to invade the underlying stroma *(16,17)*. There is further growth and invasion between wk 10 and 13, and the parenchyma undergoes branching between wk 13 and 20 to form 15–20 cords of epithelial cells that, in following weeks, canalize and eventually give rise to multiple openings at each nipple. In the remaining weeks of gestation, a small number of alveolar buds may develop at the ends of terminal ducts to form primitive TDLUs.

The development of the mesenchymal component of the human mammary gland is not well described, but it is clear that there are important differences between the human and the mouse; the result is that the TDLUs of the human breast become embedded in a relatively loose intralobular stroma, which in turn is surrounded by a more dense, collagen-rich interlobular stroma. Unlike the mouse mammary gland, in which adipocytes are immediately adjacent to epithelial cells, there appears to be no direct contact between the human mammary parenchyma and the white adipose tissue of the fat pad.

3.2. Establishment of a Stem Cell Population

The tissue-specific stem cells that allow multiple rounds of mammary gland growth, differentiation, and involution to occur in later life are presumed to be laid down at the very earliest stages of embryonic development. One suggestion is that all the ectodermally derived cells in the mammary bud are potential primitive stem cells that subsequently divide both symmetrically, to maintain and enlarge the stem cell compartment, and asymmetrically, to produce the mammary epithelial-specific cell lineages. It is not clear how many stem cells might be present at the earliest stages of mammary bud formation, but studies on chimeric mice generated by fusing the blastomeres of two different strains suggest that it is at least two *(18)*. The exact fate of each of the stem cells may well be governed by a number of factors, including interactions with surrounding epithelium, mesenchyme or extracellular matrix, and any endocrine, paracrine, or autocrine hormones or growth factors present.

The importance of the immediate environment in determining the fate of the mammary parenchyma and its stem cells has been demonstrated in tissue recombination studies. For example, experiments on rabbit embryos showed that nonmammary epidermis can form mammary buds when combined with mammary mesenchyme. Conversely, mammary buds are not formed when mammary epidermis is combined with non-mammary mesenchyme confirming that only mammary mesenchyme has the capacity to induce mammary epithelial development *(19)*. Other studies on embryonic mice indicated that the mesenchyme dictates sexual dimorphism in mammary development under the influence of testosterone *(20)*. For obvious reasons, tissue recombination studies of human embryonic tissue are not possible, but given the similarities in mammary gland development between placental animals, it seems likely that epithelial–mesenchymal interactions are critically involved in the process.

4. POSTNATAL MAMMARY GLAND DEVELOPMENT

4.1. Pubertal Mammary Gland Development

Between birth and puberty, mammary gland growth is isometric in relation to the rest of the body, but at puberty and under the influence of both ovarian and pituitary hormones, it undergoes a first phase of allometric growth *(21)*. In mice, this starts at about 21 d of age and is characterized by enlargement of the duct termini to form bulbous terminal end buds (TEBs). These TEBs are major sites of proliferation through which ductal elongation and ramification into the mammary fat pad are achieved. At the microscopic level, the outermost layer of the TEB is a single layer of pale-staining or cap cells that are not polarized, are loosely adherent, and do not have an organized cytoskeleton. It has been proposed that these cells are a stem cell population that gives rise to the different cell types of the mammary parenchyma *(21)*.

During duct extension, some of the cap cells seem to reposition along the perimeter of the TEB to become myoepithelial cells, whereas others migrate toward the lumen to become the ductal or luminal epithelium. In the body of the TEB, there is a population of highly proliferative body cells that also undergo extensive apoptosis, presumably to facilitate lumen formation. Ductal elongation and branching by bifurcation continues throughout puberty until the margins of the mammary fat pad have been reached.

The human mammary parenchyma undergoes a similar pattern of allometric growth at puberty in that the primitive ducts undergo extensive elongation and branching. The site of active epithelial proliferation is also a TEB-like structure, although this is not as bulbous as that of the mouse mammary gland, and it does not seem to share the same histological features. It is not known, at present, where a stem cell population might reside in the human TEB equivalents, although they clearly must contain stem cells because, through a process of dichotomous and sympodial branching, they give rise to new ductal branches and smaller ductules or alveolar buds *(1)*.

The alveolar buds cluster around terminal ducts to form TDLUs, which were classified by Russo and Russo according to the number of alveolar buds or ductules they contain *(17)*. The first to appear is the type 1 or *virginal lobule*, which contains about 11 alveolar buds. Once menstruation is established, there is a cyclical increase in proliferation associated with the luteal phase, and the TDLUs become more elaborate in terms of the numbers of alveoli they contain with each successive ovulatory cycle. Eventually, the breast epithelium comprises a mixture of types 1, 2, and 3 lobules, which contain an average of 11, 47, and 80 ductules, respectively.

This progressive development of the epithelium continues until the age of 35 yr. However, intervening pregnancy influences the proportions of the different types of TDLUs present, such that type 1 lobules are found more frequently in breast tissue from nulliparous women, whereas types 2 and 3 are predominant in the glands of parous women. There is some evidence that limited lobulo-alveolar (equivalent to the TDLU) development occurs in mouse mammary glands during each estrous cycle *(22)*. However, the extent of this development appears to be strain dependent and, as regression occurs at the end of the estrous cycle, the effect does not appear cumulative. A further difference between mouse and human mammary glands is that, in the human, the mammary fat pad also undergoes allometric enlargement at puberty.

4.2. Mammary Gland Development During Pregnancy and Lactation

The second phase of allometric growth in the mammary gland occurs during pregnancy. In mice, there is an initial and enormous increase in ductal proliferative activity and in the formation of alveolar buds *(23)*. In the second half of pregnancy, the alveolar buds differentiate into alveoli, which ultimately become the milk-secreting lobules. By late pregnancy, the alveoli fill most of the mammary fat pad; the luminal epithelial cells produce milk proteins and are distended because of lipid accumulation. The surrounding myoepithelial cells become discontinuous, which means that the luminal epithelium may contact the basement membrane; it has been postulated that this contact is essential for full differentiation and milk secretion *(24)*.

Just before parturition, the luminal epithelial cells accumulate rough endoplasmic reticulum and Golgi and secretory vesicles and begin secreting milk protein and lipids. The cells become flattened, intracellular milk fat globules are clearly visible, and milk accumulates in the lumen. The differentiated function of myoepithelial cells is to eject milk from the alveoli into the ducts by contraction of their dendritic processes in response to oxytocin. In mice, lactation continues in this way until weaning, a period of about 3 wk.

The changes that occur in the human mammary gland during pregnancy and lactation are not well described because there are few specimens of tissue from these stages of development. However, examination of samples obtained at surgery to remove cancers arising during pregnancy and lactation suggests that the changes are similar to those seen in rodent mammary glands; that is, there is an increase in the number of lobules and a corresponding decrease in the adipose component of the breast *(1)*. However, the human breast differs from the rodent mammary gland in its heterogeneity. Whereas the fully developed mouse mammary gland fills the mammary fat pad uniformly, it is not unusual to see completely quiescent TDLUs immediately adjacent to those undergoing a florid proliferative response in breast tissue taken from pregnant women *(1)*.

4.3. Mammary Gland Involution

Once weaning has occurred, the mammary gland undergoes involution. Milk stasis is the signal for the alveolar secretory epithelial cells to undergo apoptosis and be removed by phagocytosis. Although the ducts remain intact, the alveoli collapse until, by d 6 of involution, they have disintegrated completely, and both epithelial and stromal components are remodeled *(23)*. After 3 wk, remodeling is complete, and the mammary gland resembles the pre-pregnant state.

There is very little available information on the involution process in human breast; again, it is thought to be similar to that seen in the rodent mammary gland *(1)*. Also, it seems that, once remodeling is complete, the human gland does not revert to the virgin state, but appears more differentiated in terms of the numbers of lobules in each TDLU *(17)*. It is thought that the reduction in breast cancer risk afforded by an early first full-term pregnancy may be related to the fact that the gland is left in a more "differentiated" state following involution *(25)*.

Uniquely, the human mammary gland undergoes a second involution at the menopause. At this point, there is regression of ducts as well as lobules, and adipose tissue replaces the glandular epithelium and interlobular stroma. As a result, there is a sparse scattering of atrophic acini and ducts through the tissue *(1)*.

Review of the postnatal development of the mammary gland emphasizes the huge changes in the size and function of the epithelial compartment that are possible and how often they can occur during reproductive life. In this respect, the mammary gland is quite different from other stem cell systems, such as the bone marrow or the intestinal epithelium, in which cell turnover is constant and continuous unless the tissue has been damaged.

It has been suggested that there might not be tissue-specific stem cells in the mammary epithelium, but it is difficult to explain how the gland can go through multiple cycles of proliferation, differentiation, involution, and remodeling without invoking the presence of a long-lived stem cell population with an almost unlimited capacity to produce differentiated progeny whenever the gland needs to expand and that persists during involution. However, although for many years mammary gland-specific stem cells have been presumed to exist, it is only now that the tools needed to isolate and characterize them are becoming available.

Consequently, there are many questions about mammary gland-specific stem cells that remain to be answered, especially with respect to the human breast. These include fundamental questions such as whether there are stem cells, and if so, what are they and where are they located in the mammary epithelium. Is there a hierarchy of pluripotent stem cells that gives rise to increasingly more restricted lineage progenitors similar to that existing in the bone marrow? What are the lineages arising from the stem cells? Does the size of the stem cell population change during periods of allometric growth or, conversely, during involution? Finally, what factors, if any, control stem cell activity?

5. EVIDENCE FOR THE EXISTENCE OF MAMMARY EPITHELIAL STEM CELLS

Although stem cells have yet to be identified directly, there is ample circumstantial evidence for their existence. It has been known for years that fragments of mouse mammary epithelium can give rise to complete glands when transplanted into the fat pads of syngeneic hosts cleared of their endogenous epithelium *(26)*. More recently, limiting dilution studies in which clones derived from single cells could be identified on the basis of their patterns of mouse mammary tumor virus (MMTV) viral integration have shown that an entire fully functional mammary gland can be derived from a single cell *(27)*. Moreover, this single cell could give rise to both luminal epithelial and myoepithelial cell types, suggesting the presence of a self-renewing pluripotent stem cell in the mouse mammary gland. The study also provided evidence for the presence of more lineage-restricted stem cells as instances of ductal development in the absence of lobules and vice versa could be demonstrated in some transplants. Interestingly, even when only ducts or lobules were formed, both luminal and myoepithelial cell types were present. This implies that one lineage restriction might be at the level of ductal or lobule formation.

There are, at present, no functional assays of the repopulating capacity of human mammary epithelial cells in vivo. Attempts to transplant human breast epithelial cells into the fat pads of immunodeficient animals have been uniformly unsuccessful, probably because mouse mammary stroma is very different from that of the human. However, elegant studies on the pattern of X chromosome inactivation in microdissected human breast demonstrated that there are contiguous patches of epithelium where the same X chromosome has been inactivated, suggesting that all the cells in each of these patches are derived from a single progenitor *(28)*. Similarly, demonstration of specific patterns of loss of heterozygosity involving entire ducts or lobules suggests the presence of a common precursor *(29)*. Interestingly, luminal and myoepithelial cells isolated from the same area share patterns of genetic damage, suggesting that they have a common ancestor.

More progress has been made in developing tissue culture–based assays of human mammary stem cell activity that use cells isolated from reduction mammoplasty specimens. At low plating densities, three types of colony are formed: luminal epithelial only, myoepithelial only, or colonies containing both cell types *(30,31)*. Further analysis as to the origin of the mixed colonies found that they were derived from single cells *(32)*. The conclusion from these studies is that there are three types of progenitor present in the human mammary gland: luminal restricted, myoepithelial restricted, and stem cells that are bipotent because they can give rise to both cell types in culture.

Taken together, the results from these studies indicate that there are indeed stem cells in the mammary epithelium, and there may be a hierarchy, with a more primitive stem cell giving rise to lineage-committed progenitors.

6. LOCATION OF STEM CELLS IN THE MAMMARY GLAND

The next question is, where are these stem cells in the mammary epithelium? It has been suggested that the cap cells of the TEBs in the mouse mammary gland during pubertal development contain a population of multipotent stem cells. This suggestion is based on the observation that these cells have a high mitotic rate, and that they have two fates: They either enter the body of the TEB and become luminal epithelial cells, or they can migrate laterally along the outermost layer of the subtending duct to become myoepithelial cells *(14)*.

The suggestion that multipotent stem cells are in the TEB is supported by studies that showed that fully functional mammary glands are obtained when dissected TEBs are transplanted into cleared mammary fat pads of syngeneic hosts *(21)*. However, the TEBs and associated cap cells disappear once pubertal ductal development is complete, and they never reappear during the lifespan of the mouse. Moreover, all parts of the mammary ductal tree can regenerate entire mammary glands when transplanted into fat pads, so the best guess as to the location of stem cells is that they are distributed throughout the entire gland and are present in the TEBs during pubertal development *(21)*.

Although the human mammary gland has a TEB-like structure during pubertal development, it does not have a similar arrangement of cap and body cells. So, it is unclear whether there is a population of stem cells in the TEB-like structure, although it is thought that mammary gland-specific stem cells are distributed throughout the ductal epithelium in the human breast.

Studies in which isolated breast epithelial cells were sorted according to whether they expressed luminal- or myoepithelial-specific antigens and then placed into culture indicated that bipotent progenitors reside in the luminal population. Mixed colonies were only obtained when CK8- and CK18-positive cells were plated and never when CALLA-positive cells were used *(31)*. Thus, preliminary conclusions as to the position of stem cells in the mammary epithelium from nonpregnant, nonlactating women are that they are distributed throughout the epithelium, and that they are in the luminal and not the myoepithelial population.

7. CHARACTERIZATION OF MAMMARY EPITHELIAL STEM CELLS

Early searches for mammary-specific stem cell markers were based on the observations of Smith and Medina that mammary epithelial explants contained pale or light-staining cells, and that only these cells entered mitosis when placed into culture *(33)*. Subsequent and painstaking electron microscopic studies confirmed that these pale or undifferentiated cells undergo mitosis occasionally, and that they exist in both small and large forms *(34)*. Other cells, which are darker in appearance because they contain more organelles, are never seen to mitose, and they are assumed to be terminally differentiated.

The small light cells (SLCs) fulfill the presumed criteria for stem cells in that they are division competent and are in the luminal population, but do not contact the lumen. In addition, SLCs can be found either side by side or one above the other (in relation to the basement membrane) in heterogeneous pairs or clustered with large light cells, implying that they have divided asymmetrically. Cell counting at all stages of rodent mammary gland development indicates that the proportion of SLCs remains relatively constant and accounts for approx 3% of the total epithelial population *(34)*. This is higher than the proportion of stem cells calculated from the mammary fat pad repopulation studies (approx 1 in 2500 cells) *(27)* and has been explained by suggesting that SLCs comprise both multipotent and more lineage-restricted stem cells *(35)*. SLCs have been identified in the mammary epithelium of all mammals examined so far, including humans *(35)*. It is assumed that they are stem cells, although this has yet to be shown definitively.

Studies of other stem cell markers have built on the observations made on the position of SLCs in the rodent epithelium. Specifically, it has been postulated that, because SLCs do not contact the lumen but are situated in a suprabasal position between luminal and myoepithelial cells, they should express a general epithelial marker, but not a luminal-specific, apical membrane marker such as MUC1. Cells isolated using this strategy appear to be multipotent because both luminal and myoepithelial cell types are produced when they are placed into low-density culture. Further analysis of these cells showed that they also express $\alpha6$ integrin and CK19 *(32)*.

Further characterization of mammary epithelial stem cells has been aided greatly by cross fertilization from studies of other adult stem cell systems. For example, examination of the patterns of deoxyribonucleic acid (DNA) label retention in the study of stem cells in the skin and small intestine has now been applied to the mammary epithelium *(36,37)*. This technique involves administration of a DNA label, usually tritiated thymidine (^{3}H-dT) or a halogenated thymidine analogue such as bromodeoxyuridine (BudR), to the animal and then determination of which cells retain the label at subsequent time points. The label is taken up only by cells actively synthesizing DNA at the time of treatment. In cells that continue to divide, the DNA label will be progressively diluted such that, after a few divisions, levels are reduced to below the level of detection. However, in cells that do not continue to divide, the label will be retained. As quiescence and longevity are considered stem cell properties, these cells would be expected to retain label over long periods of time.

Accordingly, two groups have used label retention as a means of defining the stem cell population in the mouse mammary epithelium *(38,39)*. The first used a bolus of ^{3}H-dT to label the mammary epithelial cells of fully adult mice (10–12 wk old); the mice were then followed for 3 wk and showed that label-retaining cells (LRCs) comprised 0.1 to 1% of the total epithelial population *(38)*. Immunohistochemistry combined with histoautoradiography used to detect the tritium label indicated that a high proportion of the LRCs (approx 95%) also expressed the ERα.

In the second study, BudR was administered continuously for 2 wk to pubertal mice (3–5 wk old), and tissue was sampled until the mice reached 13 wk of age *(39)*. In this study, a greater proportion of LRCs was detected, probably because labeling was carried out over the period of pubertal mammary gland development, when stem cells might be expected to be more active. However, steroid receptor expression was transiently associated with retention of the DNA label because the LRCs detected at the last time point did not contain steroid receptors, although they did at earlier time points. The very long lived LRCs also appeared to be undifferentiated in that they expressed neither CK18 nor CK14, luminal and myoepithelial markers, respectively.

The data from these two studies can be interpreted as evidence for the existence of two stem cell populations in the mammary epithelium: The first is a long-lived, primary, steroid receptor-negative stem cell, and the second is a steroid receptor-positive stem cell that might be more short lived and be the more active during the estrous cycle.

Again, and for obvious reasons, in vivo DNA label retention studies cannot be carried out in women, but a method of implanting small pieces of intact normal breast tissue into immunodeficient mice to track mammary epithelial cells after administration of the label has been used *(40)*. In this study, tissue was labeled intensively with ^{3}H-dT for a period covering two S-phase durations and sampled at various time points afterward. The tissue was taken from adult premenopausal, but nonpregnant and nonlactating, women, so the study was analogous to that of Zeps et al., who used adult mice *(38)*. Accordingly, 2 wk after ^{3}H-dT injection, a population of LRCs was detectable that comprised less than 1% of the total population and expressed steroid receptors in addition to the p27^{KIP1} cyclin-dependent kinase inhibitor (CDKI), which is consistent with the conception of LRCs as a quiescent population.

The second technique that has crossed over from use in the hematopoietic system to use in the mammary gland is flow cytometric cell sorting based on exclusion of the fluorescent DNA dye Hoechst 33342 *(41)*. Hematopoietic cells that efflux the dye are called the *side population* (SP) and are able to reconstitute the bone marrow of lethally irradiated mice, suggesting that they are enriched in stem cells. Combining SP sorting with analysis of label retention indicated that, 9 wk after labeling, mammary epithelial LRCs formed 8% of the SP *(39)*.

The mouse mammary SP population is also enriched for cells expressing stem cell antigen (Sca-1, a hematopoietic stem cell marker), α6 integrin (a skin stem cell marker), and telomerase. The technique has now been applied to human breast epithelial cells isolated from reduction mammoplasty specimens, and further characterization of these human SP cells is awaited *(42)*.

It is unclear, at present, how useful SP sorting will be to the study of mammary stem cells as the ability of mouse cells isolated using this method to reconstitute mammary glands in cleared mammary fat pads is no greater than that of non-SP cells *(39,42)*. In this respect, Sca-1 appears to be a better correlate of the ability to reconstitute mammary glands, although immunohistochemical studies showed that approx 20% of mouse mammary epithelial cells express Sca-1 *(39)*. As it is unlikely that they are all stem cells, more work is needed to define exactly which cell types form the population positive for Sca-1.

Unfortunately, Sca-1 is not detectable in human tissues, which means that other mammary stem cell markers need to be found. In this respect, specific cytokeratins may be useful. Gudjonsson and colleagues *(30)* reported that the cells that give rise to mixed luminal and myoepithelial cells in culture expressed CK19, whereas Bocker et al. *(43)* suggested that CK5 positivity defines a stem cell population. Equally, the absence of lineage-specific markers could be used when defining stem cell populations. The CK5-positive stem cell population proposed by Bocker et al. cannot be stained by an anti-CK8/CK18/CK19 antibody or by one that recognizes smooth muscle actin *(43)*. The CK19-positive population identified by Gudjonsson et al. does not express MUC1 *(30)*. These findings illustrate another difficulty in defining human mammary stem cell populations: inconsistency between different research groups. This is probably because of the fact that the study of mammary stem cells in human tissue is in its infancy, and it is hoped that a consensus will be reached as investigations progress.

8. STEM CELL HIERARCHIES

Hierarchies of division-competent cells have been demonstrated in other adult tissues, such as blood and the small intestine *(44,45)*. These hierarchies comprise a primary tissue-specific stem cell that can produce all the cell types of the tissue, including new stem cells. Downstream of the primary stem cell are more restricted progenitors that give rise only to a subset of cell types, and then there are lineage-committed progenitors that produce only one particular cell type. The number of committed progenitors and the number of divisions between them and the primary stem cells varies according to the tissue. For example, there are a large number of increasingly committed progenitors in the hematopoietic system because a correspondingly large number of differentiated cell types need to be produced *(44)*. In contrast, there appears to be only one type of progenitor downstream of the primary keratinocyte stem cell in the skin *(45)*.

It is still not clear how many and what type of committed progenitors may be present in the mammary epithelium. However, the results of the studies presented above suggest that there are at least two layers of commitment downstream of the primary stem cell: The first is the commitment to produce either ducts or lobules, and the second is commitment to produce either luminal or myoepithelial cells *(see* Fig. 1).

The data derived from studies on DNA label retention, Hoechst 33342 efflux, and Sca-1 expression in the mouse mammary gland labeled during pubertal development support the suggestion that the stem cell population is heterogeneous *(39)*. The population positive for Sca-1 contains at least two different cell types, SP-positive LRCs and those that are neither LRCs nor SP cells. Welm and colleagues suggested that the Sca-1-positive population, LRCs, and SP cells represent the most primitive population that may contain the multipotent stem cells *(39)*. The steroid receptor-positive LRCs identified by Zeps and in our study in adult mouse and human mammary epithelium could be more committed or lineage-restricted stem cells *(38,40)*.

9. LIFESPAN, NUMBER, AND PROLIFERATIVE CAPACITY OF MAMMARY EPITHELIAL STEM CELLS

Primary mammary epithelial stem cells persist throughout the reproductive lifespan of the animal, allowing the gland to develop and differentiate with each pregnancy. However, serial transplantation studies indicated that mouse mammary epithelial stem cells are not immortal. Mammary epithelium can be transferred serially to host animals up to seven times before it becomes incapable of repopulating the mammary fat pad, although more than 75% of epithelial outgrowths lose their repopulating ability by the fourth passage *(46)*.

Experiments on the repopulating capacity of mammary epithelium isolated from mice of different ages and reproductive histories suggested that loss of capacity to self-renew is dictated not by chronological age, but by the number of times a stem cell divides; it has been calculated that stem cells become senescent after 40–50 divisions *(47,48)*. Interestingly, this figure agrees with the "Hayflick" number, which is the maximum number of divisions a eukaryotic cell can undergo in culture before replicative senescence

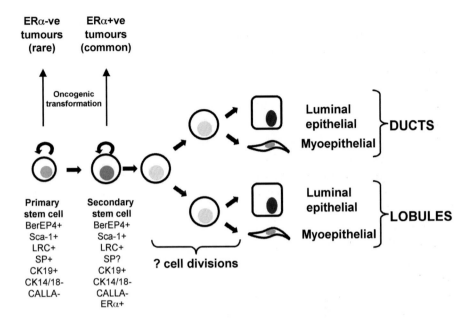

Fig. 1. A model of tissue-specific stem cells and their progeny postulated to exist in the mammary epithelium. The putative stem cells occupy an intermediate location in the epithelium in terms of their physical position and specific protein expression and are defined by their appearance (small and pale staining), ability to efflux Hoechst 33342, retention of DNA label, and in mice, expression of stem cell antigen (Sca-1). We speculate that the stem cell pool includes primary, long-lived cells with a genome that is extremely well protected against damage and a second population of more committed stem cells with a shorter lifespan that contain steroid receptors. The progeny of these stem cells may form a transit-amplifying population before commitment to the ductal and lobular lineages and then final differentiation into luminal and myoepithelial cells. We postulate that the genome of the primary stem cell population is very highly protected, whereas that of the more committed steroid receptor-containing stem cells is not. This means that the latter population may be more susceptible to transforming events, leading to loss of normal growth constraints, proliferation in response to signals such as estrogen and progesterone, and then progression to hyperplasia, to ductal carcinoma *in situ*, and ultimately to invasive breast tumors.

occurs *(49)*. However, it is probable that no stem cell actually undergoes 40–50 divisions under normal circumstances.

In the experiments that showed that just one cell can reconstitute an entire mammary gland, the stem cell was estimated to have undergone 11 self-renewing symmetrical divisions *(27)*. So, it would appear that no one stem cell survives throughout the entire lifespan of the tissue, but that the population turns over slowly by infrequent self-renewing symmetrical division.

It is still not clear whether the stem cell population size expands and contracts in response to the cyclical developmental needs of the mammary gland. Ultrastructural examination of rat mammary epithelium at all developmental stages indicated that, from nulliparity through to involution, the proportion of SLCs remains constant at about 3% of the total epithelium *(34)*. This suggests that the stem cell population increases and decreases at the same rate as the more differentiated epithelial cells. However, given what is known about stem cell properties in other tissues, it might be expected that absolute mammary stem cell numbers would remain constant; therefore, their proportion would decrease during epithelial cell growth and increase during involution.

One explanation for this disparity is that SLCs are a heterogeneous population that includes both primary mammary stem cells and lineage-committed progenitors; it is the latter population that alters in number during mammary gland growth and involution. This is supported by the results of the limiting dilution transplantation studies of Kordon and Smith, which indicated a primary stem cell frequency of approx 1 in 2500 (0.04%) *(27)*. This is not only much lower than the proportion of SLCs, but is also lower than the number of LRCs or SP cells detected in other studies *(38,39)* and suggests that label retention and Hoechst 33342 efflux are properties common to both primary stem cells and the more lineage-committed progenitors.

Another assumption is that, given that stem cells appear to divide infrequently, it might be expected that all the stem cells present in the adult mammary gland are direct descendants of those established during embryogenesis. However, Wagner and colleagues suggested that a new "adjunct" epithelial population arises during pregnancy, and that this could explain the physiological differences between nulliparous and involuted parous mammary glands in terms of sensitivity to hormones and carcinogenic agents *(50)*. Using genetic techniques to generate mammary epithelial-specific expression of the β-galactosidase gene, Wagner et al. show that this new population of cells survives postlactational involution and tissue remodeling to give rise to new alveolar cells in subsequent pregnancies. Moreover, the new adjunct cells appear to be able to contribute to both ductal and alveolar epithelial cell types when transplanted into cleared mammary fat pads. The authors concluded that the adjunct epithelial population could represent a new stem cell population derived from more differentiated cells; this population contributes to the biological changes occurring in the mammary epithelium after pregnancy and lactation. As these changes result in a decreased risk of breast cancer if they occur early enough in a woman's reproductive lifespan, it will be important to determine whether such an adjunct population of cells exists in parous human breast and, if so, what their properties are. The conclusions from these studies is that the mammary epithelial stem cell population is not as fixed in either number or characteristics as first thought.

10. MAMMARY STEM CELLS AND CANCER

As stem cells persist in the mammary gland throughout reproductive life, they must be regarded as prime targets for oncogenic transformation because they have the potential to accumulate genetic damage and to transmit it to their progeny. In this respect, it is interesting to note that susceptibility to carcinogens is greatest during pubertal breast development, when stem cells should be active and there is proliferation of the epithelium, but no lactational differentiation. This has been shown extensively in rodents, for which the gland is most sensitive to carcinogens when there are high concentrations of developing TEBs and alveolar buds containing large numbers of active stem cells *(51)*. Similar structures are present in the human breast during pubertal development, and it is not surprising that breast cancer rates are highest in women who were prepubertal or adolescent when they were irradiated as a consequence of the atomic detonations in Japan *(52)*.

The studies of dynamics of repopulation after transplantation of small numbers of mammary epithelial cells into cleared mammary fat pads may force us to reconsider the nature of "oncogenic" transformation. If mammary stem cells realize their full proliferative capacity and undergo 40–50 self-renewing symmetrical divisions, then just one cell could give rise to an additional 10^{12}–10^{13} multipotent progenitors in its lifetime, which is more than enough to kill someone of breast cancer *(27)*. However, under normal circumstances, no stem cell in the mammary gland actually realizes this potential in the lifespan of the mouse or woman, and it appears that the signals that initiate proliferation are very tightly regulated, so stem cells cannot divide uncontrollably.

These control mechanisms are poorly understood, but one possibility is that stem cell proliferative activity is influenced by its position in relation to more differentiated progeny, to the basement membrane, and to stromal cells. The role of a stem cell's immediate environment or niche is increasingly recognized, and it has been suggested that the niche acts as an organizing center that maintains "stemness" *(53)*.

In studies of skin stem cell niches, it has been shown that differentiated cells can show characteristics of stem cells if they are allowed to occupy empty stem cell niches *(54)*. The nature and location of mammary stem cell niches are, at present, unknown. It is possible that the earliest steps in tumorigenesis allow a stem cell to escape from its constraints and, once escape has been achieved, to accumulate further genetic alterations that might, for example, increase genomic instability and resistance to apoptosis, enhance sensitivity to growth promoters such as estrogen, switch on production of angiogenic factors, or confer metastatic potential.

Other questions that the study of mammary stem cells might help answer are why most malignant tumors have a luminal phenotype in terms of morphology and specific protein expression and why a large proportion expresses steroid receptors. Results of studies presented above suggested that the mammary stem cell population is heterogeneous and may comprise very long-lived primary stem cells and multipotent, but more committed and short-lived, stem cells that may express steroid receptors.

We speculate that the genome of the primary stem cells would be very highly protected against damage to ensure their continued survival and ability to proliferate, as has been shown in the small intestine, where stem cells protect their genome by a number of different mechanisms *(55,56)*. However, the genome of the more committed stem cells may not be so well protected and therefore would be more susceptible to carcinogenic transformation, with the result that one of them escapes its constraints and is able to respond to the proliferative signals of estrogen and progesterone, leading first to hyperplasia, then to ductal carcinoma in situ, and then to invasive disease.

This postulation is supported by data that shows that increased steroid receptor expression is one of the earliest changes associated with the carcinogenic process in the human breast *(57)*. In addition, genetic analysis of the proliferative lesions thought to be malignant precursors indicates that they are clonal *(58,59)*. Finally, Smith and Chepko demonstrated increased numbers of SLCs in both mouse and human breast tumors *(35)*.

The clinical implication of these findings is that successful breast cancer treatment or prevention strategies must eradicate stem cells. Current chemotherapeutic and endocrine agents reduce proliferation and induce apoptosis, but it is not clear whether mammary stem cells are susceptible to factors that signal programmed cell death. Studies of the mouse small intestine suggested that, in terms of resistance to apoptosis, there are two populations of stem cells *(56)*. One of these undergoes spontaneous apoptosis as part of the homeostatic mechanism, restricting the number of stem cells present at any one time. The other smaller population is resistant to radiation-induced apoptosis, undergoes DNA repair, and, presumably, is the primary stem cell population. We believe that a similar arrangement exists in the mammary epithelium, and if it does, new therapies need to be aimed at eliminating both stem cell populations.

11. MAMMARY STEM CELL PLASTICITY

As is made clear throughout the rest of this book, there is considerable interest in the concept of stem cell plasticity from the point of view of using cells isolated from adult tissues for therapeutic purposes. Thus, hematopoietic stem cells have been reported to give rise to differentiated liver parenchymal cells *(60)*, mesenchymal stem cells isolated from the bone marrow can give rise to chondrocytes and adipocytes *(61)*, and skin stem cells have been reported to be able to give rise to hematopoietic cells *(62)*.

At present, it is not known whether mammary stem cells can give rise to the differentiated cell types of other tissues, but as far as the mouse is concerned, the necessary experimental models are available to find this out. Until then, there can only be speculation. On one hand, the fact that mammary stem cells have features in common with other types of tissue-specific stem cells (e.g., the ability to efflux Hoechst 33342 and expression of Sca-1 in mice) suggests that they could be plastic. On the other hand, the transplantation studies that used embryonic mammary tissue *(14)* suggested that the decision to become mammary-specific stem cells is made at an early stage of development, and that reprogramming might not be possible. If plasticity was shown, the mammary gland could be a relatively accessible source of stem cells for therapeutic intervention and even cosmetic purposes such as breast reconstruction and augmentation.

12. CONCLUSIONS AND FUTURE WORK

It has been known for some time that there must be a population of self-renewing, mammary-specific stem cells that give rise to all the differentiated cell types of the epithelium and that allow the gland to undergo multiple cycles of pregnancy-associated proliferation, differentiation, and apoptosis. More recently, progress in understanding these cells has been aided by cross fertilization of ideas from studies of other types of tissue-specific stem cells. Taken together, the available data suggest that a hierarchical arrangement of primary stem cells, lineage-committed progeny, and differentiated cells exists in both human and mouse mammary epithelium (*see* Fig. 1). Tissue transplantation studies indicated that the number of progeny any primary stem cell can generate is prodigious. However, this capacity is never realized under normal circumstances, and it is believed that stem cell proliferation is highly restrained by positional cues from the immediate environment. Very little is known about these signals, but it seems important to find out what they are as they may represent therapeutic targets for breast cancer prevention and treatment.

Mammary transplantation techniques that use embryonic or adult mouse mammary epithelium are powerful tools for studying the effects of genetic alterations on stem cell function in a tissue that is well-defined developmentally. As mammary stem cells seem to have characteristics in common with stem cells in other types of tissue, at least some of the results from these studies should be more generally applicable. These common characteristics also imply that mammary stem cells may be "plastic" in that they are able to give rise to differentiated cell types of other tissues. This remains to be determined, but it is interesting to note that one adult mammary epithelial cell has shown the ultimate plasticity in that a complete cloned mammal, Dolly the sheep, was generated from it *(63)*.

REFERENCES

1. Howard, B. A., and Gusterson, B. A. (2000). Human breast development. J Mammary Gland Biol Neoplasia 5, 119–137.
2. Taylor-Papadimitriou, J., Millis, R., Burchell, J., Nash, R., Pang, L., and Gilbert, J. (1986). Patterns of reaction of monoclonal antibodies HMFG-1 and -2 with benign breast tissues and breast carcinomas. J Exp Pathol 2, 247–260.
3. Gusterson, B. A., Monaghan, P., Mahendran, R., Ellis, J., and O'Hare, M. J. (1986). Identification of myoepithelial cells in human and rat breasts by anti-common acute lymphoblastic leukemia antigen antibody A12. J Natl Cancer Inst 77, 343–349.
4. Taylor-Papadimitriou, J., Wetzels, R., and Ramaekers, F. (1992). Intermediate filament protein expression in normal and malignant human mammary epithelial cells. Cancer Treat Res 61, 355–378.
5. Taylor-Papadimitriou, J., Stampfer, M., Bartek, J., et al. (1989). Keratin expression in human mammary epithelial cells cultured from normal and malignant tissue: relation to in vivo phenotypes and influence of medium. J Cell Sci 94, 403–413.
6. Latza, U., Niedobitek, G., Schwarting, R., Nekarda, H., and Stein, H. (1990). Ber-EP4: new monoclonal antibody which distinguishes epithelia from mesothelial. J Clin Pathol 43, 213–219.
7. Emerman, J. T., Stingl, J., Petersen, A., Shpall, E. J., and Eaves, C. J. (1996). Selective growth of freshly isolated human breast epithelial cells cultured at low concentrations in the presence or absence of bone marrow cells. Breast Cancer Res Treat 41, 147–159.
8. Joshi, K., Smith, J. A., Perusinghe, N., and Monoghan, P. (1986). Cell proliferation in the human mammary epithelium. Differential contribution by epithelial and myoepithelial cells. Am J Pathol 124, 199–206.

9. Clarke, R. B., Howell, A., Potten, C. S., and Anderson, E. (1997). Dissociation between steroid receptor expression and cell proliferation in the human breast. Cancer Res 57, 4987–4991.

10. Petersen, O. W., Hoyer, P. E., and van Deurs, B. (1987). Frequency and distribution of estrogen receptor-positive cells in normal, nonlactating human breast tissue. Cancer Res 47, 5748–5751.

11. Wellings, S. R., Jensen, H. M., and Marcum, R. G. (1975). An atlas of subgross pathology of the human breast with special reference to possible precancerous lesions. J Natl Cancer Inst 55, 231–273.

12. Hovey, R. C., McFadden, T. B., and Akers, R. M. (1999). Regulation of mammary gland growth and morphogenesis by the mammary fat pad: a species comparison. J Mammary Gland Biol Neoplasia 4, 53–68.

13. Sakakura, T. (1987). Mammary embryogenesis. In: Neville, M. C., and Daniel, C. W., eds., The Mammary Gland. Development, Regulation and Function. New York: Plenum Press, pp. 37–66.

14. Daniel, C. W., and Smith, G. H. (1999). The mammary gland: a model for development. J Mammary Gland Biol Neoplasia 4, 3–8.

15. Robinson, G. W., Karpf, A. B., and Kratochwil, K. (1999). Regulation of mammary gland development by tissue interaction. J Mammary Gland Biol Neoplasia 4, 9–19.

16. Hovey, R. C., Trott, J. F., and Vonderhaar, B. K. (2002). Establishing a framework for the functional mammary gland: from endocrinology to morphology. J Mammary Gland Biol Neoplasia 7, 17–38.

17. Russo, J., and Russo, I. H. (1987). Development of the human mammary gland. In: Neville, M. C., and Daniel, C. W., eds., The Mammary Gland. Development, Regulation and Function. New York: Plenum Press, pp. 67–93.

18. Mintz, B., and Slemmer, G. (1969). Gene control of neoplasia. I. Genotypic mosaicism in normal and preneoplastic mammary glands of allophenic mice. J Natl Cancer Inst 43, 87–109.

19. Propper, A. Y. (1978). Wandering epithelial cells in the rabbit embryo milk line. A preliminary scanning electron microscope study. Dev Biol 67, 225–231.

20. Heuberger, B., Fitzka, I., Wasner, G., and Kratochwil, K. (1982). Induction of androgen receptor formation by epithelium-mesenchyme interaction in embryonic mouse mammary gland. Proc Natl Acad Sci U S A 79, 2957–2961.

21. Daniel, C. W., and Silberstein, G. B. (1987). Postnatal development of the rodent mammary gland. In: Neville, M. C., and Daniel, C. W., eds., The Mammary Gland. Development, Regulation and Function. New York: Plenum Press, pp. 3–36.

22. Andres, A. C., and Strange, R. (1999). Apoptosis in the estrous and menstrual cycles. J Mammary Gland Biol Neoplasia 4, 221–228.

23. Richert, M. M., Schwertfeger, K. L., Ryder, J. W., and Anderson, S. M. (2000). An atlas of mouse mammary gland development. J Mammary Gland Biol Neoplasia 5, 227–241.

24. Howlett, A. R., and Bissell, M. J. (1993). The influence of tissue microenvironment (stroma and extracellular matrix) on the development and function of mammary epithelium. Epithelial Cell Biol 2, 79–89.

25. Russo, J., Hu, Y. F., Silva, I. D., and Russo, I. H. (2001). Cancer risk related to mammary gland structure and development. Microsc Res Tech 52, 204–223.

26. Ormerod, E. J., and Rudland, P. S. (1986). Regeneration of mammary glands in vivo from isolated mammary ducts. J Embryol Exp Morphol 96, 229–243.

27. Kordon, E. C., and Smith, G. H. (1998). An entire functional mammary gland may comprise the progeny from a single cell. Development 125, 1921–1930.

28. Tsai, Y. C., Lu, Y., Nichols, P. W., Zlotnikov, G., Jones, P. A., and Smith, H. S. (1996). Contiguous patches of normal human mammary epithelium derived from a single stem cell: implications for breast carcinogenesis. Cancer Res 56, 402–404.

29. Lakhani, S. R., Chaggar, R., Davies, S., et al. (1999). Genetic alterations in "normal" luminal and myoepithelial cells of the breast. J Pathol 189, 496–503.

30. Gudjonsson, T., Villadsen, R., Nielsen, H. L., Ronnov-Jessen, L., Bissell, M. J., and Petersen, O. W. (2002). Isolation, immortalization, and characterization of a human breast epithelial cell line with stem cell properties. Genes Dev 16, 693–706.

31. Stingl, J., Eaves, C. J., Kuusk, U., and Emerman, J. T. (1998). Phenotypic and functional characterization in vitro of a multipotent epithelial cell present in the normal adult human breast. Differentiation 63, 201–213.

32. Stingl, J., Eaves, C. J., Zandieh, I., and Emerman, J. T. (2001). Characterization of bipotent mammary epithelial progenitor cells in normal adult human breast tissue. Breast Cancer Res Treat 67, 93–109.

33. Smith, G. H., and Medina, D. (1988). A morphologically distinct candidate for an epithelial stem cell in mouse mammary gland. J Cell Sci 90, 173–183.

34. Chepko, G., and Smith, G. H. (1997). Three division-competent, structurally-distinct cell populations contribute to murine mammary epithelial renewal. Tissue Cell 29, 239–253.

35. Smith, G. H., and Chepko, G. (2001). Mammary epithelial stem cells. Microsc Res Tech 52, 190–203.

36. Potten, C. S., and Morris, R. J. (1988). Epithelial stem cells in vivo. J Cell Sci Suppl 10, 45–62.

37. Potten, C. S., and Loeffler, M. (1990). Stem cells: attributes, cycles, spirals, pitfalls and uncertainties. Lessons for and from the crypt. Development 110, 1001–1020.

38. Zeps, N., Dawkins, H. J., Papadimitriou, J. M., Redmond, S. L., and Walters, M. I. (1996). Detection of a population of long-lived cells in mammary epithelium of the mouse. Cell Tissue Res 286, 525–536.

39. Welm, B. E., Tepera, S. B., Venezia, T., Graubert, T. A., Rosen, J. M., and Goodell, M. A. (2002). Sca-1(pos) cells in the mouse mammary gland represent an enriched progenitor cell population. Dev Biol 245, 42–56.

40. Clarke, R. B., Howell, A., Potten, C. S., and Anderson, E. (2000). P27(KIP1) expression indicates that steroid receptor-positive cells are a non-proliferating, differentiated subpopulation of the normal human breast epithelium. Eur J Cancer 36 (Suppl. 4), S28–S29.

41. Goodell, M. A., Brose, K., Paradis, G., Conner, A. S., and Mulligan, R. C.

(1996). Isolation and functional properties of murine hematopoietic stem cells that are replicating in vivo. J Exp Med 183, 1797–1806.

42. Alvi, A. J., Clayton, H., Joshi, C., et al. (2003). Functional and molecular characterisation of mammary side population cells. Breast Cancer Res 5, R1–R8.

43. Bocker, W., Moll, R., Poremba, C., et al. (2002). Common adult stem cells in the human breast give rise to glandular and myoepithelial cell lineages: a new cell biological concept. Lab Invest 82, 737–746.

44. Orkin, S. H. (2000). Diversification of haematopoietic stem cells to specific lineages. Nat Rev Genet 1, 57–64.

45. Jones, P. H. (1997). Epithelial stem cells. Bioessays 19, 683–690.

46. Daniel, C. W., De Ome, K. B., Young, J. T., Blair, K. B., and Faulkin, L. J., Jr. (1968). The in vivo lifespan of normal and preneoplastic mouse mammary glands: a serial transplantation study. Proc Natl Acad Sci U S A 61, 53–60.

47. Daniel, C. W., and Young, L. J. (1971). Influence of cell division on an aging process. Life span of mouse mammary epithelium during serial propagation in vivo. Exp Cell Res 65, 27–32.

48. Young, L. J., Medina, D., DeOme, K. B., and Daniel, C. W. (1971). The influence of host and tissue age on life span and growth rate of serially transplanted mouse mammary gland. Exp Gerontol 6, 49–56.

49. Hayflick, L. (1992). Aging, longevity, and immortality in vitro. Exp Gerontol 27, 363–368.

50. Wagner, K. U., Boulanger, C. A., Henry, M. D., Sgagias, M., Hennighausen, L., and Smith, G. H. (2002). An adjunct mammary epithelial cell population in parous females: its role in functional adaptation and tissue renewal. Development 129, 1377–1386.

51. Russo, I. H., and Russo, J. (1978). Developmental stage of the rat mammary gland as determinant of its susceptibility to 7,12-dimethylbenz[a]anthracene. J Natl Cancer Inst 61, 1439–1449.

52. McGregor, H., Land, C. E., Choi, K., et al. (1977). Breast cancer incidence among atomic bomb survivors, Hiroshima and Nagasaki, 1950–69. J Natl Cancer Inst 59, 799–811.

53. Spradling, A., Drummond-Barbosa, D., and Kai, T. (2001). Stem cells find their niche. Nature 414, 98–104.

54. Nishimura, E. K., Jordan, S. A., Oshima, H., et al. (2002). Dominant role of the niche in melanocyte stem-cell fate determination. Nature 416, 854–860.

55. Potten, C. S., Owen, G., and Booth, D. (2002). Intestinal stem cells protect their genome by selective segregation of template DNA strands. J Cell Sci 115, 2381–2388.

56. Potten, C. S., Booth, C., and Pritchard, D. M. (1997). The intestinal epithelial stem cell: the mucosal governor. Int J Exp Pathol 78, 219–243.

57. Shoker, B. S., Jarvis, C., Sibson, D. R., Walker, C., and Sloane, J. P. (1999). Oestrogen receptor expression in the normal and pre-cancerous breast. J Pathol 188, 237–244.

58. Lakhani, S. R., Collins, N., Stratton, M. R., and Sloane, J. P. (1995). Atypical ductal hyperplasia of the breast: clonal proliferation with loss of heterozygosity on chromosomes 16q and 17p. J Clin Pathol 48, 611–615.

59. Lakhani, S. R., Slack, D. N., Hamoudi, R. A., Collins, N., Stratton, M. R., and Sloane, J. P. (1996). Detection of allelic imbalance indicates that a proportion of mammary hyperplasia of usual type are clonal, neoplastic proliferations. Lab Invest 74, 129–135.

60. Petersen, B. E., Bowen, W. C., Patrene, K. D., et al. (1999). Bone marrow as a potential source of hepatic oval cells. Science 284, 1168–1170.

61. Prockop, D. J. (1997). Marrow stromal cells as stem cells for nonhematopoietic tissues. Science 276, 71–74.

62. Lako, M., Armstrong, L., Cairns, P. M., Harris, S., Hole, N., and Jahoda, C. A. (2002). Hair follicle dermal cells repopulate the mouse haematopoietic system. J Cell Sci 115, 3967–3974.

63. Wilmut, I., Schnieke, A. E., McWhir, J., Kind, A. J., and Campbell, K. H. (1997). Viable offspring derived from fetal and adult mammalian cells. Nature 385, 810–813.

11

From Marrow to Brain

Josef Priller

1. INTRODUCTION

All organisms originate from a single totipotent cell. During development, the progeny of this cell become increasingly restricted in their differentiation potential. Nevertheless, most cells retain an intact genome, and the
selective expression or repression of genes determines specific properties.
In recent years, it has become evident that adult cells can thus be reprogrammed back to totipotency (1). Moreover, stem cells that reside in adult
tissues have the capacity to self-renew and to generate several types of differentiated progeny. These stem cells do not appear to be restricted to generating only cells of their original tissue, but show remarkable plasticity when
exposed to an environment that they usually would not encounter. It has
thus been suggested that a stem cell is not necessarily a specific cellular
entity, but rather a function that can be assumed by numerous diverse cell
types (2).

One of the most intriguing examples of adult stem cell plasticity is the
recently discovered conversion of bone marrow cells to brain cells and vice
versa (3–7). Although some of the reported findings remain highly controversial (8–10), the notion of *transdifferentiation*, the conversion of one differentiated cell type to another, has challenged our concept of cell fate
determination. Moreover, the plasticity of adult stem cells has raised hopes
for future restorative treatment of debilitating disorders, particularly in the
central nervous system (CNS), which has a low regenerative capacity.

2. STEM CELLS IN THE BONE MARROW COMPARTMENT

Less than 0.1% of all nucleated cells in the bone marrow are stem cells.
Among these, hematopoietic stem cells (HSCs) provide a source of circulating erythrocytes, leukocytes, and platelets (Fig. 1). The HSC pool can be
subdivided into long-term reconstituting HSCs, which have the greatest self-
renewal capacity, and short-term HSCs, which generate hematopoietic lineages only for several weeks (11). Mesenchymal (stromal) stem cells

From: *Adult Stem Cells*
Edited by: K. Turksen © Humana Press Inc., Totowa, NJ

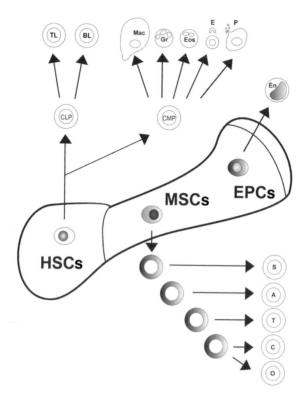

Fig. 1. Schematic view of the bone marrow stem cell compartments. Hematopoietic stem cells (HSCs) and mesenchymal (stromal) stem cells (MSCs) become gradually restricted in their differentiation potential by a succession of cell divisions. Common lymphoid progenitors (CLP) arise from HSCs and differentiate into T lymphocytes (TL) and B lymphocytes (BL). Common multipotent progenitors (CMP) derive from HSCs and give rise to monocytes/macrophages (Mac), neutrophil granulocytes (Gr), eosinophil granulocytes (Eos), erythroblasts/erythrocytes (E), and megakaryocytes/platelets (P). MSCs generate progenitor cells, which differentiate into hematopoietic-supporting stroma (S), adipocytes (A), tenocytes (T), chondrocytes (C), and osteocytes (O). Endothelial progenitor cells (EPCs) can be mobilized into the blood and give rise to vascular endothelial cells (En).

(MSCs) generate nonhematopoietic tissues, including adipocytes, tenocytes, hematopoietic-supporting stroma, chondrocytes, and osteocytes (Fig. 1) *(12)*. The bone marrow cavity also contains endothelial progenitor cells (EPCs), which can be mobilized into the peripheral blood and give rise to mature endothelial cells in vessels (Fig. 1) *(13)*.

All bone marrow stem cells derive from mesoderm, but it is unclear at present how HSCs arise and what their relationship to endothelial cells is *(14)*. Unfortunately, cell surface determinants provide only limited information on the origin of particular stem cells, and even highly enriched bone marrow stem cell populations remain heterogeneous. This has to be taken into account when plasticity is attributed specifically to HSCs or MSCs.

The expression of CD34/Stem cell antigen 1 (Sca-1), c-Kit, and CD45 antigens in the absence of lineage markers helps to differentiate HSCs from MSCs. Nevertheless, some HSCs are also found in the CD34$^-$ fraction *(15)*. HSCs also express the ABC transporter Bcrp1, which effluxes dyes such as Hoechst-33342 and allows the selection of the so-called side population of HSCs *(16,17)*. MSCs can be enriched in vitro based on their adhesive properties and the presence of CD44 *(18)*. Finally, EPCs express CD34/Sca-1, vascular endothelial growth factor receptor2 (Flk-1), and Tie-2 antigens *(13)*. They can be mobilized into the circulation by granulocyte-macrophage colony-stimulating factor, vascular endothelial growth factor, and 3-hydroxy-3-methylglutaryl coenzyme A (HMG-CoA) reductase inhibitors *(19–21)*. Selected by adherence to plastic, some of the EPCs may be of hematopoietic origin given that monocytes have been shown to give rise to endothelial-like cells in vitro *(22)*.

3. TURNING BONE MARROW INTO BRAIN

Most of the studies addressing the contribution of bone marrow-derived cells to the adult brain were performed in bone marrow chimeras, which were generated by the transplantation of genetically marked bone marrow cells into conditioned hosts (Fig. 2). Donor-derived cells were subsequently detected in the host CNS based on the expression of a transgene, retroviral tag, or the Y chromosome (when male bone marrow was transplanted into female recipients).

To identify bone marrow cells that have turned into brain cells, tissue sections were analyzed for cells showing colocalization of the bone marrow-label with tissue-specific markers and a distinctive morphology indicative of cell fate change. Unless the boundaries determined by embryologic trilaminar origin are not maintained in the adult, cells of mesodermal origin, such as microglia, brain macrophages, and endothelial cells, are more likely to be generated from the transplanted bone marrow than cells of ectodermal origin, including neurons, astrocytes, and oligodendrocytes. However, the signals regulating division and differentiation in the adult brain may be quite distinct from those in the immature brain, and the acquisition of specific cell phenotypes may therefore largely depend on local inductive signals.

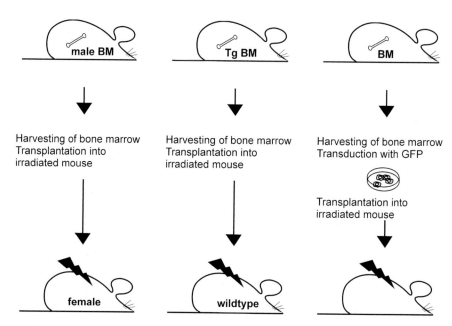

Fig. 2. Schematic view of the generation of bone marrow (BM) chimeras. Bone marrow cells are harvested from donor animals and injected into myeloblated (e.g., lethally irradiated) recipient animals. Reconstitution of hematopoiesis with donor-derived peripheral blood cell progeny occurs within several weeks after transplantation. Donor-derived cells can be identified by the presence of the Y chromosome in female animals transplanted with male bone marrow cells. When donor bone marrow is derived from a transgenic animal or is genetically marked by viral transduction, donor-derived cells can be detected in the recipient animals based on the expression of the transgene/viral-encoded protein (e.g., β-galactosidase or GFP).

3.1. Microglia and Perivascular Cells

Microglia are resident immunological effector cells that sense pathological events in the CNS *(23)*. Their origin has been one of the most contentious issues in glial research over the past decades, but it is now generally accepted that microglia are ontogenetically related to cells of the mononuclear phagocyte lineage. Resting microglia show a ramified morphology with downregulated immunophenotype; other brain macrophages, such as perivascular cells and leptomeningeal macrophages, express major histocompatibility complex (MHC) class II antigen and high levels of CD45 *(24,25)*. Microglial activation occurs in response to even subtle changes in the brain microenvironment, and activated microglia express macrophage-related antigens, such as the complement receptor 3, MHC class I and class II antigens, and a number of cytokines and cell adhesion molecules *(26)*.

In the developing brain, pial mesenchymal progenitor cells from the yolk sac seed in the brain parenchyma before its vascularization; later, circulating monocytes contribute to the population of ameboid microglia that colonize the embryonic brain *(27)*. After extensive proliferation, these ameboid microglia ultimately transform into the ramified microglia that can be found in the postnatal brain *(28,29)*. Some authors have challenged this view and suggested that microglial cells derive from neuroectodermal glioblasts *(30,31)*. Moreover, studies using irradiated adult rats transplanted with MHC-mismatched bone marrow cells strongly suggested that only perivascular cells and leptomeningeal macrophages, but not parenchymal microglia, were bone marrow derived *(32)*. Even after severe inflammatory conditions of the brain, resident microglia represented a very stable cell pool *(33)*.

In contrast, mice transplanted with bone marrow cells expressing the green fluorescent protein (GFP) showed substantial microglial engraftment up to 12 mo after bone marrow transplantation (BMT) *(7,34,35)*. In these chimeras, host perivascular cells were also completely substituted by donor-derived macrophages within 4 mo after BMT *(36)*. Although the appearance of GFP-expressing cells in perivascular and leptomeningeal sites occurred throughout the brain, preferential microglial engraftment was observed in the olfactory bulb and later in the cerebellum.

From the data of Brazelton et al. *(6)*, an estimated 5% of the microglia in the olfactory bulb were generated from bone marrow cells 8–12 wk after transplantation. By 15 wk post-BMT, up to a quarter of the cerebellar microglial population was found to be donor derived *(34)*. De Groot et al. *(37)* reported that approx 10% of the white matter microglia arose from the transplanted bone marrow, and studies using transplantation of murine bone marrow cells transgenic for β-galactosidase *(38)* suggested that 20–30% of all brain macrophages originated from the donor marrow at 4–12 mo after transplantation. In mice homozygous for a mutation in the PU.1 gene and transplanted with wild-type bone marrow cells at birth without irradiation, all microglia and macrophages throughout the brain arose from the donor bone marrow *(5)*.

Thus, it can be concluded that microglia in the postnatal murine brain may be generated in the bone marrow compartment. Because mature monocytes traffic between the blood and the brain despite an intact blood–brain barrier *(39,40)*, these cells are likely to replace the perivascular and leptomeningeal macrophage cell pools continuously and to differentiate into resi-

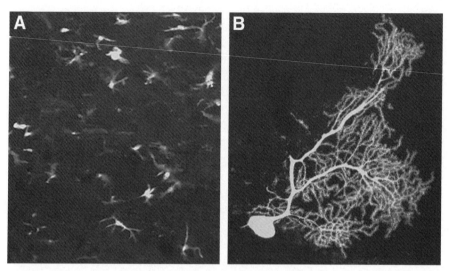

Fig. 3. Central nervous system engraftment of bone marrow–derived cells. Chimeric mice were generated by the transplantation of GFP-marked bone marrow cells into lethally irradiated wild-type mice *(7,34)*. (**A**) Four weeks after transplantation, the middle cerebral artery was occluded for 60 min, and after 14 d of reperfusion, the ischemic hemisphere of a chimera was infiltrated by donor-derived GFP-expressing microglia/macrophages. (**B**) Twelve months after transplantation, a rare GFP-expressing Purkinje neuron is seen in the cerebellum of a bone marrow chimera.

dent microglia. In vitro evidence suggests that microglial differentiation of peripheral blood monocytes may result from an interaction with astrocytes *(41)*.

Interestingly, microglial engraftment is significantly enhanced following CNS damage. Thus, focal cerebral ischemia leads to a dramatic recruitment of bone marrow-derived cells into the ischemic brain, and almost one-third of these cells develop into microglia (Fig. 3) *(34)*. Even remote lesion of the CNS by facial nerve axotomy in rodents induced microglial engraftment in proximity to the injured motoneurons *(34,42)*. In a murine model of globoid cell leukodystrophy, bone marrow-derived microglia were distributed diffusely in both gray and white matter after 100 postnatal days *(43)*. Donor-derived cells of the macrophage lineage also infiltrated the CNS in a regionally specific manner following BMT in mouse models of GM2 gangliosidosis *(44)* and Gaucher disease *(45)*. Finally, peripheral macrophages were specifically recruited in cuprizone-induced CNS demyelination *(46)*. It has thus been suggested that microglia may be used therapeutically to deliver genes of interest to the diseased brain *(34)*.

3.2. Astrocytes

Astrocytes are the most numerous glial cells of the CNS. They seem to be generated from the dorsal regions of the neural tube during development *(47)* and throughout life are continuously replenished from multipotent neuroepithelial stem cells and glial-restricted precursors *(48)*. The glial fibrillary acidic protein (GFAP) represents the major component of intermediate filaments in mature astroglia, but astrocytes may also express S100β *(49)*. Extending processes to blood vessels, astrocytes participate in the formation of the blood–brain barrier, and they contribute to tissue homeostasis *(50)*.

Astrocytes have been shown to control neuronal life directly by regulating synaptogenesis and neurogenesis *(51,52)*. Moreover, in the adult brain, GFAP-expressing astrocytes have been suggested to represent neural stem cells capable of generating macroglia and neurons *(53–55)*. It is therefore surprising that Eglitis and Mezey *(3)* found that 0.5–2% of the bone marrow-derived cells engrafting in the murine CNS after BMT expressed GFAP. Marrow stromal cells injected into the lateral ventricle of neonatal mice also differentiated into GFAP-immunoreactive astrocytes within 12 d *(56)*.

On the other hand, none of the chimeric mice transplanted with bone marrow cells expressing GFP revealed any astroglial differentiation of the donor cells *(6,10,34,35,57)*. Only when GFP-marked bone marrow cells were injected directly into the brain did they develop into GFAP-expressing astrocytes *(35)*.

In vitro, bone marrow stromal cells were induced to differentiate into astrocytes in the presence of growth factors or differentiation factors such as Noggin *(58,59)*. Multipotent adult progenitor cells (MAPCs) generated from murine MSCs differentiated into GFAP-expressing astrocytes in vitro and gave rise to astrocytes throughout the brain after injection into an early blastocyst *(60)*. However, these cells failed to generate astroglia after transplantation into adult NOD/SCID mice. In the studies of Woodbury et al. *(61)* and Deng et al. *(62)*, human and rat bone marrow stromal cells developed into neuronal phenotypes in vitro, but did not differentiate into GFAP- or S100β-expressing astroglia.

Because astrocyte proliferation is dramatically enhanced after CNS injury *(63)*, the contribution of bone marrow-derived cells to the population of reactive astroglia was studied after cerebral ischemia. In chimeric rats subjected to middle cerebral artery occlusion (MCAO), Eglitis et al. *(64)* observed that the number of bone marrow-derived astrocytes was twice as high in the ischemic hemisphere compared to the contralateral side and almost 10 times as high as in control rats within 48 h. MSCs grafted into the ischemic brains of rodents *(65,66)* or injected into the peripheral circulation after transient MCAO in rats *(67)* were reported to differentiate into GFAP-expressing astrocytes in high numbers. In contrast, Hess et al. *(57)* failed to detect donor-derived astrocytes in chimeric mice subjected to MCAO after transplantation of GFP-marked bone marrow cells. Moreover, MSCs transplanted into the rat spinal cord after contusion did not express GFAP up to 5 wk after lesion *(68)*. In contrast, MSCs administered intravenously to rats 1 d after traumatic brain injury gave rise to GFAP-expressing cells in the brain *(69)*. Thus, differentiation of bone marrow cells into astroglia remains controversial, and the switch of cell fate may depend on specific local signals.

3.3. Oligodendrocytes

In contrast to astrocytes, oligodendrocytes seem to originate more from the embryonic ventral neural tube *(70)*. Precursors of oligodendrocytes from the ventricular zone migrate to extraventricular sites during CNS development and continue to divide throughout life *(71)*. Maturation of oligodendrocytes includes the expression of specific markers, such as O4 and galactocerebroside, and the extension of endfeet toward axons, followed by the process of myelination *(72)*. In the adult, myelination is thought to occur by the recruitment of quiescent oligodendrocyte precursor cells *(73)*. Recently, adult bone marrow cells enriched in c-Kit–positive hematopoietic progenitor cells were found to differentiate into O4-immunoreactive oligodendrocytes within 6 d after intracerebral transplantation into the neonatal mouse brain *(74)*. Similarly, GFP-expressing bone marrow cells grafted directly into the adult murine brain differentiated into oligodendroglia-expressing carbonic anhydrase II within 12 wk after injection *(35)*.

In contrast, murine MAPCs transplanted into adult NOD/SCID mice did not turn into galactocerebroside-expressing oligodendroglia, whereas MSCs could be induced to adopt an oligodendroglial fate in vitro *(59,60)*. Similarly, GFP-marked bone marrow cells transplanted into myeloablated adult mice failed to generate oligodendrocytes for up to 6 mo after BMT *(35)*.

Experimental evidence suggested that GFP-expressing murine MSCs can form functional myelin on injection into a focal demyelinated lesion in the rat spinal cord *(75)*. Within 3 wk after injection, bone marrow-derived cells were found to express the myelin proteins myelin basic protein and P0. Moreover, rat MSCs predifferentiated in vitro into a Schwann cell-like phenotype were found to myelinate the regenerating fibers of the axotomized sciatic nerve 3 wk after transplantation *(76)*. As for astrocytes, the differentiation of bone marrow-derived cells into oligodendroglia therefore seems to depend largely on specific local cues.

3.4. Endothelial Cells

In the embryo, endothelial cells arise either from endothelial progenitors (angioblasts) or from stem cells, giving rise to both endothelial and hematopoietic cells (hemangioblasts) *(77)*. Even after birth, circulating EPCs generated in the adult bone marrow can be assembled into endothelial channels after *in situ* differentiation *(13)*. This *vasculogenesis* by EPCs has been reported in connection with wound healing and tumor vascularization and in response to ischemia *(78)*.

Although the cerebral vascular system is primarily developed by angiogenesis (sprouting), experiments in chimeric mice have recently revealed that bone marrow-derived cells contribute substantially to neovascularization after focal cerebral ischemia. Thus, mice transplanted with bone marrow cells expressing β-galactosidase under the control of the endothelial Tie2 promoter showed bone marrow-derived endothelia in vessels at the border of the infarct, but not in intact parenchymal cerebral vessels *(79)*. In mice transplanted with GFP-expressing bone marrow cells, 42% of the endothelial cells in the infarct were coexpressing GFP with endothelial markers, such as von Willebrand factor, CD31, and isolectin B4, at 3 d after transient MCAO. This number decreased to 26% by 14 d after MCAO *(57)*. Finally, mobilization of EPCs by statin treatment enhanced endothelial regeneration after carotid artery lesion *(80)*.

3.5. Neurons

Undoubtedly, the most intriguing example of adult stem cell plasticity is the conversion of bone marrow-derived cells into neuronal phenotypes. During development, reinforced by signals from the mesoderm, ectoderm from the dorsal side of the embryo forms neural tissue *(81)*. Anterior–posterior neural patterning occurs soon after neural induction, and signals from the underlying mesoderm and the epidermis influence dorsal–ventral patterning.

Neurogenesis occurs in defined regions of the patterned neural plate, and the developmental fate of a neural crest cell depends critically on the signals it receives from the environment through which it migrates. Neurogenesis persists throughout the life of the organism, and small populations of hippocampal, cortical, and olfactory bulb neurons continue to be born in the adult dentate gyrus and the subventricular zone *(53,82,83)*. Thus, the signals required for the neuronal differentiation of stem cells are maintained at least in some parts of the adult brain.

Interestingly, in mice transplanted with GFP–transgenic bone marrow cells, up to 0.3% of all neurons in the olfactory bulb were found to express

GFP within 4 mo after BMT *(6)*. Bone marrow-derived cells in the brain were characterized as neurons based on their expression of neuronal antigens, such as neuronal nuclei (NeuN), neurofilament, and neural nuclei class III β-tubulin. Although most of these cells did not display the morphological charac-teristics of neurons, the presence of phosphorylated cyclic adenosine *5'-monophosphate response element-binding protein suggested that the donor-derived cells responded to cues in their environment in a manner consistent with the surrounding neurons.

Similarly, rare immature bone marrow-derived cells in the mouse spinal cord and in sensory ganglia expressed NeuN, neurofilament, and class III β-tubulin at 3 mo after BMT *(84)*. When female mice homozygous for a mutation in the PU.1 gene were rescued by postnatal intraperitoneal injection of adult male bone marrow cells, 0.3–2.3% of all NeuN-immunoreactive cells throughout the brain were found to be Y chromosome-positive after 1–4 mo *(5)*. There was no overall increase in the density of donor-derived cells in the neurogenic regions.

In contrast, a subsequent study failed to detect any donor-derived neurons in the brain within 4 mo after transplantation of bone marrow cells expressing β-galactosidase into adult wild-type mice *(9)*.

In the experiments of Nakano et al. (35), none of the GFP-marked bone marrow cells transplanted by systemic infusion gave rise to cells expressing neuron-specific enolase in the brain after 6 mo. However, GFP-marked bone marrow cells injected directly into the brain were found to differentiate into neuronal phenotypes within 4 mo. Marrow stromal cells injected into the lateral ventricle of neonatal mice differentiated into neurofilament-immunoreactive cells in the brain stem *(56)*.

MAPCs generated from murine MSCs gave rise to NeuN-expressing cells throughout the brain after injection into an early blastocyst *(60)*. However, these cells failed to generate neuronal phenotypes 1–6 mo after transplantation into adult NOD/SCID mice.

Perhaps the most consistent finding is the appearance of rare GFP-expressing Purkinje cells in the cerebellum several months after transplantation of GFP-marked bone marrow cells (Fig. 3) *(7,10,85)*. Based on morphologic criteria and the expression of the γ-aminobutyric acid (GABA)-synthesizing enzyme glutamic acid decarboxylase, these newly generated neurons were considered fully developed and functionally integrated into the cerebellar cytoarchitecture *(7)*. Moreover, analysis of the brains of mice transplanted with a single GFP-marked HSC revealed only one GFP-positive nonhematopoietic cell, a Purkinje cell *(10)*.

It is thus conceivable that less than 0.1% of the Purkinje cells in the adult may arise from hematopoietic stem cells by way of transdifferentiation. Nevertheless, it has to be taken into account that intermediate stages of development have not yet been described for the bone marrow-derived Purkinje cells, and issues of potential cell fusion have to be addressed in the light of in vitro findings on stem cell fusion *(86)*. However, in the study of Corti et al. *(84)*, all the cells coexpressing GFP with neuronal markers were mononucleate and diploid.

There is also ample in vitro evidence to suggest that clonal GFP-express-ing MSCs can be induced to differentiate into mature neurons expressing markers of dopamine synthesis, serotonin and GABA *(60)*. Clones of rat and murine MSCs produced by limiting dilution were found to adopt a neu-ronal phenotype when exposed to combinations of specific growth factors or differentiation factors *(59,61)*. Human MSCs could also be induced to differentiate into cells expressing neuronal antigens in vitro *(58,62)*.

Several recent studies suggested that the injured CNS provides cues for the engraftment and neural differentiation of bone marrow-derived cells. Thus, MSCs grafted into the ischemic brains of rodents *(65,66)* or injected into the peripheral circulation after transient MCAO in rats *(67,87)* were found to differentiate into NeuN-, neurofilament-, class III β-tubulin-, or microtu-bule-associated protein 2 (MAP2)–expressing cells within 2 wk. In chimeric mice transplanted with GFP-expressing bone marrow cells, scattered donor-derived cells were immunoreactive for NeuN in the ischemic striatum 1–7 d after transient MCAO *(57)*. Almost 1% of the murine MSCs grafted into the striatum of mice treated with 1-methyl-4-phenyl-1,2,3,6-tetrahydropyridine (MPTP) acquired a dopaminergic (tyrosine hydroxylase immunoreactive) phenotype within 1 mo after transplantation *(88)*. Finally, MSCs injected into the rat spinal cord after contusion displayed NeuN immunoreactivity after 5 wk, but failed to express neurofilament or MAP2 *(68)*.

In chimeric mice transplanted with bone marrow cells expressing β-galactosidase, no donor-derived neural cells were observed in the brain up to 5 mo after stab injury *(9)*. In contrast, MSCs administered intravenously after traumatic brain injury in rats gave rise to NeuN-expressing cells in the damaged CNS *(69)*.

4. REGENERATIVE POTENTIAL OF BONE MARROW CELLS IN THE BRAIN

Despite the controversy over the differentiative potential of adult bone marrow stem cells, researchers have explored the possibilities of hastening

Table 1
Recent Studies of Mesenchymal Stem Cell
Transplantation in Rodent Models of Neurological
Disorders

Lesion	Reference
Cerebral ischemia	*65–67, 87*
MPTP	*88*
TBI/spinal contusion	*68, 69*
Spinal cord demyelination	*75, 89*
Acid sphingomyelinase deficiency	*90*

recovery of neurological deficits by the transplantation of bone marrow cells. In rodent models of stroke, Parkinson's disease, multiple sclerosis, trauma, and neurodegeneration, direct intracerebral grafting or systemic administration of bone marrow cells helped promote functional recovery (Table 1). Rats subjected to MCAO performed significantly better in motor and somatosensory behavior tests when treated with intravenous or intracarotid administration of MSCs 1 d or 7 d after ischemia *(67,87)*. Similarly, direct transplantation of MSCs into the ischemic hemispheres of mice or rats 1–7 d after MCAO improved functional recovery from the rotarod, limb placement, or modified neurologic severity score tests *(65,66)*.

In the MPTP model of Parkinson's disease, mice transplanted with MSCs by intrastriatal injection at 1 wk after MPTP administration performed significantly better on the rotarod test compared with controls *(88)*. Delayed injection of MSCs into the rat spinal cord 1 wk after contusion led to long-term improvement of locomotor function *(68)*, and MSCs injected intravenously into rats after traumatic brain injury reduced motor and neurological deficits by d 15 *(69)*. Moreover, regeneration of the axotomized sciatic nerve was accelerated by the local transplantation of MSCs predifferentiated into a Schwann cell-like phenotype *(76)*. MSCs were also found to remyelinate the rat spinal cord after focal demyelination induced by irradiation/ethidium bromide *(75,89)*. Finally, survival of a knockout mouse model of Niemann–Pick disease was enhanced after intracerebral transplantation of MSCs genetically engineered to express acid sphingomyelinase *(90)*. Thus, the use of bone marrow stem cells holds great promise for the treatment of debilitating CNS disorders.

However, it should be quite clear from the presented evidence that the differentiation of bone marrow stem cells into neurons and macroglia is a rare biological event that can only partially account for the dramatic therapeutic effects observed with bone marrow transplantation in models of neurological diseases. These effects are more likely mediated by the concomitant increase in trophic factors *(91)*. Nevertheless, as the signals instructing stem cells to adopt a particular cell fate are elucidated, the bone marrow compartment may turn out to be a valuable source of all kinds of cells destined for the CNS. As secluded as the CNS may appear, we are now beginning to realize that there may be a way from marrow to brain.

REFERENCES

1. Wilmut, I., Schnieke, A. E., McWhir, J., Kind, A. J., and Campbell, K. H. (1997). Viable offspring derived from fetal and adult mammalian cells. Nature 385, 810–813.
2. Blau, H. M., Brazelton, T. R., and Weimann, J. M. (2001). The evolving concept of a stem cell: entity or function? Cell 105, 829–841.
3. Eglitis, M. A., and Mezey, E. (1997). Hematopoietic cells differentiate into both microglia and macroglia in the brains of adult mice. Proc Natl Acad Sci U S A 94, 4080–4085.
4. Bjornson, C. R., Rietze, R. L., Reynolds, B. A., Magli, M. C., and Vescovi, A. L. (1999). Turning brain into blood: a hematopoietic fate adopted by adult neural stem cells in vivo. Science 283, 534–537.
5. Mezey, E., Chandross, K. J., Harta, G., Maki, R. A., and McKercher, S. R. (2000). Turning blood into brain: cells bearing neuronal antigens generated in vivo from bone marrow. Science 290, 1779–1782.
6. Brazelton, T. R., Rossi, F. M., Keshet, G. I., and Blau, H. M. (2000). From marrow to brain: expression of neuronal phenotypes in adult mice. Science 290, 1775–1779.
7. Priller, J., Persons, D. A., Klett, F. F., Kempermann, G., Kreutzberg, G. W., and Dirnagl, U. (2001). Neogenesis of cerebellar Purkinje neurons from gene-marked bone marrow cells in vivo. J Cell Biol 155, 733–738.
8. Morshead, C. M., Benveniste, P., Iscove, N. N., and van der Kooy, D. (2002). Hematopoietic competence is a rare property of neural stem cells that may depend on genetic and epigenetic alterations. Nat Med 8, 268–273.
9. Castro, R. F., Jackson, K. A., Goodell, M. A., Robertson, C. S., Liu, H., and Shine, H. D. (2002). Failure of bone marrow cells to transdifferentiate into neural cells in vivo. Science 297, 1299.
10. Wagers, A. J., Sherwood, R. I., Christensen, J. L., and Weissman, I. L. (2002). Little evidence for developmental plasticity of adult hematopoietic stem cells. Science 297, 2256–2259.
11. Fibbe, W. E., Mark, J., Zijlmans, J. M., and Willemze, R. (1996). Stem cells with short-term and long-term repopulating ability in the mouse. Ann Oncol 7(Suppl. 2), 15–18.
12. Minguell, J. J., Erices, A., and Conget, P. (2001). Mesenchymal stem cells. Exp Biol Med (Maywood) 226, 507–520.

13. Asahara, T., Murohara, T., Sullivan, A., et al. (1997). Isolation of putative progenitor endothelial cells for angiogenesis. Science 275, 964–967.

14. Orkin, S. H., and Zon, L. I. (2002). Hematopoiesis and stem cells: plasticity vs developmental heterogeneity. Nat Immunol 3, 323–328.

15. Bhatia, M., Bonnet, D., Murdoch, B., Gan, O. I., and Dick, J. E. (1998). A newly discovered class of human hematopoietic cells with SCID-repopulating activity. Nat Med 4, 1038–1045.

16. Goodell, M. A., Brose, K., Paradis, G., Conner, A. S., and Mulligan, R. C. (1996). Isolation and functional properties of murine hematopoietic stem cells that are replicating in vivo. J Exp Med 183, 1797–1806.

17. Zhou, S., Schuetz, J. D., Bunting, K. D., et al. (2001). The ABC transporter Bcrp1/ABCG2 is expressed in a wide variety of stem cells and is a molecular determinant of the side-population phenotype. Nat Med 7, 1028–1034.

18. Pittenger, M. F., Mackay, A. M., Beck, S. C., et al. (1999). Multilineage potential of adult human mesenchymal stem cells. Science 284, 143–147.

19. Takahashi, T., Kalka, C., Masuda, H., et al. (1999). Ischemia- and cytokine-induced mobilization of bone marrow-derived endothelial progenitor cells for neovascularization. Nat Med 5, 434–438.

20. Asahara, T., Takahashi, T., Masuda, H., et al. (1999). VEGF contributes to postnatal neovascularization by mobilizing bone marrow-derived endothelial progenitor cells. EMBO J 18, 3964–3972.

21. Dimmeler, S., Aicher, A., Vasa, M., et al. (2001). HMG-CoA reductase inhibitors (statins) increase endothelial progenitor cells via the PI 3-kinase/Akt pathway. J Clin Invest 108, 391–397.

22. Fernandez Pujol, B., Lucibello, F. C., Gehling, U. M., et al. (2000). Endothelial-like cells derived from human CD14 positive monocytes. Differentiation 65, 287–300.

23. Kreutzberg, G. W. (1996). Microglia: a sensor for pathological events in the CNS. Trends Neurosci 19, 312–318.

24. Graeber, M. B., Streit, W. J., and Kreutzberg, G. W. (1989). Identity of ED2-positive perivascular cells in rat brain. J Neurosci Res 22, 103–106.

25. Ford, A. L., Goodsall, A. L., Hickey, W. F., and Sedgwick, J. D. (1995). Normal adult ramified microglia separated from other central nervous system macrophages by flow cytometric sorting. Phenotypicdifferences defined and direct ex vivo antigen presentation to myelin basic protein-reactive CD4+ T cells compared. J Immunol 154, 4309–4321.

26. Raivich, G., Bohatschek, M., Kloss, C. U., Werner, A., Jones, L. L., and Kreutzberg, G. W. (1999). Neuroglial activation repertoire in the injured brain: graded response, molecular mechanisms and cues to physiological function. Brain Res Brain Res Rev 30, 77–105.

27. Kaur, C., Hao, A. J., Wu, C. H., and Ling, E. A. (2001). Origin of microglia. Microsc Res Tech 54, 2–9.

28. Imamoto, K., and Leblond, C. P. (1978). Radioautographic investigation of gliogenesis in the corpus callosum of young rats. II. Origin of microglial cells. J Comp Neurol 180, 139–163.

29. Ling, E. A., Penney, D., and Leblond, C. P. (1980). Use of carbon labeling to demonstrate the role of blood monocytes as precursors of the "ameboid cells" present in the corpus callosum of postnatal rats. J Comp Neurol 193, 631–657.

30. Kitamura, T., Miyake, T., and Fujita, S. (1984). Genesis of resting microglia in the gray matter of mouse hippocampus. J Comp Neurol 226, 421–433.

31. Fedoroff, S., Zhai, R., and Novak, J. P. (1997). Microglia and astroglia have a common progenitor cell. J Neurosci Res 50, 477–486.

32. Hickey, W. F., and Kimura, H. (1988). Perivascular microglial cells of the CNS are bone marrow-derived and present antigen in vivo. Science 239, 290–292.

33. Lassmann, H., Schmied, M., Vass, K., and Hickey, W. F. (1993). Bone marrow derived elements and resident microglia in brain inflammation. Glia 7, 19–24.

34. Priller, J., Flügel, A., Wehner, T., et al. (2001). Targeting gene-modified hematopoietic cells to the central nervous system: use of green fluorescent protein uncovers microglial engraftment. Nat Med 7, 1356–1361.

35. Nakano, K., Migita, M., Mochizuki, H., and Shimada, T. (2001). Differentiation of transplanted bone marrow cells in the adult mouse brain. Transplantation 71, 1735–1740.

36. Bechmann, I., Priller, J., Kovac, A., et al. (2001). Immune surveillance of mouse brain perivascular spaces by blood-borne macrophages. Eur J Neurosci 14, 1651–1658.

37. de Groot, C. J., Huppes, W., Sminia, T., Kraal, G., and Dijkstra, C. D. (1992). Determination of the origin and nature of brain macrophages and microglial cells in mouse central nervous system, using non-radioactive in situ hybridization and immunoperoxidase techniques. Glia 6, 301–309.

38. Kennedy, D. W., and Abkowitz, J. L. (1997). Kinetics of central nervous system microglial and macrophage engraftment: analysis using a transgenic bone marrow transplantation model. Blood 90, 986–993.

39. Lawson, L. J., Perry, V. H., and Gordon, S. (1992). Turnover of resident microglia in the normal adult mouse brain. Neuroscience 48, 405–415.

40. Kennedy, D. W., and Abkowitz, J. L. (1998). Mature monocytic cells enter tissues and engraft. Proc Natl Acad Sci U S A 95, 14,944–14,949.

41. Sievers, J., Schmidtmayer, J., and Parwaresch, R. (1994). Blood monocytes and spleen macrophages differentiate into microglia-like cells when cultured on astrocytes. Anat Anz 176, 45–51.

42. Flügel, A., Bradl, M., Kreutzberg, G. W., and Graeber, M. B. (2001). Transformation of donor-derived bone marrow precursors into host microglia during autoimmune CNS inflammation and during the retrograde response to axotomy. J Neurosci Res 66, 74–82.

43. Wu, Y. P., McMahon, E., Kraine, M. R., et al. (2000). Distribution and characterization of GFP(+) donor hematogenous cells in Twitcher mice after bone marrow transplantation. Am J Pathol 156, 1849–1854.

44. Oya, Y., Proia, R. L., Norflus, F., Tifft, C. J., Langaman, C., and Suzuki, K. (2000). Distribution of enzyme-bearing cells in GM2 gangliosidosis mice:

regionally specific pattern of cellular infiltration following bone marrow transplantation. Acta Neuropathol (Berl) 99, 161–168.

45. Krall, W. J., Challita, P. M., Perlmutter, L. S., Skelton, D. C., and Kohn, D. B. (1994). Cells expressing human glucocerebrosidase from a retroviral vector repopulate macrophages and central nervous system microglia after murine bone marrow transplantation. Blood 83, 2737–2748.

46. McMahon, E. J., Suzuki, K., and Matsushima, G. K. (2002). Peripheral macrophage recruitment in cuprizone-induced CNS demyelination despite an intact blood-brain barrier. J Neuroimmunol 130, 32–45.

47. Pringle, N. P., Guthrie, S., Lumsden, A., and Richardson, W. D. (1998). Dorsal spinal cord neuroepithelium generates astrocytes but not oligodendrocytes. Neuron 20, 883–893.

48. Mehler, M. F., and Gokhan, S. (1999). Postnatal cerebral cortical multipotent progenitors: regulatory mechanisms and potential role in the development of novel neural regenerative strategies. Brain Pathol 9, 515–526.

49. Pekny, M., Leveen, P., Pekna, M., et al. (1995). Mice lacking glial fibrillary acidic protein display astrocytes devoid of intermediate filaments but develop and reproduce normally. EMBO J 14, 1590–1598.

50. Kimelberg, H. K., and Norenberg, M. D. (1989). Astrocytes. Sci Am 260, 66–72, 74, 76.

51. Ullian, E. M., Sapperstein, S. K., Christopherson, K. S., and Barres, B. A. (2001). Control of synapse number by glia. Science 291, 657–661.

52. Song, H., Stevens, C. F., and Gage, F. H. (2002). Astroglia induce neurogenesis from adult neural stem cells. Nature 417, 39–44.

53. Doetsch, F., Caille, I., Lim, D. A., Garcia-Verdugo, J. M., and Alvarez-Buylla, A. (1999). Subventricular zone astrocytes are neural stem cells in the adult mammalian brain. Cell 97, 703–716.

54. Seri, B., Garcia-Verdugo, J. M., McEwen, B. S., and Alvarez-Buylla, A. (2001). Astrocytes give rise to new neurons in the adult mammalian hippocampus. J Neurosci 21, 7153–7160.

55. Heins, N., Malatesta, P., Cecconi, F., et al. (2002). Glial cells generate neurons: the role of the transcription factor Pax6. Nat Neurosci 5, 308–315.

56. Kopen, G. C., Prockop, D. J., and Phinney, D. G. (1999). Marrow stromal cells migrate throughout forebrain and cerebellum, and they differentiate into astrocytes after injection into neonatal mouse brains. Proc Natl Acad Sci U S A 96, 10,711–10,716.

57. Hess, D. C., Hill, W. D., Martin-Studdard, A., Carroll, J., Brailer, J., and Carothers, J. (2002). Bone marrow as a source of endothelial cells and NeuN-expressing cells after stroke. Stroke 33, 1362–1368.

58. Sanchez-Ramos, J., Song, S., Cardozo-Pelaez, F., et al. (2000). Adult bone marrow stromal cells differentiate into neural cells in vitro. Exp Neurol 164, 247–256.

59. Kohyama, J., Abe, H., Shimazaki, T., et al. (2001). Brain from bone: efficient "meta-differentiation" of marrow stroma-derived mature osteoblasts to neurons with Noggin or a demethylating agent. Differentiation 68, 235–244.

60. Jiang, Y., Jahagirdar, B. N., Reinhardt, R. L., et al. (2002). Pluripotency of mesenchymal stem cells derived from adult marrow. Nature 418, 41–49.
61. Woodbury, D., Schwarz, E. J., Prockop, D. J., and Black, I. B. (2000). Adult rat and human bone marrow stromal cells differentiate into neurons. J Neurosci Res 61, 364–370.
62. Deng, W., Obrocka, M., Fischer, I., and Prockop, D. J. (2001). In vitro differentiation of human marrow stromal cells into early progenitors of neural cells by conditions that increase intracellular cyclic AMP. Biochem Biophys Res Commun 282, 148–152.
63. Eddleston, M., and Mucke, L. (1993). Molecular profile of reactive astrocytes—implications for their role in neurologic disease. Neuroscience 54, 15–36.
64. Eglitis, M. A., Dawson, D., Park, K. W., and Mouradian, M. M. (1999). Targeting of marrow-derived astrocytes to the ischemic brain. Neuroreport 10, 1289–1292.
65. Li, Y., Chopp, M., Chen, J., et al. (2000). Intrastriatal transplantation of bone marrow nonhematopoietic cells improves functional recovery after stroke in adult mice. J Cereb Blood Flow Metab 20, 1311–1319.
66. Zhao, L. R., Duan, W. M., Reyes, M., Keene, C. D., Verfaillie, C. M., and Low, W. C. (2002). Human bone marrow stem cells exhibit neural phenotypes and ameliorate neurological deficits after grafting into the ischemic brain of rats. Exp Neurol 174, 11–20.
67. Chen, J., Li, Y., Wang, L., et al. (2001). Therapeutic benefit of intravenous administration of bone marrow stromal cells after cerebral ischemia in rats. Stroke 32, 1005–1011.
68. Hofstetter, C. P., Schwarz, E. J., Hess, D., et al. (2002). Marrow stromal cells form guiding strands in the injured spinal cord and promote recovery. Proc Natl Acad Sci U S A 99, 2199–2204.
69. Mahmood, A., Lu, D., Wang, L., Li, Y., Lu, M., and Chopp, M. (2001). Treatment of traumatic brain injury in female rats with intravenous administration of bone marrow stromal cells. Neurosurgery 49, 1196–1203.
70. Timsit, S., Martinez, S., Allinquant, B., Peyron, F., Puelles, L., and Zalc, B. (1995). Oligodendrocytes originate in a restricted zone of the embryonic ventral neural tube defined by DM-20 mRNA expression. J Neurosci 15, 1012–1024.
71. Fujita, S. (1965). An autoradiographic study on the origin and fate of subpial glioblasts in the embryonic chick spinal cord. J Comp Neurol 124, 51–60.
72. Warrington, A. E., Barbarese, E., and Pfeiffer, S. E. (1993). Differential myelinogenic capacity of specific developmental stages of the oligodendrocyte lineage upon transplantation into hypomyelinating hosts. J Neurosci Res 34, 1–13.
73. Wolswijk, G. (1998). Oligodendrocyte regeneration in the adult rodent CNS and the failure of this process in multiple sclerosis. Prog Brain Res 117, 233–247.
74. Bonilla, S., Alarcon, P., Villaverde, R., Aparicio, P., Silva, A., and Martinez,

S. (2002). Haematopoietic progenitor cells from adult bone marrow differentiate into cells that express oligodendroglial antigens in the neonatalmouse brain. Eur J Neurosci 15, 575–582.

75. Akiyama, Y., Radtke, C., and Kocsis, J. D. (2002). Remyelination of the rat spinal cord by transplantation of identified bone marrow stromal cells. J Neurosci 22, 6623–6630.

76. Dezawa, M., Takahashi, I., Esaki, M., Takano, M., and Sawada, H. (2001). Sciatic nerve regeneration in rats induced by transplantation of in vitro differentiated bone-marrow stromal cells. Eur J Neurosci 14, 1771–1776.

77. Pardanaud, L., Luton, D., Prigent, M., Bourcheix, L. M., Catala, M., and Dieterlen-Lievre, F. (1996). Two distinct endothelial lineages in ontogeny, one of them related to hemopoiesis. Development 122, 1363–1371.

78. Carmeliet, P., and Luttun, A. (2001). The emerging role of the bone marrow-derived stem cells in (therapeutic) angiogenesis. Thromb Haemost 86, 289–297.

79. Zhang, Z. G., Zhang, L., Jiang, Q., and Chopp, M. (2002). Bone marrow-derived endothelial progenitor cells participate in cerebral neovascularization after focal cerebral ischemia in the adult mouse. Circ Res 90, 284–288.

80. Werner, N., Priller, J., Laufs, U., et al. (2002). Bone marrow-derived progenitor cells modulate vascular reendothelialization and neointimal formation: effect of 3-hydroxy-3-methylglutaryl coenzyme a reductase inhibition. Arterioscler Thromb Vasc Biol 22, 1567–1572.

81. Chang, C., and Hemmati-Brivanlou, A. (1998). Cell fate determination in embryonic ectoderm. J Neurobiol 36, 128–151.

82. Gould, E., Reeves, A. J., Graziano, M. S., and Gross, C. G. (1999). Neurogenesis in the neocortex of adult primates. Science 286, 548–552.

83. van Praag, H., Schinder, A. F., Christie, B. R., Toni, N., Palmer, T. D., and Gage, F. H. (2002). Functional neurogenesis in the adult hippocampus. Nature 415, 1030–1034.

84. Corti, S., Locatelli, F., Donadoni, C., et al. (2002). Neuroectodermal and microglial differentiation of bone marrow cells in the mouse spinal cord and sensory ganglia. J Neurosci Res 70, 721–733.

85. Keshet, G. I., Brazelton, T., Weimann, J. M., and Blau, H. M. (2002). From marrow to brain. Paper presented at WS 7-2, Seventh European Congress of Neuropathology, Helsinki, Finland.

86. Terada, N., Hamazaki, T., Oka, M., et al. (2002). Bone marrow cells adopt the phenotype of other cells by spontaneous cell fusion. Nature 416, 542–545.

87. Li, Y., Chen, J., Wang, L., Lu, M., and Chopp, M. (2001). Treatment of stroke in rat with intracarotid administration of marrow stromal cells. Neurology 56, 1666–1672.

88. Li, Y., Chen, J., Wang, L., Zhang, L., Lu, M., and Chopp, M. (2001). Intracerebral transplantation of bone marrow stromal cells in a 1-methyl-4-phenyl-1,2,3,6-tetrahydropyridine mouse model of Parkinson's disease. Neurosci Lett 316, 67–70.

89. Akiyama, Y., Radtke, C., Honmou, O., and Kocsis, J. D. (2002). Remyelination of the spinal cord following intravenous delivery of bone marrow cells. Glia 39, 229–236.

90. Jin, H. K., Carter, J. E., Huntley, G. W., and Schuchman, E. H. (2002). Intra-cerebral transplantation of mesenchymal stem cells into acid sphingomyelinase-deficient mice delays the onset of neurological abnormalities and extends their life span. J Clin Invest 109, 1183–1191.
91. Li, Y., Chen, J., Chen, X. G., et al. (2002). Human marrow stromal cell therapy for stroke in rat: neurotrophins and functional recovery. Neurology 59, 514–523.

12

Adult Retinal Stem Cells

Monica L. Vetter and Edward M. Levine

1. INTRODUCTION

Degenerative diseases of the retina result in the loss of specific populations of retinal neurons. For example, retinitis pigmentosa is characterized by progressive loss of rod photoreceptors, macular degeneration is a common disease of the elderly in which rod and cone photoreceptors degenerate, and glaucoma is marked by a loss of retinal ganglion cells. Thus, there is considerable interest in identifying retinal stem cells with the capacity to repopulate the retina in response to disease or injury. Although this has not yet been achieved in the mammalian eye, recent results hold promise for future success in this area.

One source of evidence for a resident stem cell population in the retina is the capacity of retinal tissue to regenerate in response to injury or damage. There are considerable species differences in the potential for retinal regeneration. It has been known for some time that retinal stemlike cells exist in the eyes of fish and amphibians because classic studies demonstrated a capacity for retinal regeneration in these species (1–5). More recently, there is evidence in chick and even mammals for cell populations that have potential to differentiate into retinal cells (4,5). There are a number of different sources for cells that can undergo neural retinal differentiation; however, one common feature found thus far is that all cells capable of giving rise to retinal neurons originate from the neural ectoderm-derived structures of the optic vesicle.

Here, we review the understanding of stem cells in the adult vertebrate retina. We first describe the organization of the vertebrate retina, briefly discuss how cells of the retina are generated during development, and then present the evidence for stem cells both in the neural retina and in ocular tissues outside the retina. Finally, we discuss the prospects for therapeutic treatment of retinal disease using retinal stem cell technology.

From: *Adult Stem Cells*
Edited by: K. Turksen © Humana Press Inc., Totowa, NJ

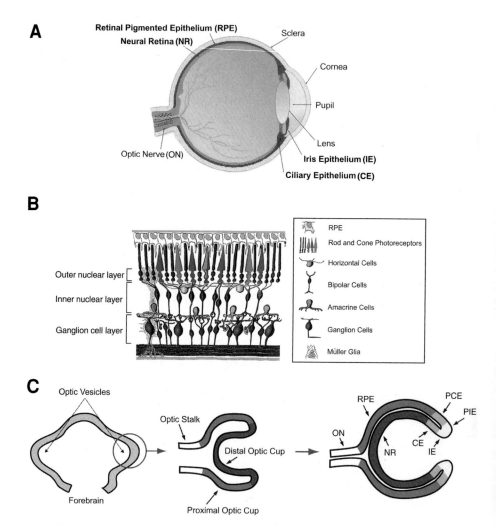

Fig. 1. Structures of the eye, neural retina, and developing optic vesicle. (**A**) The main components of the mature vertebrate eye. Ocular tissues discussed in the text and that contain cell populations with retinal stem cell properties are indicated with bold labels. (**B**) The neural retina contains six major neuronal cell types (rod and cone photoreceptors, horizontal cells, bipolar cells, amacrine cells, and ganglion cells) and one major glial cell type (Müller glia). These cell types are organized

2. ORGANIZATION OF THE VERTEBRATE OCULAR TISSUES

Several regions of the eye have been implicated in housing cell populations with stem cell properties or that have the potential to become retinal stem cells. These include the neural retina itself, the retinal pigmented epithelium (RPE), the pigmented ciliary epithelium, and the iris epithelium (Fig. 1A). In this section, we briefly describe the basic cellular organization of these tissues in the mammalian eye. For a comprehensive description of visual system anatomy, physiology, and function, see the Webvision Web site at http://webvision.med.utah.edu/index.html.

2.1. Neural Retina

The neural retina processes visual information and is subdivided into three distinct cellular layers with cellular composition that is correlated with the flow of visuosensory input (Fig. 1B). The outer nuclear layer (ONL) is composed of the rod and cone photoreceptors; the inner nuclear layer (INL) is composed of three classes of interneurons, horizontal cells, bipolar cells, and amacrine cells; and the ganglion cell layer (GCL) is composed of retinal ganglion cells and displaced amacrine cells. The Müller glia nuclei are resident in the INL, but the full extent of these cells traverses all layers and terminates to form two membranes: the external limiting membrane and the inner limiting membrane. Two synaptic layers are sandwiched between the nuclear layers. Rod and cones synapse onto the dendrites of horizontal cells and bipolar cells in the outer plexiform layer, and bipolar cells and amacrine cells synapse onto the retinal ganglion cell dendrites in the inner plexiform layer. Finally, the nerve fiber layer is composed of retinal ganglion cell axons and astrocytes. Importantly, many of these major cell classes, most notably

---

Fig. 1. *(continued)*

into three distinct layers: the ganglion cell layer, the inner nuclear layer, and the outer nuclear layer. The nonneural retinal pigment epithelium (RPE) is in close contact with the outer segments of the rod and cone photoreceptors. **(C)** Eye development begins when the optic vesicles evaginate from the walls of the forebrain at very early stages of neural development. As development proceeds, the optic vesicles invaginate to form an optic cup. The outer layer of the optic cup will become the retinal pigment epithelium (RPE) and more distally will give rise to the pigmented ciliary epithelium (PCE) and the pigmented iris epithelium (PIE). The inner layer of the optic cup will differentiate into neural retina and more distally will give rise to the nonpigmented ciliary epithelium (CE) and the nonpigmented iris epithelium (IE). The optic stalk will become the optic nerve (ON). (Panels A and B modified from Webvision, http://webvision.med.utah.edu/index.html, with the expert assistance of Mary Scriven and kind permission of Dr. Helga Kolb.)

the interneurons and ganglion cells, are further divided into many distinct subtypes.

2.2. Retinal Pigmented Epithelium

The RPE is a monolayer of epithelial cells in close contact with the outer segments of the rod and cone photoreceptors in the neural retina. RPE cells are highly differentiated in that they have an apical (toward the photoreceptors) and basal polarity, and all of the RPE cells are joined near the apical side by tight junctions (zonula occludens), which functions in part to form the blood–brain barrier. RPE cells are important for maintaining the integrity of photoreceptors by providing trophic support and by phagocytosing outer segment disks shed by the photoreceptors. RPE cells are highly enriched in the pigment melanin.

2.3. Ciliary Epithelium

The ciliary epithelium is a bilayered epithelium that extends from the periphery of the neural retina and RPE. The monolayer that is continuous with the neural retina is not pigmented, whereas the monolayer continuous with the RPE is pigmented. This tissue is an important region for attachment of several ocular structures, including the retina and lens. In addition, the nonpigmented ciliary epithelial cells secrete aqueous fluid and maintain intraocular pressure.

2.4. Iris Epithelium

The iris epithelium is similar in structure to the ciliary epithelium, and it forms the outermost margin of tissue continuous with the neural retina, RPE, and ciliary epithelium and forms the boundaries of the pupil. Like the ciliary epithelium, the iris epithelium is home to several ocular muscles that are interspersed through the epithelium.

3. DEVELOPMENT OF OCULAR TISSUES

3.1. Optic Vesicle Development

The neural ectoderm structures of the eye, including the neural retina, RPE, ciliary epithelium, and iris epithelium, are derived from the optic vesicles (Fig. 1C). These consist of bilateral evaginations from the walls of the forebrain during early nervous system development. As development proceeds, the optic vesicles expand toward the head ectoderm before invaginating to form a bilayered optic cup. The outer layer of the optic cup (proximal) will give rise to the RPE, and the inner layer of the optic cup (distal) thickens, undergoes a proliferative expansion, and will differentiate into the neural retina. The iris and ciliary epithelia arise primarily from the neuroepithelium.

3.2. Retinal Cell Fate Specification

Within the neural retina domain of the optic cup are proliferating retinal progenitors that will give rise to six major neuronal cell types (rod and cone photoreceptors, bipolar cells, amacrine cells, horizontal cells, ganglion cells), as well as to Müller glia. These retinal cell types are born in an overlapping sequence that is largely conserved across species *(6)*. Lineage analysis in a number of species has demonstrated that retinal progenitors are multipotent rather than dedicated to the generation of specific retinal cell types *(7–9)*. However, there is evidence that the competence of retinal progenitors changes over developmental time, so that at any one time, they give rise to a limited subset of retinal cell types *(6)*. In addition, there is progressive restriction in developmental potential of retinal progenitors. Thus, the ultimate fate of differentiating retinal cells depends on the intrinsic competence of the retinal progenitors and extrinsic signals that provide specific differentiation cues *(6,10,11)*.

4. DEFINITION OF A RETINAL STEM CELL

As any reader of the stem cell literature can attest, it is essentially impossible to assign a precise and all-inclusive definition of what constitutes a stem cell or a stem cell population. A major reason for this is that the application of the term *stem cell* has become context dependent. This has occurred in large part because of the experimental interventions often used to identify a stem cell, and because different tissues have distinct requirements in maintaining cell numbers and cell-type diversity. Moreover, what is considered a primary characteristic of a stem cell in one tissue may be considered a minor characteristic in another. Thus, to evaluate the fast-growing field of stem cell research critically, the context in which the term stem cell is applied must be understood.

Potten and Loeffler (1990) proposed an adult stem cell lineage model that is based on the robust and continuous renewal of the crypt epithelium in the small intestine *(12)*. Primary characteristics of the adult stem cell are that they are undifferentiated, have a long cell cycle time, and are self-renewing. A subset of stem cell progeny rapidly proliferate (transit-amplifying cells) and produce all of the differentiated progeny that make up the complement of cell types in a given structure. Although the transit-amplifying cell can generate many progeny, it is not capable of long-term self-renewal under normal conditions. Finally, with appropriate stimuli such as tissue damage, an adult stem cell population may have the propensity to regenerate the damaged tissue. The remainder of this chapter considers putative sources of adult retinal stem cells with these characteristics in mind.

In the vertebrate neural retina, two predominant precursor cell populations are recognized: stem cells and progenitor cells *(2,4,5)*. During embryonic development, as yet there is no evidence for a self-renewing retinal stem cell population. Rather, the cells of the mature retina born during the developmental period are derived from multipotent progenitors. These progenitors are essentially a transit-amplifying cell population because they proliferate rapidly, give rise to the complement of neural cell types, and are ultimately depleted. As described in the following sections, there are now several putative sources of adult retinal stem cells, some originating in the neural retina and some originating from other ocular tissues. These candidate stem cell populations were discovered in vivo as well as by in vitro approaches.

5. SOURCES OF STEM CELLS IN THE RETINA

There are several cell populations in the neural retina that have the potential to contribute to retinal regeneration, although considerable species differences exist. Fish, amphibians, and chicks have a population of stem cells at the margins of the retina that contribute to the normal growth of the eye and can repopulate cells in response to damage or injury *(4,5)*. In adult fish, there is also a more specialized progenitor population in the retina that generates rod photoreceptors selectively *(2)*. Evidence also suggests that, in response to neurotoxic damage, Müller glia in the chick retina can reenter the cell cycle and give rise to subsets of retinal neurons *(13)*. These different cell populations are reviewed below.

5.1. Ciliary Marginal Zone of the Retina

In fish and amphibians, growth of the retina and RPE after the initial embryonic period is achieved by the addition of cells from a proliferative zone at the retinal margins known as the ciliary marginal zone (CMZ). This was demonstrated by showing that a pulse of [^{3}H]-thymidine in the postembryonic period labels cells at the margins of the retina, and these labeled cells ultimately differentiate into retinal neurons and glia *(14–17)*. Thus, new cells are added in rings as the retina grows. In the CMZ, the most peripheral cells are slowly dividing and only become labeled with prolonged [^{3}H]-thymidine labeling *(2)*. This is consistent with the low rate of proliferation characteristic of stem cells. Cells that are more centrally located divide more rapidly and can be labeled with short pulses of [^{3}H]-thymidine, suggesting that these are rapidly cycling progenitors.

Injury or damage to the retina can stimulate proliferation of cells in the CMZ. Elegant experiments showed that selective ablation of amacrine and bipolar cells by intraocular injections of kainate in *Rana pipiens* tadpoles can stimulate increased cell production from the margins, which results in selective replacement of the ablated cells *(18)*. Thus, the cells of the CMZ not only contribute to the normal growth of the retina, but also can be a source for regeneration of retinal cells.

To determine whether cells in the CMZ are multipotent or instead represent committed precursors of specific retinal cell types, the lineage tracer rhodamine dextran was injected into individual precursor cells at the margins of the *Xenopus laevis* retina *(19)*. It was shown that cells in the CMZ are multipotent and can give rise to all major retinal cell types as well as to nonneural pigment epithelial cells. Analysis of clone size suggested that, in the margins, there are both self-renewing stem cells and progenitors with more limited proliferative potential *(19)*. This has been confirmed by *in situ* analysis of gene expression at the margins of fish and *Xenopus* retinas *(20–24)*. These studies showed that there is a peripheral-to-central sequence of gene expression that recapitulates the temporal sequence of gene expression during retinal histogenesis *(25)*. In fish, the cells of the CMZ give rise to all retinal cell types except rods, which are instead derived from a dedicated rod precursor population distributed throughout the INL of the retina (*see* Section 5.2.).

Until recently, it was believed that the CMZ was a feature unique to fish and amphibians. However, it has now been shown that in chicks a proliferative population of cells exists at the margins of the retina for up to 3 wk after hatching *(26)*. These dividing cells differentiate and give rise to amacrine, bipolar, and Müller cells, but were not found to differentiate into ganglion cells, horizontal cells, or photoreceptors. However, injection of insulin and fibroblast growth factor 2 (FGF-2) into the vitreous chamber of posthatch chickens could stimulate the production of retinal ganglion cells, suggesting that it is either the absence of appropriate signals or the presence of inhibitory signals that normally prevents these cells from differentiating into all retinal cell types *(27)*. Additional experiments will be required to determine whether cells in the CMZ of the chick are fully multipotent. Interestingly, unlike fish and amphibians, neurotoxic lesions to the retina do not provoke an increase in proliferation of cells in the chicken CMZ *(26)*.

There is no evidence yet for CMZ-like cells in the mammalian retina, although it has been suggested that cells in the ciliary body of the eye are analogous to cells of the CMZ found in other vertebrates (*see* Section 6.1.). It will be interesting to determine why there are such intriguing species differences in the development and maintenance of a CMZ in the vertebrate retina.

5.2. Rod Precursor Lineage of Fish

In addition to the CMZ, another adult neurogenic progenitor cell population was initially discovered in retinas of teleost fish over 20 yr ago *(28–30)*. These cells were termed *rod precursor cells* because they were observed to give rise exclusively to rod photoreceptors. In contrast to the CMZ cells, rod precursors are distributed along the entire central-to-peripheral plane of the mature retina and are found in the ONL interspersed with mature photoreceptors.

Several characteristics of the rod precursor cell suggest that it is a transit-amplifying cell rather than a true stem cell. First, its rate of cell division is quite rapid. Second, pulse labeling with BrdU or [^{3}H]-thymidine followed by long survival times showed that the rod precursor cells do not self-renew, but rather differentiate into rods. Third, the cell output appears lineage restricted. Thus, even by minimal criteria, rod precursor cells do not qualify as bona fide adult stem cells. However, because this cell population persists throughout most of the life of the fish, it has been postulated that, in the retina, a stem cell population exists with an output that is the rod precursor population.

The identification of the putative rod precursor stem cell population has been elusive *(1)*. For some time, the location of the rod precursor stem cell has been postulated to reside in the INL. This was demonstrated by [^{3}H]-thymidine injections followed by successively longer survival times to trace the fate of the labeled cells. These proliferative INL cells divide and form clusters of labeled cells termed *neurogenic clusters*. Subsequently, individual cells migrate into the ONL, where they give rise to the rod precursor cell population.

A study by Otteson and colleagues *(31)* suggests that the neurogenic clusters may be maintained by yet another population of INL cells, and they propose that this population may be the stem cells. As opposed to a relatively short pulse of BrdU or [^{3}H]-thymidine, goldfish were exposed to BrdU continuously for 9 d. This long labeling period would allow for the detection of a slow-dividing cell population. Using this approach, they observed two morphologically and spatially distinct populations of BrdU cells in the INL. One population had cell bodies of a fusiform morphology and was positioned toward the outer portion of the INL; this population is most likely the neurogenic clusters. The other population had a spherical morphology, was much fewer in number, and was positioned immediately adjacent to the neurogenic cluster at the inner boundary of the INL. In experiments in which the fish were allowed to survive for varying periods following BrdU treatment, it was observed that the BrdU-labeled spherical cells were maintained, the BrdU-labeled fusiform cells decreased, and the BrdU-labeled cells in the ONL increased. These observations suggest that the spherical cells self-renew and give rise to the neurogenic clusters, which in turn give rise to the rod precursor cells. Consistent with this, these different cell populations were organized into radial arrays.

Several issues still need to be resolved, however, before it can be concluded that the spherical cells are stem cells. First, it needs to be determined that the neurogenic clusters are lineal descendants of the spherical cells. Second, can the spherical cells repopulate neurogenic clusters in a manner similar to that shown in depletion–replacement experiments done in the hematopoietic lineage and more recently in the subventricular zone of adult rodent brains?

Another interesting feature of the rod precursor lineage is its potential to become multipotent following injury to the retina. On physical ablation or severe neurotoxic damage, retinal neurons and glia are replaced, and it is well established that the source of new cells arises from the mature retina *(1–3)*. Interestingly, it is suggested that the rod precursors in the ONL and the neurogenic clusters and spherical cells in the INL all contribute to the regenerative process *(31)*. Thus, it appears that the healthy retinal environment restricts the rod precursor lineage.

5.3. Müller Glia

It is now well established that adult neurogenesis occurs in the subventricular zones lining the lateral ventricles in the cerebrum *(32)*. Two cell populations have been suggested as the neural stem cell: the astrocytes of the subependyma and the ependymal cells that line the ventricles *(33,34)*. Although the precise identity of the stem cell is still controversial *(35,36)*, in either case, the candidate populations are both glial cells. This is intriguing because it suggests that adult neural stem cells may be glia that dedifferentiate and then transdifferentiate into neurons.

As described in Section 2.1., the Müller glia are derived from the retinal progenitor population *(7)*. Because they are lineally related to retinal neurons, it is tempting to speculate that the Müller glia may have the potential to behave as retinal stem cells. To date, there is no evidence that Müller glia are an intrinsic source of retinal neurons under normal conditions.

A study by Fischer and Reh *(13)*, however, suggested that, in response to neurotoxic damage, Müller glia reenter the cell cycle, and new neurons are generated in the avian retina. N-Methyl-D-aspartate was administered to the retinas of juvenile chicks at a dose sufficient to kill amacrine cells and possibly other retinal cell types. After 2 d, a robust induction of proliferation was observed by BrdU incorporation in the INL. This proliferative response was transient, however, suggesting that the proliferating cells may have exited the cell cycle and differentiated.

To address this possibility, the fate of the BrdU-labeled cells were followed by examining expression of cell class-restricted proteins over a period of days to weeks. Initially, the overwhelming majority of BrdU-labeled cells expressed glutamine synthetase (GS), a marker of Müller glia. In days, however, many of the BrdU-labeled cells were GS negative, but expressed the homeodomain proteins Pax6, Chx10, and the basic helix-loop-helix protein CASH1, all of which are transcription factors coexpressed exclusively in retinal progenitors. Furthermore, some BrdU-positive cells also expressed Hu, a marker of differentiated ganglion cells and amacrine cells, or cellular retinoic acid binding protein (CRABP), a marker of differentiated amacrine and bipolar cells. These observations suggest that, given the appropriate stimulus, Müller glia have the potential to dedifferentiate into a progenitor-like state and then differentiate into retinal neurons.

Although the above study is certainly provocative, several unanswered questions still remain and warrant consideration. For instance, because clonal analysis of the damaged-induced proliferative cells was not done, it remains to be demonstrated that the BrdU-positive retinal neurons arose from dedifferentiated Müller glia. Second, the authors did not observe expression of neurotransmitters or of a photoreceptor phenotype in the BrdU-positive neurons. It is therefore important to determine whether these neurons have the capacity to mature fully or adopt a photoreceptor fate and what the signals are that can promote these fates. Finally, it is not yet known whether adult avian Müller glia retain this neurogenic potential, or whether mammalian Müller glia have the propensity to dedifferentiate into a progenitor-like cell that can generate retinal neurons.

6. SOURCES OF STEM CELLS OUTSIDE THE RETINA

Outside the neural retina itself, there exist cell populations that have the potential to contribute to regeneration of retinal neurons. In a number of species, including chicks and amphibians, the nonneural RPE can transdifferentiate into retinal tissue *(37)*. In mammals, there are cells from both the pigmented epithelium of the ciliary body and the iris epithelium that, under certain conditions, can differentiate into retinal neurons in culture *(38–40)*.

6.1. Pigmented Ciliary Epithelium in Mammals

In mammals, neither the neural retina nor the RPE show any evidence for regenerative potential in adults. This raises the question of whether there are any cells in the adult eye with retinal stem cell-like properties. Long-term labeling of 4-wk-old rats with BrdU revealed a small population of proliferative cells in the pigmented ciliary body *(38)*. The number of cells that incorporated BrdU could be stimulated in explant culture by treatment with FGF-2, a known mitogen for neural stem cells. Under no conditions was BrdU incorporation detected in the retina, RPE, or nonpigmented ciliary body. The presence of proliferative cells in the ciliary body and the fact that this structure shares embryological origins with the neural retina raised the possibility that these proliferative cells could represent stemlike cells in the adult mammalian eye. In addition, the ciliary body may be related to the ciliary marginal zone of the fish, amphibian, and chick neural retina, which contains resident retinal stem cells that contribute to retinal growth.

Dissociated cells from the pigmented ciliary epithelium of both mouse and rat could be grown in culture, in which, at very low frequency, they formed neurospheres, a proliferative colony of neural stemlike cells *(38,39)*. Proliferative cells in the neurospheres were positive for nestin, a marker expressed by neural stem cells. The neurospheres had the capacity for self-renewal because a subset of single pigmented cells from dissociated neurospheres gave rise to new neurosphere colonies when recultured *(38,39)*, and this could be repeated for at least six generations *(39)*. Treatment with FGF-2 in serum-free culture enhanced neurosphere formation; however, there appeared to be production of endogenous FGF-2, permitting neurospheres to grow in the absence of exogenous FGF-2 *(39)*. Neurosphere colonies did not arise from cultures of adult neural retina, RPE, iris epithelium, or ciliary muscle or from nonpigmented ciliary process cells. Pigmentation was not required for neurosphere formation because colonies could be generated from albino tissue. Neurospheres could not be cultured from adult neural retina or from adult RPE. Even neurospheres derived from E14 neural retina did not show a capacity for self-renewal. Thus, cells derived from the ciliary epithelium were unique in having stem cell-like properties. Interestingly, the ability of ciliary margin tissue to give rise to neurospheres is conserved across species because colonies could also be derived from postmortem adult bovine and human ciliary margin tissue *(39)*.

Neurospheres appeared to arise from single pigmented ciliary body cells that proliferated in culture and gave rise to mixtures of pigmented and nonpigmented cells. Nonpigmented cells in the neurosphere expressed markers of retinal progenitors such as the homeodomain transcription factor Chx10. This suggests that pigmented cells of the ciliary body may be transdifferentiating into nonpigmented neural retinal progenitors *(38,39)*.

To determine whether neurosphere colonies could differentiate into retinal neurons or glia, the cells were grown under conditions that promote differentiation. Cells from both primary and secondary neurosphere colonies differentiated and expressed markers for rods, bipolar cells, and Müller glia; however, markers for retinal ganglion cells, horizontal cells, and amacrine cells were either not detected or were extremely rare *(38,39)*. This may be because of the culture conditions under which the assay was performed because amacrine cell differentiation could be enhanced by high-density pellet culture conditions *(39)*. Thus, neurospheres derived from the pigmented ciliary epithelium not only self-renew, but are multipotential and give rise to cells with retinal progenitorlike properties that can differentiate into neurons and glia that express markers of differentiated retinal cells.

This work has obviously generated real excitement over the promise for regenerating retinal neurons lost to disease or injury; however, a number of questions remain. For example, can the retinal stemlike cells from the pigmented ciliary epithelium give rise to all classes of retinal neurons? Will these cells (or progenitors derived from these cells) survive and differentiate when transplanted in vivo? And, can we define conditions to promote the differentiation of selected retinal cell types for replacement of retinal neurons lost to diseases such as retinitis pigmentosa or glaucoma? In addition, the role of these stemlike cells in vivo is not yet clear. Will it be possible to stimulate these cells in vivo to transdifferentiate to retinal progenitors and differentiate into retinal neurons in reponse to disease or injury? Much work remains to be done before the therapeutic potential of this work can be realized.

6.2. Iris Epithelium

The iris epithelium is contiguous with the pigmented ciliary epithelium, raising the possibility that it may share some of the stem cell-like properties of its neighbor (*see* Chapter 13). Traditionally, the iris epithelium has been associated with lens regeneration, but the potential for these cells to generate retinal tissue had not been carefully examined. Iris tissue does not give rise to neurosphere colonies *(38,39)*; however, iris cells grown in monolayer culture in the presence of FGF proliferate and could be stimulated to differentiate into neurofilament 200-positive cells *(40)*. However, unlike differentiated cells derived from pigmented ciliary epithelium, the differentiated cells derived from the iris epithelium did not express rhodopsin, a marker of differentiated rod photoreceptors. No other retinal markers were examined in these experiments.

To determine whether iris-derived cells could be induced to differentiate into rodlike cells, the photoreceptor-specific homeobox gene *Crx* was overexpressed. Differentiated cells expressing *Crx* expressed rhodopsin as well as recoverin, another gene expressed in photoreceptors and a subset of bipolar neurons, suggesting that Crx expression in iris-derived cells was sufficient to promote photoreceptor differentiation *(40)*. Similar results were obtained by overexpression of *Crx* in pigmented ciliary epithelium cells grown in monolayer culture (which does not normally result in photoreceptor differentiation). Interestingly, overexpression of *Crx* in a neural stem cell line derived from the adult rat hippocampus did not promote expression of photoreceptor-specific markers, suggesting that tissues derived from the optic vesicle may have a unique ability to differentiate into neurons with retinal properties.

In summary, these experiments demonstrated that cells derived from the iris epithelium have the potential to differentiate into photoreceptors, but only when *Crx* is expressed. However, because the self-renewing potential of these cells was not examined and their multipotential properties were not tested, there is no evidence yet that true retinal stemlike cells exist in the iris epithelium.

6.3. RPE Transdifferentiation

Urodele amphibians (newts and salamanders) have a remarkable capacity for adult regeneration, and this is also true in the eye. Removal of the neural retina results in complete retinal regeneration from cells of the RPE. After removal of the retina, the cells of the RPE proliferate, dedifferentiate, and lose their pigmentation, then form a second layer of cells that differentiates into neural retina *(37)*. Similar transdifferentiation of RPE to neural retina is possible in anuran amphibians until metamorphosis, but does not occur in adults. In other vertebrate species, such as chicks and rodents, RPE transdifferentiation is restricted to embryonic periods (up to E4.5 in chick and E13 in rats) and diminishes as development proceeds *(41,42)*.

Fibroblast growth factors, such as FGF-1, FGF-2, or FGF-8, can stimulate transdifferentiation of embryonic RPE to neural retina in vitro in multiple species *(43–47)*. In addition, RPE transdifferentiation can also be promoted by overexpression of cell intrinsic factors. For example, expression of the basic helix–loop–helix transcription factor NeuroD in chick RPE cells isolated at E6 promoted neuronal differentiation and expression of the photoreceptor markers, including visinin *(48,49)*. Similarly, adult human RPE cells expressing the oncogenic form of the ras signaling molecule adopted a neuronal phenotype and expressed several neuronal markers, including neurofilament and neuron-specific enolase, although retinal-specific markers were not examined *(50)*. Thus, RPE cells in homeothermic vertebrates may retain some capacity to transdifferentiate beyond the embryonic period, but the full potential of these cells remains to be examined. Again, clues may lie with the intriguing species differences in the ability of RPE cells to transdifferentiate to neural retina.

7. CONCLUSIONS AND FUTURE PROSPECTS

Tissues derived from the optic vesicle have a unique ability to generate retinal neurons. This neurogenic capacity, however, diminishes in most vertebrate organisms as they reach adulthood, with the exception of teleost fishes and urodele amphibians. The continued neurogenic capacity found in these organisms is due in large part to the presence of active adult stem cells. Interestingly, recent studies demonstrated that most, if not all, vertebrate classes have cell populations in the eye that retain neurogenic potential and, if given the appropriate stimulus, can actively differentiate into retinal neurons. Whether these cell populations can be coaxed into becoming productive retinal stem cells and used for therapeutic purposes is still an open question.

To utilize a stem cell therapy to replace dying retinal neurons, many significant hurdles need to be overcome. With respect to the findings presented in this chapter, it has not been demonstrated that the newly identified neurogenic cell populations produce fully differentiated and functional neurons. Thus, further studies are needed to determine how these cell populations can be manipulated into producing sufficient progeny without introducing deleterious changes in their genomes and then differentiating into functional neurons of the cell class desired (i.e., photoreceptors for retinitis pigmentosa and macular degeneration, ganglion cells for glaucoma). An important approach in this regard is to continue to identify and understand the regulatory pathways that promote retinal development and regeneration and develop strategies to activate these pathways in cell populations with neurogenic potential.

REFERENCES

1. Raymond, P. A., and Hitchcock, P. F. (1997). Retinal regeneration: common principles but a diversity of mechanisms. Adv Neurol 72, 171–184.
2. Reh, T. A., and Levine, E. M. (1998). Multipotential stem cells and progenitors in the vertebrate retina. J Neurobiol 36, 206–220.

3. Raymond, P. A., and Hitchcock, P. F. (2000). How the neural retina regenerates. Results Probl Cell Differ 31, 197–218.

4. Perron, M., and Harris, W. A. (2000). Retinal stem cells in vertebrates. Bioessays 22, 685–688.

5. Reh, T. A., and Fischer, A. J. (2001). Stem cells in the vertebrate retina. Brain Behav Evol 58, 296–305.

6. Cepko, C. L., Austin, C. P., Yang, X., Alexiades, M., and Ezzeddine, D. (1996). Cell fate determination in the vertebrate retina. Proc Natl Acad Sci U S A 93, 589–595.

7. Turner, D. L., and Cepko, C. L. (1987). A common progenitor for neurons and glia persists in rat retina late in development. Nature 328, 131–136.

8. Holt, C. E., Bertsch, T. W., Ellis, H. M., and Harris, W. A. (1988). Cellular determination in the *Xenopus* retina is independent of lineage and birth date. Neuron 1, 15–26.

9. Wetts, R., and Fraser, S. E. (1988). Multipotent precursors can give rise to all major cell types of the frog retina. Science 239, 1142–1145.

10. Fuhrmann, S., Chow, L., and Reh, T. A. (2000). Molecular control of cell diversification in the vertebrate retina. Results Probl Cell Differ 31, 69–91.

11. Livesey, F. J., and Cepko, C. L. (2001). Vertebrate neural cell-fate determination: lessons from the retina. Nat Rev Neurosci 2, 109–118.

12. Potten, C. S. and Loeffler, M. (1990). Stem cells: attributes, cycles, spirals, pitfalls, and uncertainties. Lessons for crypt. Development 110, 1001–1020.

13. Fischer, A. J., and Reh, T. A. (2001). Muller glia are a potential source of neural regeneration in the postnatal chicken retina. Nat Neurosci 4, 247–252.

14. Straznicky, K., and Gaze, R. (1971). The growth of the retina in *Xenopus laevis*: an autoradiographic study. J Embryol Exp Morphol 26, 67–79.

15. Hollyfield, J. (1971). Differential growth of the neural retina in *Xenopus laevis* larvae. Dev Biol 24, 264–286.

16. Johns, P. (1977). Growth of the adult goldfish eye. III. Sources of the new retinal cells. J Comp Neurol 176, 343–357.

17. Meyer, R. (1978). Evidence from thymidine labeling for continuing growth of retina and tectum in juvenile goldfish. Exp Neurol 59, 99–111.

18. Reh, T. A. (1987). Cell-specific regulation of neuronal production in the larval frog retina. J Neurosci 7, 3317–3324.

19. Wetts, R., Serbedzija, G. N., and Fraser, S. E. (1989). Cell lineage analysis reveals multipotent precursors in the ciliary margin of the frog retina. Dev Biol 136, 254–263.

20. Levine, E. M., Hitchcock, P. F., Glasgow, E., and Schechter, N. (1994). Restricted expression of a new paired-class homeobox gene in normal and regenerating adult goldfish retina. J Comp Neurol 348, 596–606.

21. Hitchcock, P. F., Macdonald, R. E., VanDeRyt, J. T., and Wilson, S. W. (1996). Antibodies against Pax6 immunostain amacrine and ganglion cells and neuronal progenitors, but not rod precursors, in the normal and regenerating retina of the goldfish. J Neurobiol 29, 399–413.

22. Levine, E. M., Passini, M., Hitchcock, P. F., Glasgow, E., and Schechter, N.

(1997). Vsx-1 and Vsx-2: two Chx10-like homeobox genes expressed in over-lapping domains in the adult goldfish retina. J Comp Neurol 387, 439–448.

23. Sullivan, S. A., Barthel, L. K., Largent, B. L., and Raymond, P. A. (1997). A goldfish Notch-3 homologue is expressed in neurogenic regions of embryonic, adult, and regenerating brain and retina. Dev Genet 20, 208–223.

24. Perron, M., Kanekar, S., Vetter, M. L., and Harris, W. A. (1998). The genetic sequence of retinal development in the ciliary margin of the *Xenopus* eye. Dev Biol 199, 185–200.

25. Harris, W. A., and Perron, M. (1998). Molecular recapitulation: the growth of the vertebrate retina. Int J Dev Biol 42, 299–304.

26. Fischer, A. J., and Reh, T. A. (2000). Identification of a proliferating marginal zone of retinal progenitors in postnatal chickens. Dev Biol 220, 197–210.

27. Fischer, A. J., Dierks, B. D., and Reh, T. A. (2002). Exogenous growth factors induce the production of ganglion cells at the retinal margin. Development 129, 2283–2291.

28. Johns, P. R., and Fernald, R. D. (1981). Genesis of rods in teleost fish retina. Nature 293, 141–142.

29. Johns, P. R. (1982). Formation of photoreceptors in larval and adult goldfish. J Neurosci 2, 178–198.

30. Julian, D., Ennis, K., and Korenbrot, J. I. (1998). Birth and fate of proliferative cells in the inner nuclear layer of the mature fish retina. J Comp Neurol 394, 271–282.

31. Otteson, D. C., D'Costa, A. R., and Hitchcock, P. F. (2001). Putative stem cells and the lineage of rod photoreceptors in the mature retina of the goldfish. Dev Biol 232, 62–76.

32. Alvarez-Buylla, A., and Garcia-Verdugo, J. M. (2002). Neurogenesis in adult subventricular zone. J Neurosci 22, 629–634.

33. Doetsch, F., Caille, I., Lim, D. A., Garcia-Verdugo, J. M., and Alvarez-Buylla, A. (1999). Subventricular zone astrocytes are neural stem cells in the adult mammalian brain. Cell 97, 703–716.

34. Johansson, C. B., Momma, S., Clarke, D. L., Risling, M., Lendahl, U., and Frisen, J. (1999). Identification of a neural stem cell in the adult mammalian central nervous system. Cell 96, 25–34.

35. Barres, B. A. (1999). A new role for glia: generation of neurons! Cell 97, 667–670.

36. Morshead, C. M., and van der Kooy, D. (2001). A new "spin" on neural stem cells? Curr Opin Neurobiol 11, 59–65.

37. Zhao, S., Rizzolo, L., and Barnstable, C. (1997). Differentiation and transdifferentiation of the retinal pigment epithelium. Int Rev Cytol 171, 225–266.

38. Ahmad, I., Tang, L., and Pham, H. (2000). Identification of neural progenitors in the adult mammalian eye. Biochem Biophys Res Commun 270, 517–521.

39. Tropepe, V., Coles, B. L., Chiasson, B. J., et al. (2000). Retinal stem cells in the adult mammalian eye. Science 287, 2032–2036.

40. Haruta, M., Kosaka, M., Kanegae, Y., et al. (2001). Induction of photo-receptor-specific phenotypes in adult mammalian iris tissue. Nat Neurosci 4, 1163–1164.

41. Coulombre, J., and Coulombre, A. (1965). Regeneration of neural retina from the pigmented epithelium in the chick embryo. Dev Biol 12, 79–92.

42. Zhao, S., Thornquist, S., and Barnstable, C. (1995). In vitro transdifferentiation of embryonic rat pigment epithelium to neural retina. Brain Res 677, 300–310.

43. Park, C., and Hollenberg, M. (1989). Basic fibroblast growth factor induces retinal regeneration in vivo. Dev Biol 134, 201–205.

44. Pittack, C., Jones, M., and Reh, T. A. (1991). Basic fibroblast growth factor induces retinal pigment epithelium to generate neural retina in vitro. Development 113, 577–588.

45. Guillemot, F., and Cepko, C. (1992). Retinal cell fate and ganglion cell differentiation are potentiated by acidic FGF in an in vivo assay of early retinal development. Development 114, 743–754.

46. Pittack, C., Grunwald, G. B., and Reh, T. A. (1997). Fibroblast growth factors are necessary for neural retina but not pigmented epithelium differentiation in chick embryos. Development 124, 805–816.

47. Vogel-Hopker, A., Momose, T., Rohrer, H., Yasuda, K., Ishihara, L., and Rapaport, D. H. (2000). Multiple functions of fibroblast growth factor-8 (FGF-8) in chick eye development. Mech Dev 94, 25–36.

48. Yan, R. T., and Wang, S. Z. (2000). Differential induction of gene expression by basic fibroblast growth factor and NeuroD in cultured retinal pigment epithelial cells. Vis Neurosci 17, 157–164.

49. Yan, R. T., and Wang, S. Z. (2000). Expression of an array of photoreceptor genes in chick embryonic retinal pigment epithelium cell cultures under the induction of neuroD. Neurosci Lett 280, 83–86.

50. Dutt, K., Scott, M., Sternberg, P. P., Linser, P. J., and Srinivasan, A. (1993). Transdifferentiation of adult human pigment epithelium into retinal cells by transfection with an activated H-ras proto-oncogene. DNA Cell Biol 12, 667–673.

13

Multipotentiality of Iris Pigment Epithelial Cells in Vertebrate Eye

Mitsuko Kosaka, Guangwei Sun, Masatoshi Haruta, and Masayo Takahashi

1. INTRODUCTION

The discovery of adult stem cells indicated a previously unrecognized degree of plasticity in stem cell function (1–3). Recent extensive studies have suggested that mammalian stem cells residing in one tissue may have the capacity to produce differentiated cell types for other tissues and organs (4–6). However, more recent reports raised questions about some of the earlier results, proposing that transdifferentiation consequent to cell fusion could underlie many observations otherwise attributed to an intrinsic plasticity of tissue stem cells (7,8). Thus, cell transdifferentiation is of great interest, albeit a poorly understood process invoked to explain how tissue-specific adult stem cells can lose their properties and generate new cells of other tissues.

The fact that differentiated adult cells can change their fate has been known for over a century. The phenomenon of Wolffian lens regeneration in newts (9) has attracted the interest of developmental biologists for long time because it is the clearest and most representative example of trans-differentiation naturally occurring in adult vertebrates: Melanin-producing iris pigment epithelial (PE) cells become crystallin-producing lens cells. A number of studies on the phenomenon of newt lens regeneration were published (10–17), but the molecular basis of this switch in the phenotype of PE cells is mostly unknown. At present, revisiting and rethinking the old phenomenon of Wolffian lens regeneration in adult newts could provide a useful opportunity for obtaining a real idea of somatic cell plasticity in vertebrates.

In this chapter, the historical background of the studies on trans-differentiation using PE cells is briefly reviewed. Current knowledge about the differentiation potency of iris PE cells in postnatal and adult vertebrates, including mammals, is summarized.

From: *Adult Stem Cells*
Edited by: K. Turksen © Humana Press Inc., Totowa, NJ

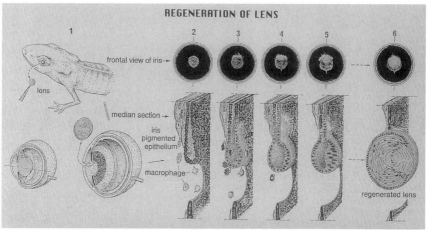

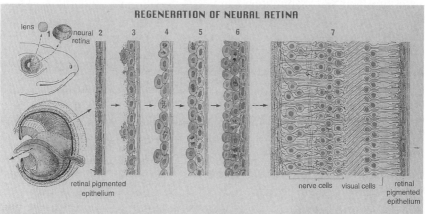

Fig. 1. Regeneration of lens and neural retina in the newt. The newt has a strong ability to regenerate lost parts of the body even after the individual has grown into an adult. In addition to the regeneration of limbs, remarkable examples are found in the eye. When the lens is surgically removed through an opening in the cornea, cells of the pigmented epithelium of the iris become depigmented and proliferate to make a new lens, which grows into the size of an adult lens with a morphology indistinguishable from the normal one. When the neural retina is removed, the retinal PE cells proliferate and regenerate a complete retina.

2. TRANSDIFFERENTIATION INTO LENS CELLS

2.1. Lens Regeneration in Urodeles

The newt has a strong ability to regenerate lost parts of the body throughout its lifetime. In addition to the regeneration of limbs, remarkable examples are found in the eye. When the lens is surgically removed from an adult newt eye, a structurally and functionally complete lens always regenerates from the PE of the dorsal papillary margin of the iris (Fig. 1). This case clearly demonstrates that fully differentiated cells can switch their type of differentiation and be reprogrammed into another differentiative pathway.

Cell culture studies showed that both dorsal and ventral dissociated newt iris PE cells can transdifferentiate into lens cells in vitro *(18,19)*. In addition to iris PE cells, it was found that retinal PE cells of adult newt can transdifferentiate into lens cells in vitro *(20)*.

Through studies to clarify the cellular origin of lens regeneration in the newt, the retinal PE cells of many vertebrate species have been shown to possess a dormant potency to transdifferentiation into the lens *(13,14)*. Recent observations proposed that retinal PE cells isolated even from adult human cells could also transdifferentiate into lenslike cells in vitro *(21)*.

2.2. A Model System of Lens Transdifferentiation Using Chick Retinal PE Cells

Cell culture of PE cells was attempted as a modern approach to establish an experimental system and to analyze transdifferentiation at the cellular and molecular levels. The introduction of phenylthiourea (PTU) and hyaluronidase (HUase) in PE cell cultures has permitted exact control of lens transdifferentiation of PE cells *(22)*. When retinal PE cells from chick embryos (E9) were dissociated and cultured in standard medium containing dialyzed FCS, PTU, and testicular HUase, they dedifferentiated rapidly and grew vigorously. By frequent passage before reaching confluence, it is possible to maintain the undifferentiated state in which cells express neither PE type- nor lens type-specific cell markers. Interestingly, the dedifferentiated PE cells can rapidly reexpress the differentiated PE cell phenotype after withdrawal of PTU and HUase. Furthermore, when seeded at high cell density in medium with dialyzed FBS, PTU, HUase, and ascorbic acid, the dedifferentiated PE cells transdifferentiated into lens cells.

Using this unique system of retinal PE cells, the biochemical and molecular studies have been extensively demonstrated and the results summarized *(13,14,23)*. It was shown that the effect of crude HUase on transdifferentiation was because of fetal growth factor 2 (FGF-2) contamination in the commercial preparations of the enzyme, and that FGF-2 promoted growth and lens transdifferentiation of retinal PE cells. In addition, the cell–cell contact and cell–substratum interactions were suggested to be important for the stabilization of retinal PE cells.

Although valuable observations were obtained from the culture system of embryonic retinal PE cells, the attempts to elucidate the molecular mechanisms of transdifferentiation in vitro have met with limited success. During long-term experience with their cell cultures, it was revealed that retinal PE cells from chick embryos were sensitive to culture conditions. In addition, utilization of the chemical agent PTU, an inhibitor of melanin synthesis that modifies cell surface properties *(22,24)*, has made it difficult to analyze lens transdifferentiation systematically. It remains unclear whether the PE cells can be induced to dedifferentiate without PTU. Moreover, it is still obscure whether the undifferentiated state is necessary for lens transdifferentiation to occur in vitro. So, trials for the establishment of an improved culture system led to the use of newt eye iris PE cells, in which lens transdifferentiation naturally occurs.

It has been difficult to prepare a pure culture of the iris PE cells of the chick embryo because of the tight adhesion of the epithelium with the stroma of the iris; therefore, transdifferentiation of the iris PE cells has scarcely been studied. A simple method was established to prepare pure iris PE cells from postnatal chick *(25)*. Interestingly, the iris PE cells isolated from postnatal chicks were much more stably maintained than the retinal PE cells, and their cells transdifferentiated and dedifferentiated efficiently and reproducibly through methods similar to those utilized for the chick retinal PE cells *(25)*.

2.3. Analysis for Lens Transdifferentiation Using Iris PE Cells of Postnatal Chick

In an effort to delineate regulatory factors in lens trandifferentiation of iris PE cells, the effects of known growth factors on iris PE cells were tested. FGF-2 was shown to promote cell proliferation and transdifferentiation of iris PE cells similar to retinal PE cells *(25)*. Further analysis showed epidermal growth factor (EGF) also had a significant effect on iris PE cells, just like FGF-2; the addition of EGF and FGF-2 to pure cell cultures of the iris PE cells synergistically induced the phenotypic change of lens transdifferentiation *(26)*. This finding contributes greatly to the simplification of the humoral requirements for the induction of lens transdifferentiation and provides a powerful system for the molecular analysis of lens transdifferentiation. Furthermore, it was found that the addition of EGF alone as well as FGF-2 alone could also induce in vitro lens transdifferentiation of embryonic retinal PE cells from early chick embryos (E5) (Fig. 2).

Mitogen-activated protein (MAP) kinase is the central component of a signal transduction pathway that is activated by growth factors interacting

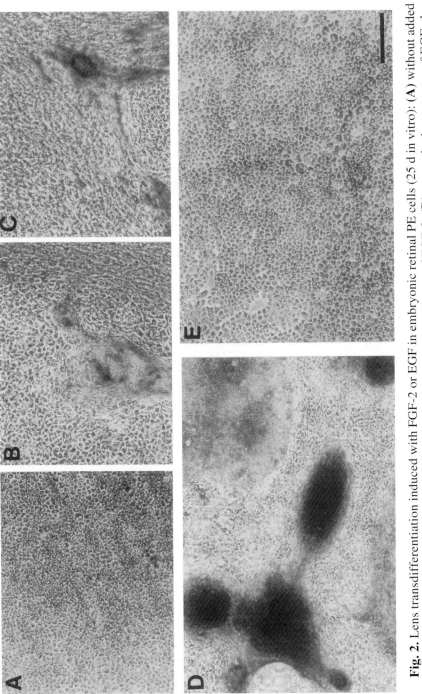

Fig. 2. Lens transdifferentiation induced with FGF-2 or EGF in embryonic retinal PE cells (25 d in vitro): (**A**) without added growth factors; (**B**) grown in the presence of EGF; (**C**) grown in the presence of FGF-2; (**D**) grown in the presence of EGF plus FGF-2; (**E**) with PD098059 for 1 h before the addition of EGF plus FGF-2. Scale bar, 200 μm.

with receptors that have protein tyrosine kinase activity. To evaluate the specific role of the MEK (MAP kinase kinase)–MAP kinase pathway in the transdifferentiative action of EGF and FGF-2, PD098059, a specific inhibitor of MEK, was used in culture. The compound completely blocked the dedifferentiation, rather than the transdifferentiation, of PE cells, providing the first evidence of the requirement of MAP kinase pathway activated by EGF and FGF-2 in the early process of lens transdifferentiation (Fig. 2).

2.4. Lens Regeneration and Lens Development

Eye development in vertebrates has been an excellent model system to investigate fundamental processes in developmental biology, from tissue induction to the formation of highly specialized structures such as the lens and the retina. This complex system develops primarily from three embryonic parts: the optic vesicle (OV), which is a lateral evagination from the wall of the diencephalons; the surrounding mesenchyme; and the overlaying surface ectoderm (SE). The OV contacts the SE and triggers a response that leads to a thickening of the SE and triggers a response that leads to a thickening of the ectoderm, the lens placode, which later develops into the mature lens. The lens placode internalizes to form the lens vesicle; the distal OV invaginates to form the optic cup, with the inner layer developing into the neuroretina, and the outer layer forming the retinal PE.

The expression and function of numerous genes have been correlated with defined cell types and stages of eye development. In particular, the study of the transcription factor Pax6 promoted understanding of eye development. Pax6 is a member of the Pax family of transcription factors. It contains two DNA-binding motifs: the paired domain and paired-type homeodomain *(27)*.

Reports have described the temporal and spatial functions of the transcription factor Pax6 in the developing vertebrate eye. Pax6 is shown to play essential roles in successive steps triggering lens differentiation, and in the retina, it junctions to maintain multipotency and proliferation of retinal progenitor cells *(28,29)*.

It was tested whether the regulatory genes in eye development similarly perform an important function during the lens regeneration process through trandifferentiation of iris PE cells. RNA blot analysis has shown that the transcription of the pax6 gene was rapidly activated on induction of lens transdifferentiation of chick iris PE cells after the addition of EGF and FGF-2 in vitro *(26)*. Furthermore, other regulatory genes in lens development, such as six3 *(30)* and l-maf *(31)*, were also induced during lens transdifferentiation of iris PE cells in vitro *(26)*. Our data led to the suggestion

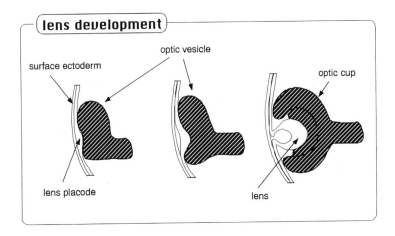

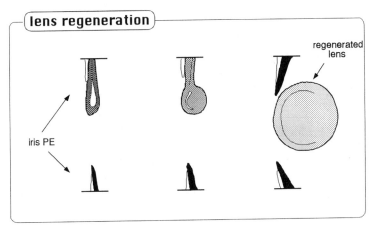

Fig. 3. Lens "development" and "regeneration." During early development of the vertebrate eye, when the optic vesicle appears near the surface ectoderm, the ectoderm cells begin to differentiate from the lens. When mesenchymal cells derived from the neural crest cover the optic cup, cells in the posterior pole of the outer layer of the optic cup begin to synthesize melanosomes and differentiate from the PE cells. When the lens is surgically removed from an adult newt eye, a structurally and functionally complete lens always regenerates from the PE of the dorsal papillary margin of the iris.

that these genes could be master regulators in lens regeneration and normal lens development in vivo, although the developmental origins of cells forming the lens were clearly distinct from each other (Fig. 3).

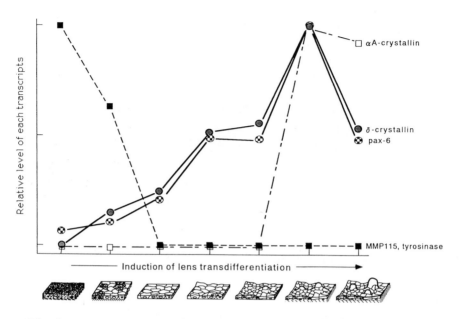

Fig. 4. Transcriptional activities of pax6, PE-specific, and lens-specific genes in the process of transdifferentiation. Total RNA was isolated and examined by Northern blot analysis. Change of relative mRNA levels of each gene was shown at each step during the lens transdifferentiation of iris PE cells from postnatal chick. Morphological changes of the cells were illustrated and observed in culture conditions permissive for lens transdifferentiation *(bottom)*. The cells gradually lost their PE phenotype, continued to proliferate through the loss of contact growth inhibition, and formed multicellular layers. Typical lentoid bodies developed in the multilayered portion.

2.5. Gene Expressions During Lens Transdifferentiation of Iris PE Cells In Vitro

In the process of lens transdifferentiation, expression of PE-specific or lens-specific genes for differentiation markers is strictly regulated at the transcriptional level *(26,32)*. When iris PE cells could give rise to additional PE cells in control growth medium with serum, the transcripts of PE-specific genes were easily detected, and those of lens-specific genes were absent. In contrast, when the cells were maintained for about 3 wk in conditions permissive for lens transdifferentiation, the transcripts of PE-specific marker genes were not detected; however, the expression of lens-specific crystallin genes were observed (Fig. 4).

The transcript of the pax6 gene was detected at a low level in iris PE cells in control growth medium. After the addition of EGF and FGF-2 in conditions permissive for lens trandifferentiation, the expression of the pax6 gene was rapidly upregulated. Expression of the pax6 gene in the process of lens transdifferentiation was analyzed by RNA blot. The elevated expression of the pax6 gene and the induction of δ-crystallin gene expression were similarly observed in the early stage of lens transdifferentiation, when the levels of other differentiation marker genes hardly changed (Fig. 4). This result suggests that the state of dedifferentiated PE cells expressing no differentiation markers is not necessary for the transdifferentiation process.

δ-Crystallin is the major lens protein in the chick and appears first in the lens placode in chick embryos. Recent findings showed that Pax6 binds cooperatively with another transcriptional factor, Sox2, to the δ-crystallin enhancer, forming a ternary complex that mediates δ-crystallin expression in the lens placode *(33)*. During the lens transdifferentiation process in culture, similar interactions may occur in the iris PE cells as in the lens placode during development. This culture system could provide a model system for further functional studies to determine the roles of various important genes in triggering lens differentiation regulatory interactions.

3. TRANSDIFFERENTIATION INTO NEURONAL CELLS

3.1. Transdifferentiation of Retinal PE Cells Into Neuronal Cells

In the newt, the retinal PE has retained the capacity to form a new and complete neural retina, including a new optic nerve (Fig. 1). In anuran species, retinal regeneration after complete retinectomy has not been observed *(34,35)*. It is known, however, that transplantation of retinal PE sheets into the posterior eye chamber of *Rana* or *Xenopus* leads to the production of a new retina through the transdifferentiation of the retinal PE. Furthermore, the embryonic retinal PE of many species of vertebrate can also be induced to dedifferentiate and transdifferentiate by altering its environment. In birds and mammals, this retinal transdifferentiation of the retinal PE can occur (only over a narrow period during early eye development) in fetal or embryonic stages, but this capacity is lost during development *(36)*. Thus, it has been thought that the ability of retinal PE cells to produce retinal neurons decreases as embryonic development proceeds.

3.2. Multipotentiality in the Iris PE Cells of Postnatal Chicks

As mentioned, dissociated iris PE cells from the postnatal chick can be expanded and transdifferentiate into lens cells in culture. It is unknown whether the iris PE cells can transdifferentiate into neuronal cell types just as embryonic retinal PE cells can. Our observations have shown that the iris PE isolated from postnatal chick cells can be induced into neuronal cells expressing panneural, glial, and specific retinal neuron markers under certain culture conditions (G. W. Sun and M. Kosaka, unpublished data). These results raise the possibility that some iris PE cells from the postnatal chick have the capacity to transdifferentiate into multiple cell types (not only lens, but also retinal neurons) in vitro. Furthermore, our preliminary observations suggested that the postnatal chick iris PE cells retain a population of neural stem cells similar to that found in the embryonic eye (G. W. Sun and M. Kosaka, unpublished data). Taken together, it can be concluded that fully differentiated iris PE cells from the postnatal chick have the ability to transdifferentiate into multiple cell types, which was classically observed in newts.

3.3. Transdifferentiation Into Neuronal Cells in Mammalian Iris PE Cells

In contrast to the data on neuronal transdifferentiation of retinal PE in amphibians, fish, and birds, very few studies have been carried out with mammalian PE. The embryonic rat retinal PE could transdifferentiate into neuronal cells, but this could occur only during a narrow period of development (E12–E13) *(36,37)*.

To know whether the fully differentiated adult mammalian iris PE cells possess any ability to transdifferentiate into neuronal cell types, we purified and cultured iris tissue cells from adult rats. The results demonstrated that adult iris-derived cells generate neuronal cells expressing panneural marker proteins, but not specific markers for retinal neurons as in response to treatment with FGF-2 *(38)*.

As mentioned, it is proposed that regulatory factors in the developmental process may also play an important role for the transdifferentiation of other cell types. Crx is the homeobox gene specifically expressed in the photoreceptors of the mature retina and is crucial in photoreceptor differentiation. Crx binds to and transactivates genes for several photoreceptor cell-specific proteins *(39,40)*.

In an attempt to obtain photoreceptor cells from adult iris cells, it was examined whether iris-derived cells from adult rats could acquire photoreceptor-specific phenotypes as a result of the ectopic expression of Crx using replication-defective recombinant virus vectors. The iris-derived cells infected with Crx became small and round, characteristic of the rod photoreceptors in monolayer culture, and most of the cells expressed rhodopsin protein, a specific marker of photoreceptors (Fig. 5). In addition, preliminary

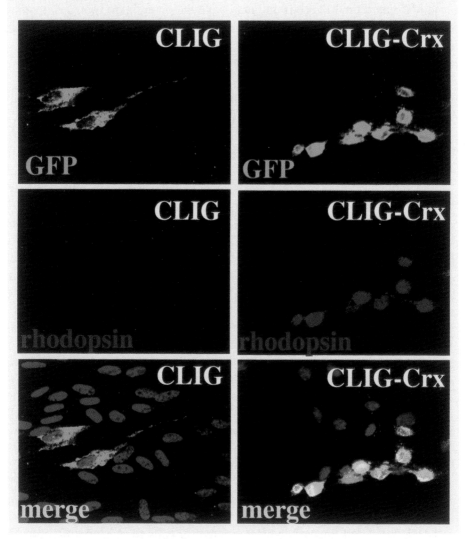

Fig. 5. Induction of rhodopsin expression from adult iris tissue cells. Iris tissue cells were isolated from adult rat eye and were infected by the recombinant retroviruses CLIG and CLIG-Crx. The CLIG virus-infected cells (enhanced green fluorescent protein [EGFP] positive) were flat and large, and none of them expressed rhodopsin. The CLIG-Crx virus-infected cells (EGFP positive) were small and round, and most expressed rhodopsin. Nuclei in cells stained with 4'–6-diamidino-2-phenylindole DAPI; scale bar, 50 μm.

results showed some of the iris PE cells purely isolated from adult rat eyes were capable of proliferation under certain culture conditions. The results suggest that iris PE cells from adult mammalian eyes possess higher flexibility than retinal PE cells and may have the potential to transdifferentiate into neuronal cells similar to PE cells of the ciliary body (*see* Section 3.4.).

3.4. Source for Neural Retina Generation

Several sources of neural regeneration are also known in the eyes of vertebrates in addition to the newts. They include neural stem cells in the ciliary marginal zone (CMZ) of fish, frogs, and salamanders *(41,42)*; Müller glia of fish and birds *(43,44)*; and rod progenitors of fish *(45)* in addition to the retinal PE of frogs, salamanders, embryonic chicks, and embryonic rodents *(46–48)*.

Although retinal regeneration has never been observed in adult mammals, reports indicate the presence of retinal stem cells in adult mammals *(49,50)*. Single pigmented ciliary margin (PCM) cells can proliferate in vitro to form spherical colonies of cells that can differentiate into rod photoreceptors. Its extreme peripheral position makes it topologically analogous to the CMZ of amphibians and fish. It is not known whether the fact that mammalian stem cells are localized in PCM, the most peripheral part of the CMZ, is related to the ability of the iris PE cells to transdifferentiate. The iris and the ciliary body contain PE cells, which are commonly derived from the ventral diencephalons during embryonic development, similar to neural retina.

Preliminary results also suggest that iris PE cells isolated from adult rat and expanded in culture express molecular markers for undifferentiated retinal progenitor cells (M. Kosaka, unpublished). It raises a possibility that the dissociated PE cells derived from iris and ciliary body might have stem cell characteristics similar to the ones found in the PCM of mammals. We are currently testing this possibility using adult mammalian iris PE cells.

By transfecting primary cell cultures with E1A (viral deoxyribonucleic acid [DNA]) as well as H-ras and c-myc (proto-oncogenes), cell lines were established from adult human retinal PE *(51)*. The cell lines derived from H-ras transfection contained cells with a neuronal phenotype. Recently established clonal cell lines of dedifferentiated PE cells from retinal PE cells of an 80-yr-old man were reported. In some cell lines, neuronlike cells expressing neuron-specific neurofilament protein subunits appeared in culture (G. Eguchi, unpublished). These observations suggest that adult mammalian PE cells, including human cells, have the capacity for transdifferentiation into neuronal cells.

4. SUMMARY AND FUTURE PERSPECTIVES

Iris PE cells are entirely postmitotic and stably maintained in the adult vertebrate eye. Once dissociated from eye tissue, however, some of the iris PE cells of vertebrate species, including adult mammals, can efficiently expand and show remarkable plasticity to transdifferentiate into multiple cell types in vitro. When exposed to growth factors such as FGF-2, these cells were induced to proliferate and lose their PE phenotype, such as the case for embryonic retinal PE cells. It remains unclear whether the dedifferentiated cells from iris PE could be retinal stem cells with the same potential as the ciliary body-derived cells in adult mammals and whether these iris-derived cells could transdifferentiate into any other lineage except for the eye. Further information derived from the analysis of the transdifferentiation in iris PE cells may deepen the understanding of the mechanisms of stabilization and destabilization of the cell phenotype in differentiated tissues.

In spite of the plasticity of iris PE cells in vitro, lens or retinal transdifferentiation of iris PE cells was never observed in postnatal chick or adult mammals. This inconsistency is likely because of an inhibitory environment in the eye of such animals. Once freed from inhibition (or if inhibitory factors can be overcome) in the eye, regeneration of the lens or retina may occur though the transdifferentiation of PE cells, even in adult mammals. Our analysis suggested that dissociated iris PE cells isolated from fully differentiated tissue may be almost as plastic as embryonic cells, at least in vitro, and that these cells could be a source of lens or neural regeneration under appropriate conditions.

In addition to basic approaches, it should also be possible and useful for future clinical applications to attempt to reconstruct functional lens or retinal neurons. A remarkable clinical advantage of using dissociated iris PE cells is that iris tissues can be easily and safely obtained by the peripheral iridectomy, unlike the PE cells in the ciliary body and retina. The multipotent cells in iris PE cells may provide a potential source for autologous retinal transplantation.

ACKNOWLEDGMENTS

We gratefully acknowledge the help and suggestions of Prof. Goro Eguchi (president of Kumamoto University) and many colleagues in his previous laboratory at the National Institute of Basic Biology in Japan. M. K. is thankful to Dr. Tamotsu Yoshioka (chairman of Kurashiki Medical Center) for encouragement. Our work described in this chapter was mainly supported by the Japan Science and Technology Corporation (JST).

REFERENCES

1. Morrison, S. J. (2001). Stem cell potential can anything make anything? Curr Biol 11, R7–R9.
2. Weissman, I. L. (2001). The evolving concept of a stem cell: entity or function? Cell 105, 829–841.
3. Temple, S. (2001). The development of neural stem cells. Nature 414, 112–1127.
4. Clarke, D. L., Johansson, C. B., Wilbertz, J., et al. (2000). Generalized potential of adult neural stem cells. Science 288, 1660–1663.
5. Bernstein, I. D., Singer, J. W., Andrews, R. G., et al. (1987). Treatment of acute myeloid leukemia cells in vitro with a monoclonal antibody recognizing a myeloid differentiation antigen allows normal progenitor cells to be expressed. J Clin Invest 79, 1153–1159.
6. Galli, R., Borello, U., Gritti, A., et al. (2000). Skeletal myogenic potential of human and mouse neural stem cells. Nat Neurosci 10, 986–991.
7. Terada, N., Hamazaki, T., Oka, M., et al. (2002). Bone marrow cells adopt the phenotype of other cells by spontaneous cell fusion. Nature 416, 542–545.
8. Ying, Q. L., Nichols, J., Evans, E. P., and Smith, A. G. (2002). Changing potency by spontaneous fusion. Nature 416, 545–548.
9. Wolff, G. (1895). Entwicklungsphysiologische Studien. I. Die Regeneration der Urodelenlinse. Arch Mikrosk Anat Entwickl Org 1, 380–390.
10. Sato, T. (1940). Vergleichende Studien uber die Geschwindigkeit der Wolffschen Linsenregeneration bei Triton taeniatus und bei Diemyctylus pyrrhogaster. Wilhelm Roux' Arch. Entw-Mech Org 140, 570–613.
11. Eguchi, G. (1964). Electron microscopic studies on lens regeneration. II. Formation and growth of lens vesicle and differentiation of lens fibers. Embryologia 8, 247–287.
12. Eguchi, G. (1986). Instability in cell commitment of vertebrate pigmented epithelial cells and their transdifferentiation into lens cells. Curr Top Dev Biol 20, 21–37.
13. Eguchi, G., and Kodama, R. (1993). Transdifferentiation. Curr Opin Cell Biol 5, 1023–1028.
14. Eguchi, G. (1998). Transdifferentiation as the basis of eye lens regeneration. In: Ferretti, P. and Geraudie, G., eds., Cellular and Molecular Basis of Regeneration: From Invertebrates to Humans. New York: Wiley, pp. 207–228.
15. Yamada, T., Roesel, M. E., and Beauchamp, J. J. (1975). Cell cycle parameters in dedifferentiating iris epithelial cells. J Embryol Exp Morphol 34, 497–510.
16. Yamada, T. (1977). Control mechanisms in cell-type conversion in newt lens regeneration. Monogr Dev Biol 13, 1–126.
17. Yamada, T., and McDevitt, D. S. (1984). Conversion of iris epithelial cells as a model of differentiation control. Differentiation 27, 1–12.
18. Abe, S., and Eguchi, G. (1977). An analysis of differentiative capacity of pigmented epithelial cells of adult newt iris in clonal cell culture. Dev Growth Differ 19, 309–317.
19. Eguchi, G. (1988). Cellular and molecular background of Wolffian lens regeneration. Cell Differ Dev Suppl., 147–158.

20. Eguchi, G. (1979). Transdifferentiation in pigmented epithelial cells of vertebrate eyes in vitro. In: Evert, J. D., and Okada, T. S., eds., Mechanisms of Cell Change. New York: Wiley, pp. 273–291.
21. Tsonis, P. A, Jang, W., Rio-Tsonis, K. D., and Eguchi, G. (2001). A unique aged human retinal pigmented epithelial cell line useful for studying lens differentiation in vitro. Int J Dev Biol 45, 753–758.
22. Itoh, Y., and Eguchi, G. (1986). In vitro analysis of cellular metaplasia from pigmented epithelial cells to lens phenotypes: a unique model system for studying cellular and molecular mechanisms of "transdifferentiation." Dev Biol 115, 353–362.
23. Kodama, R., and Eguchi, G. (1995). From lens regeneration in the newt to in vitro transdifferentiation of vertebrate pigmented epithelial cells. Semin Cell Biol 6, 143–149.
24. Masuda, A., and Eguchi, G. (1984). Phenylthiourea enhances Cu cytotoxicity in cell cultures: its mode of action. Cell Struct Funct 9, 25–35.
25. Kosaka, M., Kodama, R., and Eguchi, G. (1998). In vitro culture system for iris-pigmented epithelial cells for molecular analysis of transdifferentiation. Exp Cell Res 245, 245–251.
26. Kosaka, M., Mochii, M., and Eguchi, G. EGF and FGF-2 synergistically induce in vitro lens transdifferentiation of pigmented epithelial cells through activating several master genes for lens development. In press.
27. Walther, C., and Gruss, P. (1991). Pax-6, a murine paired box gene, is expressed in the developing CNS. Development 113, 1435–1449.
28. Marquardt, T., Ashery-Padan, R., Andrejewski, N., Scardigli, R., Guillemot, F., and Gruss, P. (2001). Pax6 is required for the multipotent state of retinal progenitor cells. Cell 105, 43–55.
29. Ashery-Padan, R., and Gruss, P. (2001). Pax6 lights-up the way for eye development. Curr Opin Cell Biol 13, 706–714.
30. Oliver, G., Loosli, F., Koster, R., Wittbrodt, J., and Gruss, P. (1996). Ectopic lens induction in fish in response to the murine homeobox gene Six3. Mech Dev 60, 233–239.
31. Ogino, H., and Yasuda, K. (1998). Induction of lens differentiation by activation of a bZIP transcription factor, L-Maf. Science 280, 115–118.
32. Agata, K., Kobayashi, H., Itoh, Y., Mochii, M., Sawada, K., and Eguchi, G. (1993). Genetic characterization of the multipotent dedifferentiated state of pigmented epithelial cells in vitro. Development 118, 1025–1030.
33. Kamachi, Y., Uchikawa, M., Tanouchi, A., Sekido, R., and Kondoh, H. (2001). Pax6 and SOX2 form a co-DNA-binding partner complex that regulates initiation of lens development. Genes Dev 15, 1272–1286.
34. Stone, L. S. (1959). The role of retina pigment cells in regenerating neural retina of adult salamander eyes. J Exp Zool 113, 9–31.
35. Okada, T. S. (1980). Cellular metaplasia or transdifferentiation as a model for retinal cell differentiation. Curr Top Dev Biol 16, 349–380.
36. Zhao, S., Thornquist, S. C., and Barnstable, C. J. (1995). In vitro transdifferentiation of embryonic rat retinal pigment epithelium to neural retina. Brain Res 677, 300–310.

37. Stroeva, O. G., and Brodskii, V. (1968). A comparative study of the relationship between the ploidy of the nuclei of the pigmented epithelium of the retina and its growth and differentiation in tritons and rats. Zh Obshch Biol 29, 177–185.

38. Haruta, M., Kosaka, M., Kanegae, Y., et al. (2001). Induction of photorecep-tor-specific phenotypes in adult mammalian iris tissue. Nat Neurosci 4, 1163–1164.

39. Furukawa, T., Morrow, E. M., and Cepko, C. L. (1997). Crx, a novel otx-like homeobox gene, shows photoreceptor-specific expression and regulates pho-toreceptor differentiation. Cell 91, 531–541.

40. Chen, S., Wang, Q. L., Nie, Z., et al. (1997). Crx, a novel Otx-like paired-homeodomain protein, binds to and transactivates photoreceptor cell-specific genes. Neuron 19, 1017–1030.

41. Reh, T. A., and Nagy, T. (1989). Characterization of Rana germinal neuroepi-thelial cells in normal and regenerating retina. Neurosci Res Suppl 10, S151–S161.

42. Wetts, R., Serbedzija, G. N., and Fraser, S. E. (1989). Cell lineage analysis reveals multipotent precursors in the ciliary margin of the frog retina. Dev Biol 136, 254–263.

43. Braisted, J. E., Essman, T. F., and Raymond, P. A. (1994). Selective regenera-tion of photoreceptors in goldfish retina. Development 120, 2409–2419.

44. Fischer, A. J., and Reh, T. A. (2001). Muller glia are a potential source of neural regeneration in the postnatal chicken retina. Nat Neurosci 4, 247–252.

45. Hitchcock, P. F., and Raymond, P. A. (1992). Retinal regeneration. Trends Neurosci 15, 103–108.

46. Reh, T. A., and Levine, E. M. (1998). Multipotential stem cells and progenitors in the vertebrate retina. J Neurobiol 36, 206–220.

47. Raymond, P. A., and Hitchcock, P. F. (1997). Retinal regeneration: common principles but a diversity of mechanisms. Adv Neurol 72, 171–184.

48. Raymond, P. A., and Hitchcock, P. F. (2000). How the neural retina regener-ates. Results Probl Cell Differ 31, 197–218.

49. Tropepe, V., Coles, B. L., Chiasson, B. J., et al. (2000). Retinal stem cells in the adult mammalian eye. Science 287, 2032–2036.

50. Ahmad, I., Tang, L., and Pham, H. (2000). Identification of neural progenitors in the adult mammalian eye. Biochem Biophys Res Commun 270, 517–521.

51. Dutt, K. C., Scott, M., Sternberg, P. P., Linser, P. J., and Hjelmeland, L. M. (1993). Transdifferentiation of adult human pigment epithelium into retinal cells by transfection with an activated H-ras proto-oncogene. DNA Cell Biol 12, 667–673.

14

Stem Cell Biology of the Inner Ear and Potential Therapeutic Applications

Thomas R. Van De Water, Ken Kojima, Ichiro Tateya, Juichi Ito, Brigitte Malgrange, Philippe P. Lefebvre, Hinrich Staecker, and Mark F. Mehler

1. INTRODUCTION

The mammalian inner ear is composed of two sensory receptor areas: the cochlea, responsible for translating auditory stimuli, and the vestibule, responsible for a sense of balance. The bilateral inner ears of mammals develop from a pair of thickened branchial ectodermal placodes that invaginate for sequential formation of otic pits, otocysts, and then complete membranous labyrinths (1). During development of inner ear sensory epithelium, several phenotypes of sensory and nonsensory cells differentiate (i.e., hair cells, support cells, glia, and neurons).

All of these cellular phenotypes are derived from the otic placode epithelium through a series of cell fate decisions and segregation into specialized sensory areas (i.e., sensory receptors and the statoacoustic ganglion complex of the cochlea and vestibule) (2–4) (Fig. 1). However, the lineage relationships among the cellular phenotypes in inner ear sensory receptors (i.e., hair cells, support cells, neurons, and glia) are not completely understood.

In contrast to the inner ear sensory receptors, neural stem cells that generate neurons or glia have been isolated in vitro from neural crest and central nervous system (CNS) tissues of either developing or mature mammals (5–7), and the potential of stem cell therapy for the treatment of neurological diseases has been raised (see reviews in refs. 8 and 9). In the inner ear, it has not been determined whether adult mammalian inner ear sensory organs possess multipotent stem cells that generate all the major cellular phenotypes in the tissues that compose the inner ear sensory receptors.

Dissociated cell cultures of adult inner ear sensory organs are difficult to establish because of the small size of the tissue samples that can be isolated

From: *Adult Stem Cells*
Edited by: K. Turksen © Humana Press Inc., Totowa, NJ

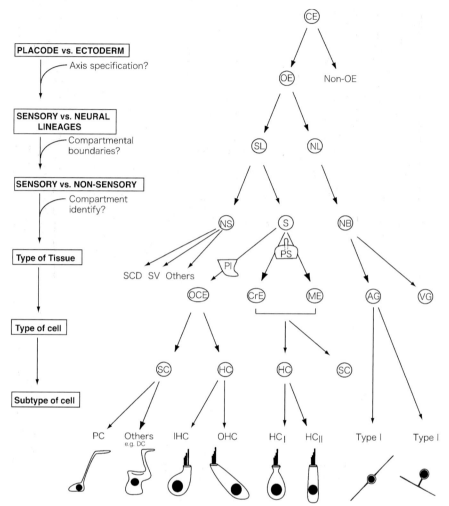

Fig. 1. Schematic of cell fate decisions that underlie the pattern of cell specification in the developing inner ear of mammals. (Modified from ref. *4.*) *Abbr:* AG, auditory ganglion neurons; CE, cephalic ectoderm; CrE, crista epithelium; DC, Deiters' cells; HC, hair cell; HC_I, type I hair cell; HC_{II}, type II hair cell; IHC, inner hair cell; ME, macular epithelium; NB, neuroblasts; NL, neural lineage; NS, nonsensory cells; OCE, organ of Corti epithelium; OE, otic placode epithelium; OHC, outer hair cell; PC, pillar cell; PI, pars inferior; PS, pars superior; S, sensory cells; SC, support cell; SCD, semicircular duct epithelium; SL, sensory lineage; SV, stria vascularis; VG, vestibular ganglion neurons.

and the complications to surgical approaches that the protective bony labyrinth presents as it encases and protects the membranous labyrinth of the inner ear. To overcome these major experimental limitations imposed both by the inaccessible nature of mammalian mechanosensory epithelium in adult animals and by the small number of sensory cells available, dissociated culture systems of mammalian sensory epithelial cells have been established using growth factors *(10)* or immortalizing oncogenes *(11–16)*. Immortalized cell lines that have been established using oncogenes have most likely undergone some genetic alterations during the immortalization process *(9)*. However, cells transformed with oncogenes should be able to proliferate continuously through an unlimited number of mitotic divisions and to retain a potential to express several characteristics of differentiated cells that derive from the tissue of their original origin *(17)*.

2. CELL CULTURE EXPERIMENTS

Four research groups have established immortalized cell lines from sensory epithelia of the rodent inner ear using the temperature-sensitive simian virus 40 tumor antigen *ts*A58, which codes for a thermolabile variant of the large T antigen (Tag) *(18,19)*. Expression of Tag is controlled by the *H2kb* promoter, which is normally active in most cell types and can be upregulated in the presence of γ-interferon (γIF). The thermolabile Tag epitope is expressed stably in transformed cells incubated at the permissive temperature of 33°C in the presence of γIF, but Tag degenerates rapidly when the transformed cells are incubated at 37°C. When the transformed cells with *H2kb–ts*A58 are cultured at 33°C in the presence of γIF (proliferative culture condition [PC condition]), the cells maintain the ability to proliferate in vitro. However, after removal of γIF from the culture media and an increase in temperature of incubation to 39°C (differentiation culture condition [DC condition]), the cells cease proliferation and begin to differentiate. Cell culture studies using transgenes regulated conditionally by a shift of temperature and removal of γIF can provide insights into mechanisms of cell differentiation and the lineage relationships of inner ear sensory cells of mammals.

Immortalized cell lines that express features of hair cell progenitors were established from cochlea sensory epithelium by Rivolta and colleagues *(13)* and from vestibular sensory epithelia by Zheng and colleagues *(14)*. Zheng et al. established the clonal cell line utricalar epithelium cell line four (UEC-4) by transfection of primary cell cultures derived from postnatal days 3 and 4 (P3 and P4) rat utricular epithelia with a retrovirus encoding the *ts*A58/U19T antigen. Under the PC condition, clone UEC-4 generated epithelial cell phenotypes that expressed polygonal cell shapes and epithelial cell marker epitopes (i.e., tight junction protein zona occludens protein one (ZO1) and a lectin molecule convacalin A (ConA) that binds selectively to

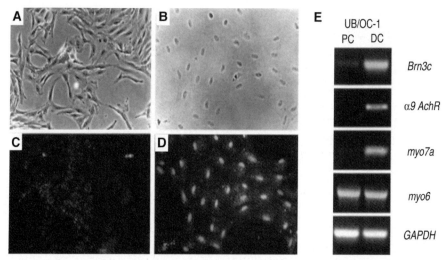

Fig. 2. Immortomouse clone UB/OC-1 cultures increased their expression of hair cell-related genes under the differentiation culture (DC) condition. Phase contrast images of clonal cell line UB/OC-1 derived from E13 auditory sensory epithelia under **(A)** PC and **(B)** DC conditions and immunofluorescence images of the cells labeled with an antibody to Brn3c under **(C)** PC and **(D)** DC conditions. Most nuclei in UB/OC-1 were unlabeled for Brn3c under PC conditions (C), and all nuclei were labeled after 14 d under DC conditions (D). Scale bar is 100 μm. **(E)** Reverse transcriptase polymerase chain reaction analysis showed that UB/OC-1 expressed low levels of transcripts of *Brn3c,* α9AchR, and *myos7a* under PC conditions when compared with cultures after 14 d in vitro under DC conditions.

epithelial cells) *(20)*. When the clonal cell line UEC-4 was incubated under a DC condition in the presence of basic fibroblast growth factor (bFGF), most of cells derived from this clone differentiated into hair cell-like phenotypes (i.e., expressed two kinds of calcium-binding proteins, calretinin and calmodulin). Inner ear hair cells express these calcium-binding proteins, but not the other epithelial cells of the inner ear *(21)*. The results from this study indicate that the clone UEC-4 may contain hair cell progenitor.

Rivolta et al. *(13)* established clonal cell lines University of Bristol organ of Corti cell line out (UB/OC-1) and UB/OC-2 from presumptive organ of Corti epithelium of an embryonic d 13 (E13) *H-2Kb–ts*A58 transgenic mouse (immortomouse). Under the PC condition, these two cell lines expressed an epithelial cell marker protein (cytokeratin), and hair cell marker epitopes were absent or present only at a low level *(13)*. Under the DC condition, UB/OC1 and UB/OC-2 expressed higher levels of hair cell marker epitopes and transcripts, that is, fibrin, α9 subunit of the nicotinic acetylcholine receptor (α9-*AchR*), *Brn3c, myo7a,* and *myo6* (Figs. 2 and 3).

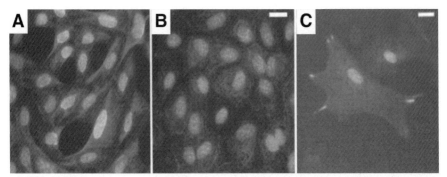

Fig. 3. The DC condition induced cells in clone UB/OC-2 cultures to express fimbrin. Immunofluorescence images of cell line UB/OC-2 derived from E13 auditory sensory epithelia. **(A,B)** Cultured cells were labeled with anticytokeratin rhodamine isothiocyanate (TRITC) and anticadherin fluorescein isothiocyanate (FITC) antibodies. The nuclei stained blue with DAPI. Both epithelial makers are detected under (A) PC conditions as well as under (B) DC conditions. However, cell boundaries are more evident under DC conditions. Note the reorganization of cytokeratin filaments under DC conditions. **(C)** After 14 d in vitro under DC conditions, UB/OC-2 expressed fimbrin epitopes (FITC) localizing primarily to focal contacts. Nuclei were stained blue with 4', 6-diamino-2-phenylindole DAPI. Scale bar is 25 μm. (*See* color plate 6 in the insert following p. 82.)

Fimbrin is an actin cross-linking protein that occurs only in the hair cell stereocilia in the cochlea *(21)*. The POU domain transcription factor *Brn3c* is: (1) expressed specifically by differentiating hair cells in the inner ear from about stage E14; (2) essential for hair cell differentiation; and (3) able to generate hearing impairments in humans when mutated *(22,23)*. Transcripts for *α9AChR* are expressed specifically by the hair cells of the inner ear and are believed to be involved in synaptic transmission between efferent nerves and hair cells *(24)*. Myosin VI and myosin VIIa epitopes are expressed in hair cells of the inner ear and are involved in the organization and maturation of stereocilia and participate in the functioning of hair cells *(25)*.

These results indicate that the immortalized cell lines UB/OC-1 and UB/OC-2 contain hair cell progenitors. To investigate the origin of clone UB/OC-1, Rivolta et al. *(13)* profiled the pattern of gene expression by the cells of the UB/OC-1 cultures. UB/OC-1 cell cultures express the transcription factor GATA3, which is also expressed in sensory and nonsensory epithelial cells of the developing mouse cochlear duct *(26)*.

This cell line also expresses a negative basic helix–loop–helix (bHLH) transcription factor *Hes1* and a glycoprotein of the extracellular matrix of the tectorial membrane β-*tectorin*. Transcripts of *Hes1* are expressed mainly in the cells of the greater epithelial ridge (GER), which represents the medial

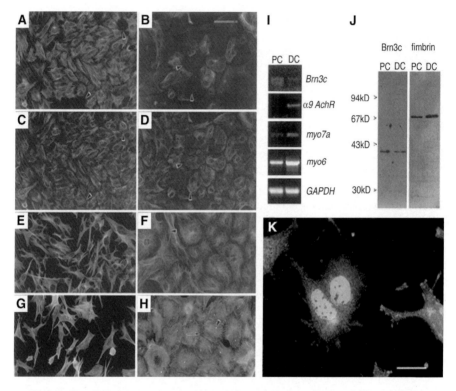

Fig. 4. The DC condition induced cells in clonal cell line UB/UE-1 to differentiate into either hair cell or support cell phenotypes. (**A–H**) Expression of cytoskeletal proteins in the cell line UB/UE-1, **A, C, E,** and **G** under PC and **B, D, F,** and **H** under DC conditions. The cultured cells were labeled with antibodies to cytokeratin (**A, B**), vimentin (**C, D**), or tublin (**E, F**), or with rhodamine-phalloidin (**G, H**) to label actin. Most cells expressed vimentin, so the antibody to this protein was used to identify cells that did not express cytokeratin. Thus, A and C (PC condition) and B and D (DC condition) show the same cells double labeled for cytokeratin and vimentin. Up to 10% of cells were unlabeled for cyto-keratin under the PC condition, shown by the arrowhead in A and C, but up to 60% were labeled under DC conditions, as shown by the arrowheads in B and D. Although vimentin was expressed in most cells at both culture conditions, the labeling was less intense under DC conditions. The changes in cell morphology under DC conditions are clearly illustrated in cells labeled for microtubules and actin filaments. The distribution of microtubules was more even and symmetrical in cells under DC conditions, as seen in E and F. The more intensely labeled filaments may be flagellar axonemes, as shown at the arrowhead in F. Under DC condition, cells possessed strong, punctate labeling for actin (arrowhead in H) and numerous well-developed structures resembling stress fibers (arrow in H). Scale bar, shown in B for A–H, is 100 μm. (**I**) Reverse transcriptase polymerase chain reaction showing the relative levels of expression of different hair cell markers between cells under PC and DC

aspect of the developing organ of Corti, and the lesser epithelial ridge (LER), which represents the lateral aspect of the developing organ of Corti *(27,28)*. The expression of β-*tectorin* messenger RNA (mRNA) is present only in the cells of the GER and the pillar cells during development *(29)*. On the other hand, UB/OC-1 does not express transcripts of another inhibitory bHLH gene, *Hes5*, which is normally expressed in the cells of the LER, under either PC or DC conditions in vitro.

Transcripts of a neural bHLH transcription factor, *Math1,* are expressed in the differentiating hair cells at early embryonic stages, but absent in the other cells of the GER *(30)*. Organotypic cultures of organ of Corti explants from an E13 immortomouse under the PC condition demonstrated that incorporation of BrdU and expression of large T antigen could be detected in GER cells, but not in developing sensory epithelial cells. Rivolta et al. *(13)* concluded that the cell line UB/OC-1 was derived from a GER cell of the developing immortomouse cochlea.

Some of the GER cells can be induced to differentiate into hair cells by overexpression of *Math1 (31)*. Histological studies have shown that inner hair cells derive from progenitor cells located in the GER, and that outer hair cells are derived from progenitor cells located in the LER *(32)*. The results of UB/OC-1 cell culture studies support the concept that inner hair cell progenitors derive from GER cells during development of the mammalian inner ear.

Lawlor et al. derived clonal cell line UB/UE1 from support cells of P2 immortomouse utricular macula sensory epithelium isolated by treatment with thermolysin so only epithelial cells were present *(15)*. Under the PC condition, over 90% of the progeny of this clone expressed the support cell marker, cytokeratin. However, under the DC condition, there was a decrease in the percentage of support cell phenotypes expressing cytokeratin to 40% in the clone UB/UE-1, and these cultures generated hair cell phenotypes that express fimbrin epitopes and transcripts of *myo7a* and *α9AchR* (Fig. 4A–J).

In a subsequent study of clonal cell line UB/UE-1, Rivolta et al. showed that the DC condition induced this cell line to differentiate into hair cell and support cell phenotypes through asymmetrical cell divisions detected by the tracing of mitochondrial localization *(33)* (Fig. 4K). Mammalian stem cell populations may undergo either symmetric or asymmetric cell divisions,

Fig. 4. *(continued)* conditions. **(J)** Immunoblots of UB/UE-1 using anti-Brn3c and fimbrin antibodies. A slight decrease in both mRNA and protein was observed for Brn3c under DC conditions. Fimbrin was upregulated under DC conditions. **(K)** BrdU labeling of a UB/UE-1 cell dividing asymmetrically after 1 d under DC conditions. BrdU is green; mitochondria are red. Scale bar is 25 μm. (*See* color plate 7 in the insert following p. 82.)

depending on their developmental stage and regional environmental cues. Symmetric cell divisions promote expansion of the progenitor pool, whereas asymmetric cell divisions allow the subsequent elaboration of neurons and glia *(34,35)*. These results suggest that the vestibular sensory epithelial cell line UB/UE-1 is a progenitor that has the potential to differentiate into both support cell and hair cell phenotypes through the process of asymmetric cell division.

Malgrange et al. established dissociated cell cultures from a newborn (P0) rat cochlea without resorting to the use of oncogenes. They utilized epidermal growth factor (EGF), bFGF, or a combination of both to stimulate nestin-positive cells isolated from P0 organ of Corti sensory epithelial cells to proliferate in vitro and to form otospheres *(10)*. Previous reports described that both of these growth factors possess the ability to induce sensory epithelial cells to proliferate in organotypic cultures of the inner ear *(36,37)*. After 2 d in culture, most of cells that formed the otospheres showed immature cell features (i.e., incorporation of BrdU and expression of the neural stem cell marker nestin) (Fig. 5A,B).

After 2 wk in vitro, the cell colonies obtained from multiple cell passages of the original otosphere colony cells began to express a hair cell marker epitope (myosin VIIa) or a support cell marker epitope (p27^{kip1}) (Fig. 5C, D). In the mature inner ear sensory organs, support cells express a cyclin-dependent kinase inhibitor p27^{kip1}, which inhibits G1-S cell cycle progression *(38)*. After 2 wk in vitro, these cultured cells underwent cellular differentiation, and the cells derived from P0 rat sensory epithelia differentiated into hair cell phenotypes with morphological features similar to those of mature hair cells (i.e., cuticular plate and stereociliary bundles), and their nuclei became polarized (Fig. 6).

This study is the first report of a progenitor cell culture obtained using low-density dissociated cell cultures without the additional requirement of an immortalizing oncogene, indicating that common progenitors of hair cells and support cells exist in the developing mammalian inner ear sensory epithelium of the organ of Corti of the newborn rat *(10)*. Additional cell culture studies using oncogenes or growth factors for the generation of clonal populations have demonstrated that hair cells and support cells have common cell lineages in the mammalian inner ear, as described in studies using retroviruses in the chick inner ear *(39,40)*.

Barald et al. *(11)* generated five clonal cell lines from otocysts obtained from E9 immortomouse embryos. One of the five clonal cell lines, IMO-6, differentiated into neuronal phenotypes under DC conditions when brain-derived neurotrophic factor (BDNF) was added to the medium. Under this DC condition with BDNF supplementation, most of cultured cells in the

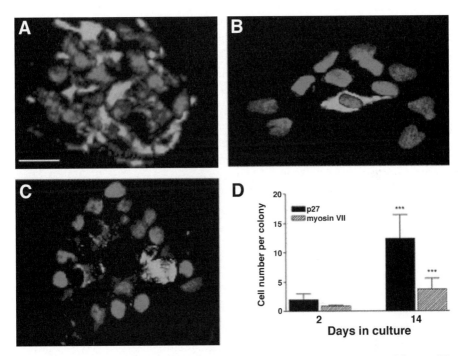

Fig. 5. Nestin (+) cells isolated from the P0 rat organ of Corti (1) proliferate, (2) form otospheres, (3) generate hair cell and support cell phenotypes in response to EGF. (**A**) Confocal laser scanning images of double immunostaining for BrdU (red) and nestin (green). After 2 d in vitro (DIV) in the presence of EGF, a large number of cells proliferated to form otospheres with incorporation of BrdU in the nuclei and expressed nestin epitopes in their cytoplasm. (**B**) Double immunofluorescence staining for BrdU (red) and myoin VIIa (green). After 2 DIV in the presence of EGF, a small number of cells expressed myosin VIIa epitopes in otospheres. (**C**) After 14 DIV in the presence of EGF, cells isolated from P0 rat organ of Corti generated cells expressing epitopes of either p27^{kip1} (red) or myosin VIIa (green) in otospheres. (**D**) Counts of the number of hair cells (i.e., expressed myosin VIIa epitopes) and support cell phenotypes (i.e., expressed epitopes of p27^{kip1}) per otosphere at 2 and 14 DIV in the presence of EGF. Scale bar is 15 μm in (A–C). (*See* color plate 8 in the insert following p. 82.)

clone expressed neurofilaments and glutamylated tubulin epitopes. In mice, the statoacoustic ganglion complex of the inner ear begins to form at E9 by delamination of neuronal progenitors from several sites in the otic cup neuroepithelium and then separates to form both the auditory (spiral) and vestibular (Scarpa's) ganglia of the inner ear's VIIIth nerve ganglion complex. This result indicates that clone IMO-6 originated from a neural progenitor cell lineage area in the otic epithelium.

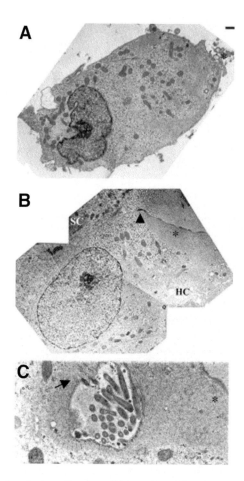

Fig. 6. Developing hair cells that differentiated in the otosphere cell cultures contained cuticular plate material and stereocilia. (**A**) A developing hair cell in an otosphere after 2 d in vitro (DIV) in the presence of epidermal growth factor. The nucleus is polarized. Cuticular platelike material is present (*). No stereocilia can be observed. (**B**) A hair cell that developed in a 14-d, EGF-stimulated otosphere. In the hair cell, the nucleus shows polarity, and an organized cuticular plate structure is present (*). There are many cell-to-cell contacts with a clearly defined apical junctional complex (arrowhead) between this hair cell and a joining cell that has abundant surface microvilli. (**C**) The cuticular plate area (*) of a developing hair cell from a 14-d, EGF-stimulated otosphere showing the organization of a stereociliary bundle with several stereociliary rootlets (arrow) projecting into the matrix of the cuticular plate. Scale bar is 1.6 µm in A and B and 0.6 µm in C.

It has been suggested from the results of chick/quail chimeric tissue explants in the avian inner ear that most glial cells in the statoacoustic ganglion complex of the chick inner ear ganglia originate from nearby cephalic neural crest cells *(41)*. However, Barald et al. *(11)* showed that polyclonal cell cultures derived from fetal otic epithelia of an immortomouse embryo can generate glial cell-like phenotypes possessing morphological features similar to glia. Kalinec et al. *(16)* derived a clonal cell line Kalinec organ of Corti cell line one (OCK-1) from the mature sensory epithelia of a 2-wk-old immortomouse organ of Corti. Under the PC condition, most of the OCK-1 cells expressed nestin and did not express either neuronal or glial cell marker epitopes. However, under the DC condition, this clonal cell line exhibited the potential to differentiate into cells expressing sensory epithelial cell (OCP2), neuronal (neurofilament 200 kD), and glial (glial fibrillary acidic protein [GFAP] and 2',3'-cyclic nucleotide-3'-phosphodiesterase [CNPase]) cell marker epitopes. Hair cell and support cell progenitors of the mouse inner ear proliferate and differentiate in the areas of presumptive sensory receptor epithelium with peak terminal mitoses occurring between E13 and E18 *(42)*. Kalinec et al. established clonal cell lines from the organ of Corti of a 2-wk-old immortomouse. There is a possibility that quiescent bipotential neuronal/glial progenitor cells exist in the sensory epithelium after terminal mitosis, as well as the additional possibility that, after cell cycle exit, oncogene-induced mature sensory epithelial cells possess the capacity to dedifferentiate into an immature cellular phenotype.

Long-term cultures of auditory sensory epithelial cells were established from adult guinea pigs using EGF to stimulate proliferation in vitro. These cells expressed a hair cell marker Brn3c and epithelial cell markers cytokeratin and ZO-1 *(43)*. It has been assumed that, after terminal cell cycle arrest, cells in mammalian sensory epithelium do not proliferate *(42)*. However, Kuntz and Oesterle demonstrated that transforming growth factor-α (TGF-α) combined with insulin can induce adult rat vestibular sensory epithelial cells to proliferate and differentiate, forming new support cells and hair cells as shown by their incorporation of radiolabeled *(3H)*-thymidine *(44)*.

Lefebvre and colleagues performed an in vitro study that suggests retinoic acid combined with TGF-α and insulinlike growth factor type I stimulate ototoxin-treated auditory sensory epithelium to generate replacement auditory hair cells in organ of Corti explants derived from immature, 1-d-old rat pups *(45)*. The hair cells generated in the ototoxin-damaged organ of Corti explants during the course of this study could be blocked by cytosine arabinoside, suggesting renewed cell division may be required for the generation of replacement auditory hair cells in this system. The results from organotypic and dissociated cell cultures of inner ear sensory organs suggest that hair cell progenitor cells may be present in immature sensory organs of the mammalian inner ear. However, the abilities of hair cell progenitors to proliferate and differentiate into mature sensory phenotypes (i.e., hair cells and support cells) may be suppressed by signals (e.g., p27^{kip1}) in the immature mammalian inner ear sensory organs.

The results obtained from studies using organotypic and dissociated cell cultures of the mammalian inner ear obtained with or without the use of oncogenes demonstrated that the developing and mature inner ear sensory organs possess both sensory epithelial progenitors (i.e., have the potential to generate both hair cells and support cells) and inner ear ganglion neural progenitors (i.e., have the potential to generate both neurons and glia). However, these studies have not shown the presence of multipotent progenitor cells with the capacity to differentiate directly into both a sensory epithelial cell lineage (i.e., hair cells and support cells) and a neural cell lineage (i.e., neurons and glia). The sensory epithelial cells and neurons are thought to derive from the otic placode during inner ear development of mammals. Furthermore, clonal analyses of insect sense organs in which all types of sensory cells are homologous to those of the mammalian inner ear (i.e., neurons, glial, hair cells, and support cells) clearly demonstrated that multipotent progenitor cells do indeed exist (46). In the near future, there is anticipation that stem cells that have the capacity to differentiate into sensory epithelial cells and the different neural cell subtypes will be identified in the sensory epithelial tissues of developing and mature mammalian inner ears.

3. GROWTH FACTOR THERAPY

Hair cell death caused by aging, ototoxins, or excessive sound exposure results in an irreversible loss of sensory cells and sensorineural hearing loss because spontaneous regeneration of auditory hair cells has not been demonstrated in the mammalian inner ear (47,48). However, Kopke and colleagues have shown that retinoic acid in combination with growth factors can stimulate ototoxin-treated vestibular sensory epithelia to produce replacement and repaired vestibular hair cells with a recovery of function in the maculae and cristae of adult guinea pigs (i.e., vestibulo-ocular reflex [VOR]) (49).

These results indicate that existing support cells may have the capability to act as hair cell progenitors within adult vestibular sense organs, but that any residual progenitor cells will need to be stimulated by specific regional environmental cues, such as retinoic acid, to express their innate potential to differentiate into new hair cell phenotypes. Several experimental reports have described attempts to induce hair cell regeneration or replacement of lost hair cells by treating with growth factors, transfection of genes (e.g.,

Math1), and the transplantation of neural stem cells into injured auditory receptors.

4. POTENTIAL THERAPIES

Ito and colleagues proposed three strategies for neural repair of inner ear sensory epithelia after hair cell loss caused by ototoxic or additional insults that initiate cell death programs in oxidative stress-injured hair cells *(50)*. The first strategy is the induction of transformation from inner ear nonsensory epithelial cells to hair cell phenotypes. Kelley and colleagues showed that support cells participate in the replacement of damaged hair cells lost after laser microbeam irradiation in vitro *(51)*. Zheng and Gao demonstrated that overexpression of *Math1* induced some of the cells in the GER to differentiate into juvenile hair cells in vitro *(31)*.

The second strategy is the induction of hair cell progenitors to differentiate into hair cell phenotypes following cell proliferation. The results of in vitro experiments by Lefebvre and colleagues indicated that treatment with retinoic acid can induce organ of Corti explants derived from immature (P0) mammalian inner ears to differentiate and thus supply replacement hair cells *(45)*, and the in vivo results of Kopke and colleagues showed that growth factor treatment of adult animals with vestibular deficits can result in a return of function *(49)*.

The third strategy is to attempt the transplantation of stem cells with the capacity to undergo self-renewal, and for these transplanted stem cells to undergo differentiation into hair cells. Next, we demonstrate and discuss the feasibility of this third approach, that is, transplantation therapy using multipotent neural stem cells to facilitate stem cell repair of damaged inner ears.

Undifferentiated cells that exhibit multipotency (e.g., embryonic stem cells and neural stem cells) are candidates for donor material to be transplanted into injured inner ears to replace damaged hair cells. These specific cell types were selected because of their known potential to differentiate into a range of appropriate mature recipient cell phenotypes (e.g., neural stem cells have been shown to have the potential to differentiate into not only neural lineages, but also blood cell lineages, when introduced into the bone marrow microenvironment) *(52)*.

To examine whether the developing mammalian inner ear has the requisite recipient environmental potential to induce transplanted neural stem cells to differentiate into sensory epithelial cells, Ito and colleagues grafted neural stem cell lines derived from an adult rat hippocampus into the developing inner ears of neonatal rats *(53)*. This clonal neural stem cell line was labeled with a β-*Gal* gene to identify the grafted (donor) cells from the other cells (recipient) in the transplanted donor inner ears. Four weeks after transplantation, the recipient animals were sacrificed, and immunohistochemical analysis was performed to localize the transplanted cells in the recipient's inner ear tissues. The immunolocalization of the transplanted donor cells in this study demonstrated that neural stem cells could survive in the inner ear tissues of the recipient ear for at least 1 mo; that is, β-galactosidase-positive cells migrated on top of or into the hair cell layer of the transplanted inner ears. The grafted donor cells that survived in the hair cell layer of the recipient inner ear strongly expressed actin filament structures, identified by staining with rhodamine-labeled phalloidin and with a staining pattern similar to that of normal inner ear hair cells.

Nishida and colleagues demonstrated that a larger number of transplanted neural stem cells could survive in a damaged retina as opposed to an uninjured retina, indicating that the microenvironment created by tissue damage contains both trophic and tropic signals that permit transplanted neural stem cells to migrate and integrate in the sensory epithelia of the eye *(54)*.

Tateya and colleagues transplanted neural stem cells derived from the forebrains of E11.5-enhanced transgenic mice expressing green fluorescent protein (GFP) into the damaged inner ears of adult mice (Fig. 7) *(55)*. The neural stem cells (donor) were transplanted into the inner ears of adult mice (8 wk old; recipient); the inner ears were damaged by an injection of neomycin into the posterior semicircular canal. Three days after the injection of neomycin, suspensions of neural stem cells were introduced into the middle turn of the cochleae of the recipient mice.

Twenty five days after the transplantation, neural stem cells migrated into and integrated in the organ of Corti at the level of the outer hair cells. In addition, a subset of the transplanted neural stem cells also migrated and survived in the hair cell layer of the utricle in the vestibule of the recipient mouse (Fig. 8). The integrated donor neural stem cells could be stained with rhodamine-labeled phalloidin, indicating that these transplanted neural stem cells had migrated into the hair cell layers of the inner ear sensory epithelia damaged by the injected ototoxin and had differentiated into hair cell phenotypes.

These results demonstrate that neural stem cells have the potential to differentiate into inner ear sensory cell lineages. These results support the hypothesis that the microenvironment in the area of damaged inner ear sensory epithelia possesses instructive cues that can induce donor multipotent stem cell (e.g., neural stem cell) to begin to differentiate into hair cell phenotypes. With further study and functional testing, stem cell transplantation

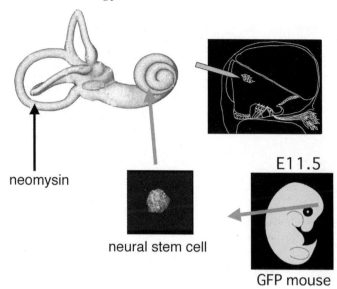

Fig. 7. Donor embryonic neural stem cells can be transplanted into the scala media of ototoxic-damaged cochleae of neonatal rats. Ten µL of neomycin were injected into the posterior semicircular canal. Three d after the injection of neomycin, a cell suspension of neural stem cells forming spheres was injected into the scala media of the cochlea. The neural stem cell spheres were obtained from forebrain tissues of embryonic d 11.5 transgenic mice with GFP gene controlled by actin promotor using serum-free medium containing basic fibroblast growth factor and epidermal growth factor. (*See* color plate 9 in the insert following p. 82.)

therapy may become an essential candidate therapy for the restoration of both auditory and vestibular function in newly damaged inner ears. Thus, the composite experimental observations obtained from in vivo transplantation studies and from in vitro stem cell culture studies may soon lead to future success in the functional neural repair of the inner ear.

Studies have shown that neural stem cells are present in multiple regions of the adult mammalian central nervous system, although ongoing neurogenesis has been documented in only a small subset of these brain regions *(9,56)*. In addition, constitutive adult neurogenesis appears to affect local circuit neurons preferentially rather than the long-relay projection neurons that are often damaged in neurodegenerative diseases *(57–58)*. However, recent reports suggested that the scope of ongoing neurogenesis during adult life may have been seriously underestimated, in part because of the absence of sensitive measures of neural stem cell activation and expansion over extended time frames *(60,61)*.

Further, recent studies have shown that a variety of neural parenchymal insults may significantly potentiate this endogenous stem cell response,

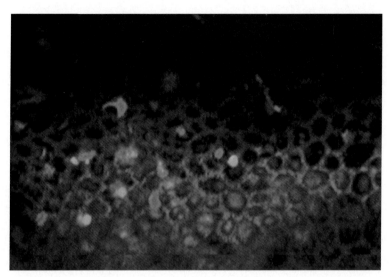

Fig. 8. Donor neural stem cells integrate into the inner ear sensory receptor epithelium of the recipient's inner ear. Ten d after the transplantation of neural stem cells, grafted cells (green) migrated and integrated into the utricle at the level of hair cells, which was identified by rhodamine-labeled phalloidin (red) staining. (*See* color plate 10 in the insert following p. 82.)

extend the spectrum of responsive stem/progenitor cell domains, and enhance the variety of differentiated neural cell types elaborated in response to pathological stimuli *(62–64).* More important, this regenerative response to neural cell injury may be profoundly augmented using various classes of cytokines and additional agents that block the receptor for myelin components involved in inhibiting neuroregenerative responses *(65–69).* These multidisciplinary approaches have resulted in the reestablishment of appropriate neural network connections that exhibit signal integration, synaptic plasticity, and behavioral recoveries, including the formation of new memory traces *(70,71).*

These cumulative observations suggest that similar approaches directed toward the reactivation of endogenous quiescent stem cells in the "otic" microenvironment may represent an attractive therapeutic option for future repair of components of the inner ear damaged by a variety of neurotoxic insults.

REFERENCES

1. Sher, A. E. (1971). The embryonic and postnatal development of the inner ear of the mouse. Acta Otolaryngol Suppl 285, 1–77.

2. Torres, M., and Giraldez, F. (1998). The development of the vertebrate inner ear. Mech Dev 71, 5–21.
3. Represa, J., Frenz, D. A., and Van De Water, T. R. (2000). Genetic patterning of embryonic inner ear development. Acta Otolaryngol 120, 5–10.
4. Fekete, D. M., and Wu, D. K. (2002). Revisiting cell fate specification in the inner ear. Curr Opin Neurobiol 12, 35–42.
5. Stemple, D. L., and Anderson, D. J. (1992). Isolation of a stem cell for neurons and glia from the mammalian neural crest. Cell 71, 973–985.
6. Reynolds, B. A., Tetzlaff, W., and Weiss, S. (1992). A multipotent EGF-responsive striatal embryonic progenitor cell produces neurons and astrocytes. J Neurosci 12, 4565–4574.
7. Reynolds, B. A., and Weiss, S. (1992). Generation of neurons and astrocytes from isolated cells of the adult mammalian central nervous system. Science 255, 1707–1710.
8. Svendsen, C. N., and Smith, A. G. (1999). New prospects for human stem-cell therapy in the nervous system. Trends Neurosci 22, 357–364.
9. Gage, F. H. (2000). Mammalian neural stem cells. Science 287, 1433–1438.
10. Malgrange, B., Belachew, S., Thiry, M., et al. (2002). Proliferative generation of mammalian auditory hair cells in culture. Mech Dev 112, 79–88.
11. Barald, K. F., Lindberg, K. H., Hardiman, K., et al. (1997). Immortalized cell lines from embryonic avian and murine otocysts: tools for molecular studies of the developing inner ear. Int J Dev Neurosci 15, 523–540.
12. Holley, M. C., Nishida, Y., and Grix, N. (1997). Conditional immortalization of hair cells from the inner ear. Int J Dev Neurosci 15, 541–552.
13. Rivolta, M. N., Grix, N., Lawlor, P., Ashmore, J. F., Jagger, D. J., and Holley, M. C. (1998). Auditory hair cell precursors immortalized from the mammalian inner ear. Proc R Soc Lond B Biol Sci 265, 1595–1603.
14. Zheng, J. L., Lewis, A. K., and Gao, W. Q. (1998). Establishment of conditionally immortalized rat utricular epithelial cell lines using a retrovirus-mediated gene transfer technique. Hear Res 117, 13–23.
15. Lawlor, P., Marcotti, W., Rivolta, M. N., Kros, C. J., and Holley, M. C. (1999). Differentiation of mammalian vestibular hair cells from conditionally immortal, postnatal supporting cells. J Neurosci 21, 9445–9458.
16. Kalinec, F., Kalinec, G., Boukhvalova, M., and Kachar, B. (1999). Establishment and characterization of conditionally immortalized organ of Corti cell lines. Cell Biol Int 23, 175–184.
17. Freshney, R. I. (2000). Culture of Animal Cells. New York: Wiley-Liss.
18. Jat, P. S., and Sharp, P. A. (1989). Cell lines established by a temperature-sensitive simian virus 40 large-T-antigen gene are growth restricted at the nonpermissive temperature. Mol Cell Biol 9, 1672–1681.
19. Noble, M., Groves, A. K., Ataliotis, P., and Jat, P. S. (1992). From chance to choice in the generation of neural cell lines. Brain Pathol 2, 39–46.
20. Zheng, J. L., and Gao, W. Q. (1997). Analysis of rat vestibular hair cell development and regeneration using calretinin as an early marker. J Neurosci 17, 8270–8282.

21. Pack, A. K., and Slepecky, N. B. (1995). Cytoskeletal and calcium-binding proteins in the mammalian organ of Corti: cell type-specific proteins displaying longitudinal and radial gradients. Hear Res 91, 119–135.

22. Erkman, L., McEvilly, R. J., Luo, L., et al. (1996). Role of transcription factors Brn-3.1 and Brn-3.2 in auditory and visual system development. Nature 381, 603–606.

23. Xiang, M., Gan, L., Li, D., et al. (1997). Essential role of POU-domain factor Brn-3c in auditory and vestibular hair cell development. Proc Natl Acad Sci U S A 94, 9445–9450.

24. Elgoyhen, A. B., Johnson, D. S., Boulter, J., Vetter, D. E., and Heinemann, S. (1994). Alpha 9: an acetylcholine receptor with novel pharmacological properties expressed in rat cochlear hair cells. Cell 79, 705–715.

25. Hasson, T., Gillespie, P. G., Garcia, J. A., et al. (1997). Unconventional myosins in inner-ear sensory epithelia. J Cell Biol 137, 1287–1307.

26. Lawoko-Kerali, G., Rivolta, M. N., and Holley, M. (2002). Expression of the transcription factors GATA3 and Pax2 during development of the mammalian inner ear. J Comp Neurol 442, 378–391.

27. Zheng, J. L., Shou, J., Guillemot, F., Kageyama, R., and Gao, W. Q. (2000). Hes1 is a negative regulator of inner ear hair cell differentiation. Development 127, 4551–4560.

28. Zine, A., Aubert, A., Qiu, J., et al. (2002). Hes1 and Hes5 activities are required for the normal development of the hair cells in the mammalian inner ear. J Neurosci 21, 4712–4720.

29. Rau, A., Legan, P. K., and Richardson, G. P. (1999). Tectorin mRNA expression is spatially and temporally restricted during mouse inner ear development. J Comp Neurol 405, 271–280.

30. Bermingham, N. A., Hassan, B. A., Price, S. D., et al. (1999). Math1: an essential gene for the generation of inner ear hair cells. Science 284, 1837–1841.

31. Zheng, J. L., and Gao, W. Q. (2000). Overexpression of Math1 induces robust production of extra hair cells in postnatal rat inner ears. Nat Neurosci 3, 580–586.

32. Lim, D. J., and Rueda, J. (1992). Structural development of the cochlea. In: Romand, R., ed., Development of Auditory and Vestibular System 2. New York: Elsevier, pp. 33–58.

33. Rivolta, M. N., and Holley, M. C. (2002). Asymmetric segregation of mitochondria and mortalin correlates with the multi-lineage potential of inner ear sensory cell progenitors in vitro. Brain Res Dev Brain Res 133, 49–59.

34. Chenn, A., and McConnell, S. K. (1995). Cleavage orientation and the asymmetric inheritance of notch1 immunoreactivity in mammalian neurogenesis. Cell 82, 631–641.

35. Morrison, S. J., Shah, N. M., and Anderson, D. J. (1997). Regulatory mechanisms in stem cell biology. Cell 88, 287–298.

36. Yamashita, H., and Oesterle, E. C. (1995). Induction of cell proliferation in mammalian inner-ear sensory epithelia by transforming growth factor alpha and epidermal growth factor. Proc Natl Acad Sci U S A 92, 3152–3155.

37. Zheng, J. L., Helbig, C., and Gao, W. Q. (1997). Induction of cell proliferation by fibroblast and insulin-like growth factors in pure rat inner ear epithelial cell cultures. J Neurosci 17, 216–226.
38. Chen, P., and Segil, N. (1999). p27(Kip1) links cell proliferation to morphogenesis in the developing organ of Corti. Development 126, 1581–1590.
39. Fekete, D. M., Muthukumar, S., and Karagogeos, D. (1998). Hair cells and supporting cells share a common progenitor in the avian inner ear. J Neurosci 18, 7811–7821.
40. Lang, H., and Fekete, D. M. (2001). Lineage analysis in the chicken inner ear shows differences in clonal dispersion for epithelial, neuronal, and mesenchymal cells. Dev Biol 234, 120–137.
41. Noden, D. M. (1983). The role of the neural crest in patterning of avian cranial skeletal, connective, and muscle tissue. Dev Biol 96, 144–165.
42. Ruben, R. J. (1967). Development of the inner ear of the mouse: a radioautographic study of terminal mitosis. Acta Otolaryngol Stockh 220, 1–44.
43. Zhao, H. B. (2001). Long-term natural culture of cochlear sensory epithelia of guinea pigs. Neurosci Lett 315, 73–76.
44. Kuntz, A. L., and Oesterle, E. C. (1998). Transforming growth factor alpha with insulin stimulates cell proliferation in vivo in adult rat vestibular sensory epithelium. J Comp Neurol 399, 413–423.
45. Lefebvre, P. P., Malgrange, B., Staecker, H., Moonen, G., and Van De Water, T. R. (1993). Retinoic acid stimulates regeneration of mammalian auditory hair cells. Science 260, 692–695.
46. Hartenstein, V., and Posakony, J. W. (1989). Development of adult sensilla on the wing and notum of *Drosophila melanogaster*. Development 107, 389–405.
47. Forge, A., Li, L., Corwin, J. T., and Nevill, G. (1993). Ultrastructural evidence for hair cell regeneration in the mammalian inner ear. Science 259, 1616–1619.
48. Warchol, M. E., Lambert, P. R., Goldstein, B. J., Forge, A., and Corwin, J. T. (1993). Regenerative proliferation in inner ear sensory epithelia from adult guinea pigs and humans. Science 259, 1619–1622.
49. Kopke, R. D., Jackson, R. L., Li, G., et al. (2001). Growth factor treatment enhances vestibular hair cell renewal and results in improved vestibular function. Proc Natl Acad Sci U S A 98, 5886–5891.
50. Ito, J. (2003) Recent advances in regenerative medicine in otology. Acta Otolaryngol Suppl., in press.
51. Kelley, M. W., Talreja, D. R., and Corwin, J. T. (1995). Replacement of hair cells after laser microbeam irradiation in cultured organs of Corti from embryonic and neonatal mice. J Neurosci 15, 3013–3026.
52. Bjornson, C. R., Rietze, R. L., Reynolds, B. A., Magli, M. C., and Vescovi, A. L. (1999). Turning brain into blood: a hematopoietic fate adopted by adult neural stem cells in vivo. Science 283, 534–537.
53. Ito, J., Kojima, K., and Kawaguchi, S. (2001). Survival of neural stem cells in the cochlea. Acta Otolaryngol 212, 140–142.
54. Nishida, A., Takahashi, M., Tanihara, H., et al. (2000). Incorporation and differentiation of hippocampus-derived neural stem cells transplanted in injured adult rat retina. Invest Ophthalmol Vis Sci 41, 4268–4274.

55. Tateya, I., Nakagawa, T., Iquchi, F., et al. (2003). Fate of neural stem cells grafted into injured inner ears of mice. Neuroreport 14, 1667–1681.

56. Panchision, D. M., and Mckay, R. D. (2002). The control of neural stem cells by morphogenic signals. Curr Opin Genet Dev 12, 478–487.

57. Kruger, G. M., and Morrison, S. J. (2002). Brain repair by endogenous progenitors. Cell 110, 399–402.

58. Mehler, M. F., and Gokhan, S. (2001). Developmental mechanisms in the pathogenesis of neurodegenerative diseases. Prog Neurobiol 63, 337–363.

59. Mehler, M. F., and Gokhan, S. (2000). Mechanisms underlying neural cell death in neurodegenerative diseases: alterations of a developmentally-mediated cellular rheostat. Trends Neurosci 23, 599–605.

60. Rietze, R., Poulin, P., and Weiss, S. (2000). Mitotically active cells that generate neurons and astrocytes are present in multiple regions of the adult mouse hippocampus. J Comp Neurol 424, 397–408.

61. Gould, E., and Gross, C. G. (2002). Neurogenesis in adult mammals: some progress and problems. J Neurosci 22, 619–623.

62. Yamamato, S., Yamamato, N., Kitamura, K., and Nakafuku, M. (2001). Proliferation of parenchymal neural progenitors in response to injury in the adult rat spinal cord. Exp Neurol 172, 115–127.

63. Yamamato, S., Nagao, M., Sugimori, M., et al. (2001). Transcription factor expression and Notch-dependent regulation of neural progenitors in the adult rat spinal cord. J Neurosci 21, 9814–9823.

64. Parent, J. M., Valentin, V. V., and Lowenstein, D. H. (2002). Prolonged seizures increase proliferating neuroblasts in the adult rat subventricular zone-olfactory bulb. J Neurosci 22, 3174–3188.

65. Brittis, P. A., and Flanagan, J. G. (2001). Nogo domain and a Nogo receptor: implications for axon regeneration. Neuron 30, 11–14.

66. Domeniconi, M., Cao, Z., Spencer, T., et al. (2002). Myelin-associated glycoprotein interacts with the Nogo66 receptor to inhibit neurite outgrowth. Neuron 35, 283–290.

67. Wang, K. C., Koprivica, V., Kim, J. A., et al. (2002). Oligodendrocyte-myelin glycoprotein is a Nogo receptor ligand that inhibits neurite outgrowth. Nature 417, 941–944.

68. Nakatomi, H., Kuriu, T., Okabe, S., et al. (2002). Regeneration of hippocampal pyramidal neurons after ischemic brain injury by recruitment of endogenous neural progenitors. Cell 110, 429–441.

69. GrandPre, T., Li, S., and Strittmatter, S. M. (2002). Nogo-66 receptor anatagonist peptide promotes axonal regeneration. Nature 417, 547–551.

70. Gates, M. A., Fricker-Gates, R. A., and Macklis, J. D., (2000). Reconstruction of cortical circuitry. Prog Brain Res 127, 115–156.

71. Arlotta, P., Magari, S. S., and Maklis, J. D., (2003). Induction of adult neurogenesis: molecular manipulation of neural precursors *in sity*. Ann NY Acad Sci 991, 229–236.

15

Engineering the In Vitro Cellular Microenvironment for the Control and Manipulation of Adult Stem Cell Responses

Ali Khademhosseini and Peter W. Zandstra

1. INTRODUCTION

Stem cells have generated a great deal of excitement as a potential source of cells for transplantation because of their ability to self-renew and differentiate into functional cells of various tissues *(1–3)*. Stem cells can be derived from multiple stages of development as well as numerous adult tissues. Adult tissues are an attractive and readily accepted source of stem cells because such cells have demonstrated efficacy in multiple types of cellular therapeutics *(4,5)* and can be directly obtained from individual patients, thereby eliminating the difficulties associated with tissue rejection. Despite this enormous potential, the use of adult stem cells has been limited, primarily because of the inability to identify these rare cells from the heterogeneous tissue populations *(6)* and to expand populations of cells that retain stem cell properties in vitro.

Historically, many adult tissues were thought incapable of regeneration. However, cells with regenerative capability have been detected in most adult tissues, including liver *(7–9)*, intestine *(10)*, retina *(11)*, skin *(12)*, muscle *(13)*, neural *(14)*, mammary glands *(15)*, and others. Although extensive documentation of the properties of many of these cells with respect to their stem cell characteristics (i.e., individual cells with the capacity for extended self-renewal and multilineage differentiation) is under way, taken together, it is clear that adult tissues may provide an untapped source of cells for cellular therapies.

Numerous studies suggest that the proliferative and differentiative potential of tissue-specific stem cells changes during ontogeny *(16–18)* and is dependent on intrinsic factors such as telomere shortening *(19)* and genetic stability *(20)*. The ability to measure changes in the developmental potential of stem cells is limited by the inability to fingerprint such cells genetically

From: *Adult Stem Cells*
Edited by: K. Turksen © Humana Press Inc., Totowa, NJ

(21) and by the properties of the assays used to detect them (such as tissue homing *[22]*). Despite these limitations, it is well recognized that adult stem cells in vivo have a proliferative potential much beyond the lifespan of the organism. For example, a single hematopoietic stem cell (HSC) not only can reconstitute hematopoiesis in primary recipients by contributing to both lymphoid and myeloid cells *(23,24)*, but also can reconstitute secondary and tertiary hosts *(25–27)*. Although less-rigorously analyzed (to a large extent because of the lack of transplantation assays based on tissue repopulation), adult stem cells from other tissues also have extensive regenerative capacities *(10,12,13)*. Despite the apparent intrinsic capability of adult stem cells for extensive self-renewal, efforts to grow these cells in culture have failed to recapitulate their in vivo potential.

The interest in adult stem cells has been elevated by recent reports that some adult stem cells, or stem cell populations, may be capable of crossing lineage boundaries by differentiating into cells with unexpected developmental properties. For example, bone marrow-derived cells have been reported to give rise to different types of muscle cells, such as from unfractionated bone marrow *(28–30)* or enriched HSC-like cells *(31,32)*; liver cells, such as from unfractionated rodent *(33,34)* and human bone marrow *(35,36)*, or enriched HSC-like cells *(34,37)*; lung *(38)* and neuronal cells, including neurons detected in vivo *(39,40)* and in vitro *(41,42)*; and astroglia and microglia *(43,44)*. Strong evidence of the multiorgan generating capability of bone marrow-derived stem cells has been demonstrated in the ability of a single cell to reconstitute hematopoiesis in primary and secondary recipients as well as to differentiate into apparently functional epithelial cells of the liver, lung, intestine, and skin *(45)*.

Stem cells not derived from bone marrow may also have developmental capacities outside their tissue of origin *(46,47)*, although recent reports have led investigators to question many of these early results *(48–51)*. Even with the uncertainties regarding the intrinsic potential of adult stem cells, the ability of cells of one tissue to give rise to differentiated cells of another tissue (either because of broader differentiation capacity or because the tissue in question contains multiple types of stem cells) may be of great therapeutic potential and could provide alternative adult stem cell sources that are readily accessible (i.e., peripheral blood, skin, or fat-derived stem cells).

This chapter highlights some of the main bioengineering challenges in the development of adult stem cell-based therapies also; methods to control the self-renewal and differentiation of adult stem cells and to create clinically relevant bioprocesses are discussed. Particular emphasis is given to analyzing these techniques in the context of well-established adult stem cell

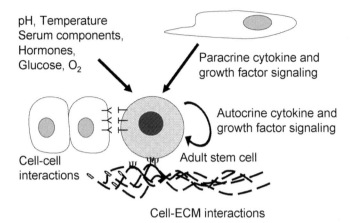

Fig. 1. Adult stem cell niche. The microenvironment of adult stem cells is regulated through a complex network of paracrine and autocrine soluble signals as well as cell–cell and cell–ECM bound signals. Physicochemical parameters such as pH, temperature, oxygen, perfusion, and mechanical stimuli can also influence cell fate decisions.

systems and the effect of the cellular microenvironment on the responses of such cells.

2. STEM CELL MICROENVIRONMENT

In vivo, stem cells reside in a complex microenvironment characterized by their local geometry (structural and physicochemical), by specific types of surrounding tissue cells, and by soluble and extracellular matrix (ECM) components *(52,53)*. The properties of this microenvironment are dynamic, depend on the specific tissue, and are affected by factors such as vascularization and loading (Fig. 1). Importantly, the analysis and understanding of the role of the microenvironment on stem cell responses should be motivated by more than the desire simply to mimic the in vivo milieu. The in vivo microenvironment dynamically exposes cells to positive and negative regulators of specific stem cell responses; the selective application of these regulatory mechanisms during in vitro culture will ultimately depend on the type of cell response to be elicited and the ability to control dominant (i.e., response-determining) culture parameters.

2.1. Cytokines and Growth Factors

Cytokines and growth factors are important regulators of the tissue microenvironment. They are produced by stem cells or their neighboring cells in an autocrine or paracrine manner and often combine with other microenvironmental components to elicit nonlinear responses, for instance, threshold-based *(54)* or synergistic responses. Due in part to difficulties in the quantitative identification of most tissue-specific stem cells and their immediate derivatives, defining cytokine networks that allow for controlled self-renewal and differentiation of these cells has been challenging. So far, most experiments have studied the response of putative stem cells to individual or limited numbers of combinations of cytokines. Few studies have rigorously analyzed and optimized the effect of multiple cytokines, as well as interactions between these cytokines, on stem cell responses.

Factorial experiments are one approach to overcome some of these limitations and quantitatively analyze the effect of cytokine interactions on stem cell responses *(18,54–58)*. By analyzing complex interactions between various cytokines, relationships commonly missed by conventional dose–response approaches are detectable. For example, with respect to HSCs, this type of analysis has been helpful in defining self-renewal and differentiation factors *(59)*, a threshold cytokine concentration effect on self-renewal and differentiation *(54)*, and changes in cell's cytokine responsiveness with ontogeny *(18)*.

These and other results (*see* ref. *60* for a review) have led to the identification of molecules thought important for the regulation of HSC fate and have led to the development of feeder-free (and serum-free) culture systems. Feeder-free cultures provide more control in studying the effects of specific signals and are desirable in clinical applications. Briefly, these studies have identified a "cocktail" of cytokines containing stem cell factor (SCF), flt-3 ligand (FL), and interleukin 11 (IL-11) family of cytokines *(61)* (with or without the addition of thrombopoietin, TPO) *(62)* and revealed much about the mechanisms of cytokine action on stem cells.

For example, even though SCF has been shown to be critical for the maintenance and expansion of HSCs *(63)*, it cannot by itself maintain HSCs in vitro *(64)*. SCF acts synergistically with various growth factors, including TPO, FL, and IL-11, to induce proliferation and maintenance of myeloid and lymphoid progenitors *(65)*. FL has also been shown to synergize with a wide variety of hematopoietic cytokines (in particular, SCF and the IL-11 family of cytokines) to stimulate the proliferation, self-renewal, and differentiation of HSCs *(66)*.

Despite these results, a definitive cocktail that leads to reproducible expansion of HSCs has not yet been developed; illustrating the underlying complexity in cytokine networks *(60)* and pointing to the need to develop more effective in vitro culture technologies.

Significant progress has recently been made with respect to growing other types of adult stem cells in cytokine-supplemented media. For example, cells with characteristics of neural stem cells (NSCs) have been expanded as neurospheres in well-defined culture conditions *(67)*. Typically, epithelial growth factor (EGF) *(68,69)* and fibroblast growth factor 2 (FGF-2) *(70–72)* are used to propagate the early tissue dissociates containing NSCs. The presence of these growth factors seems to prevent the differentiation of NSCs and allows their continual proliferation. These properties have allowed the creation of bioreactors, which have been used to expand neurosphere-forming cells *(73)*.

Multipotential adult stem cells have been isolated from human and murine skin *(74)*. Cultures derived from these cells plated clonally and maintained for many passages at low cell densities generated both neuroectodermal (neurons and glia cells) and mesodermal (smooth muscle and adipocytes) tissues. Interestingly, the propagation of these cells seems to depend, among other things, on the addition of EGF and FGF to the culture, conditions similar to those defined for NSCs *(68,75)*. It remains to be seen whether these so-called multipotent adult progenitor cells can be isolated directly from specific tissues or arise as a product of in vitro culture.

In addition to identifying the types of growth factors and cytokines important in the control of adult stem cell growth and differentiation, it is becoming well recognized that their (relative) concentrations *(56)*, mode of presentation *(76)*, and order of application *(77)* also play an important role in eliciting particular responses from stem cells. The significant increase in the complexity of experiments investigating these parameters clearly indicates the need for quantitative, systems biology approaches as a tool in analyzing such interactions. These types of approaches have been used to analyze the cross-talk between two independent ligand-activated signaling pathways *(78)* and may one day be useful for the analysis of individual candidate stem cells *(79,80)*.

To study the "effective concentration" of growth factors present in the microenvironment of an individual cell, it is necessary to understand important mediating steps, such as the transport properties of the ligand in the cellular vicinity, the complexities of ligand–receptor binding interactions, the role of ECM binding on ligand availability and ligand–receptor complex internalization *(81)*, and the downstream consequences of signaling activation *(82)*. Clearly, to signal through a receptor, a ligand must reach its receptor on the cell membrane. This step could provide a significant barrier to signaling and is dependent on a number of parameters, such as the degradation and diffusion rates of the soluble ligand, ligand interactions with the

ECM molecules, mixing and turbulence of the surrounding fluid, and other parameters in the microenvironment. Bioengineering approaches can be used to engineer proteins with modified stability and diffusive properties *(83,84)* to optimize transport to the cells. Protein engineering approaches have also been used to design and select for *(85)* proteins with modified affinity for their particular receptor *(81,86)*.

Significant deviations in supplemented growth factor concentrations can occur throughout the culture period as a result of receptor–ligand complex internalization and degradation by cells *(54,81,87)*. For example, a cell-associated depletion of growth factors and cytokines in both hematopoietic *(54)* and embryonic stem (ES) cell *(88)* cultures has been observed. Significant progress has been made in understanding and manipulating the mechanisms that underlie ligand possession by cells *(89)*. This information has been applied to optimize biological responses of T cells to engineered and mutant IL-2 proteins *(87,90)*. These strategies, along with protein engineering techniques *(91,92)*, can also be used to develop receptor–ligand complexes that dissociate in the acidic environment of the endosome, thus allowing for higher ligand recycling rates *(93)*, an approach that has been shown to enhance the "effective" concentration of the ligand in the vicinity of the cell greatly *(89,93)* and may lead to decreased exogenous requirements of cytokines and growth factors.

Another approach to growth factor supplementation that holds significant promise for the modulation of stem cell responses is the design of ligand–receptor complexes that cannot be internalized and thus may allow the delivery of controlled and sustained stimulation to the cells. This can be achieved by immobilizing proteins to various surfaces and scaffolds. To achieve this goal, techniques ranging from direct protein adsorption to covalent linkage of aldehyde-containing surface groups to amine base side chains, as utilized in protein–protein interaction arrays *(94)*, can be used. These simple strategies, however, may be limited because nonspecific adsorption of serum proteins may "mask" the immobilized proteins. To overcome this potential limitation, a linking molecule (such as polyethylene oxide, PEO) may be used to tether the growth factor to a surface or ECM. This approach has been successful in covalently binding EGF to PEO *(95)*, for which the tethered EGF elicited the deoxyribonucleic acid (DNA) synthesis of hepatocytes at rates similar to that of its soluble counterpart and significantly greater than that of adsorbed EGF at comparable surface concentrations. Significant challenges exist in the implementation of these technologies, both in terms of the biomaterial design strategies and in terms of the underlying biological mechanisms that need to be mimicked *(96–101)*.

2.2. Cell–Cell and Cell–Extracellular Matrix Interactions

In vivo cells are typically in direct contact with surrounding cells and ECM. ECM is a dynamic assembly of interacting molecules that recognizes and regulates cell function in response to endogenous and exogenous stimuli *(102)*. ECM is produced by cells and consists of collagens, proteoglycans, adhesive glycoproteins, and glycoasaminoglycans and associated bound protein modulators of cell function. Along with providing a framework in which cells form tissues, ECM directly modulates cell attachment, shape, morphology, migration, orientation, and proliferation. ECM also serves as a reservoir for various growth factors. It has been proposed that the existence of matrix is essential for the activity of many growth factors (such as hepatocyte growth factor [HGF], transforming growth factor-β [TGF-β], and acidic and basic FGF) *(103)*. The complex combination of signals provided by the ECM to adult stem cells likely provides the cell with information unique to the tissue of origin and is important for the regulation of stem cell self-renewal, differentiation, and homing.

In the bone marrow, HSC interactions with adhesion molecules (e.g., CD34, stem cell antigen 1 [Sca-1], selectins, and various integrins) on the vascular endothelium have been reported to aid in cell homing during hematopoietic reconstitution experiments and to regulate cellular trafficking during homeostasis (for review, see ref. *104*). Cell adhesion molecules have also been shown to be present in other types of stem cells. For example, mesenchymal stem cells (MSCs) were first isolated based on their adherence (albeit poorly characterized) to tissue culture surfaces *(105)*.

Cell–cell and cell–ECM interactions have been shown to greatly influence the self-renewal and differentiation of stem cells *(106,107)*. An example of the importance of cell–cell and cell–matrix interactions in the modulation of adult stem cell fate is the maturation of intestinal crypt stem cells *(10,108–113)*. These cells give rise to epithelial cells that line the gastrointestinal tract and typically lie in the base of test-tube-like structures *(114)*. As cells move toward the luminal pole, they go through a series of differentiation and proliferation steps so that the pole is occupied by short-lived functional cells. Mathematical *(115)* and experimental *(108)* studies suggested that as few as four to six stem cells are sufficient in maintaining homeostasis for each crypt. Once these cells leave a stem cell niche within the base of the structure, they are induced to differentiate. This differentiation is thought to be regulated by a variety of signals, including cell–cell and cell–ECM signals.

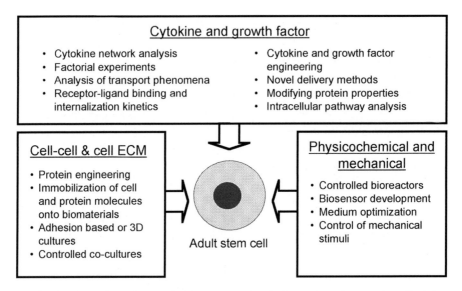

Fig. 2. Bioengineering methods to control, mimic, and analyze stem cell niche. A number of engineering approaches have been developed to control the immediate microenvironment of adult stem cells. These techniques aim to control the interactions between tissue stem cells and stimuli selected for their potential role in the design of stem cell-based bioprocesses.

A number of ECM proteins and receptors have been reported to be expressed differentially in the stem cell region of the crypt *(116)*. However, even though numerous molecules (such as Notch) mediated by cell–cell and cell–ECM interactions are suspected, the molecular mechanisms that induce this behavior remain elusive (for review, *see* ref. *117*). Further insight into understanding such signals will facilitate the design of culture technologies that mimic critical aspects of the in vivo microenvironment and facilitate better control over stem cell responses in vitro (Fig. 2).

2.3. Physicochemical Parameters

Physicochemical properties such as pH and oxygen and glucose concentrations are another important aspect of the stem cell niche. Changes in such parameters have been shown to be critical in both embryonic development and ES cell differentiation *(118,119)*, as well as adult stem cell regulation *(120)*. Low oxygen concentration has been linked to the activation of transcriptional factors such as hypoxia-inducible factor *(121)*, which in turn regulates the expression of signaling molecules, such as erythropoietin and vascular endothelial growth factor (VEGF), cytokines that influence stem and progenitor cell behavior. In hematopoietic precursors, low oxygen tension increases the size and frequency of hematopoietic colonies in semisolid media *(122,123)*.

However, the exact role of oxygen in the development of HSCs is yet to be determined, let alone the role of oxygen in the function of other tissue-derived stem cells. For example, it has been observed that, with low oxygen concentrations, HSC and progenitor numbers seem to be maintained *(124,125)*, whereas HSC expansion has been shown to occur at higher oxygen concentrations *(126,127)*. For neurosphere-forming cells, low oxygen concentrations inhibit cellular proliferation *(73)*.

In addition, it has been demonstrated that oxygen tension is also important in the regulation of MSCs *(128,129)*. These studies suggested that MSC proliferation and myogenic and bone differentiation are enhanced in physiological oxygen concentrations; higher oxygen concentrations induce adipocyte differentiation.

Clearly more research is required to completely understand the effects of oxygen on progenitor vs more differentiated cell populations. Different tissue microenvironments are comprised of widely varying physicochemical properties, and these may play a direct or indirect role in the observed functional differences between cells seeded in different tissues.

Another microenvironmental cue that may influence the in vivo and in vitro responses of stem cells is mechanical stimuli. It has long been known that mechanical forces play an important role in the development and maintenance of vascular, muscle, and bone tissues *(130–132)*. Mechanical stimuli may initiate mechanotransductive signaling pathways that are still largely unresolved *(133)*. The effect of these forces on the differentiation of stem cells is under study. For example, compressing marrow-derived stromal cells thought to contain MSCs encourages bone development; stretching MSCs immobilized in a matrix encourages tendon and cartilage formation *(134)*. However, the mechanisms by which mechanical stimuli affect the differentiation and self-renewal of mesenchymal or other adult stem cells largely remain to be determined.

3. MODELS OF STEM CELL BEHAVIOR

The design and implementation of models predictive of cell responses should facilitate the rapid investigation of a large number of "experimental" conditions and lead to a more in-depth understanding of the biological mechanisms that control stem cell behavior. Numerous models have been developed to describe the behavior of stem cells and to predict self-renewal and differentiation of these cells. In fact, mathematical modeling of stem cell behavior is as old as the concept of stem cells *(135)*.

Most established models have typically tried to develop an understanding of the way that stem cells respond to changes in the cellular microenvironment by imposing either stochastic or deterministic constraints onto the results of in vitro and in vivo experiments. Significant evidence exists that both stochastic and directive mechanisms play important roles in the regulation of stem cell responses, and that the particular mechanism (i.e., threshold-based responses; see ref. *136*) used may be dependent on the tissue system/cellular microenvironment.

For example, in hematopoiesis, several studies suggest that exposure to growth factors may not be obligatory for the differentiation of primitive cells, and that at least under certain conditions, the identity of the differentiated cell population may be intrinsically determined *(137)*. Particularly interesting in this regard is the recent demonstration that coexpression of multiple lineage-restricted genes precedes commitment in multipotent progenitors *(138–141)*. This multilineage "priming" process is consistent with the flexibility in the gene expression profiles seen during osteoprogenitor development and implies that the commitment of a multipotent cell to a particular pathway may reflect the stabilization of a particular subset of expressed genes *(142)*. The stabilization process may occur in a stochastic manner in the absence of a particular instructive signal. Conversely, the commitment of an undifferentiated cell may proceed through an instructive signal that stabilizes a particular set or subset of expressed transcription factors. Although it is not clear whether all adult stem cell types utilize the same underlying mechanisms for the control of their responses, many in vitro studies supported the existence of this two-level (stochastic in the absence of signal, directive in the presence of signal) regulatory mechanism *(138,143–146)*.

The above-described low-level expression of multiple transcription factors may also be at the root of at least some stem cell plasticity phenomena *(147)*. In this case, plastic stem cells may also express (either "stochastically" or as a result of ligand-mediated upregulation) transcription factors associated with cells of other tissues, and exposure to particular tissue microenvironments may directly or indirectly (through survival mechanisms) elicit this novel differentiation capacity.

Mathematical descriptions of stochastic differentiation mechanisms have typically utilized Monte Carlo simulations to mimic the probabilistic nature of stem cell responses. For example, Till et al. found that their experiments on colony-forming unit spleen (CFU-S) cell self-renewal were best described in computer simulations when the probability of self-renewal was fixed at 0.6 *(135)*. They also explained the colony size distributions using the same approach, although others have suggested that heterogeneity in the transplanted population is also consistent with such differences *(146)*. Since then, the stochastic models have been extended to the hematopoietic progenitor cells *(148,149)*, NSCs *(150–152)*, and intestinal crypt stem cells *(112,115)*.

Deterministic models typically incorporate external conditions to derive kinetic data that describe the growth (and differentiation) rates of populations of cells. These models have been used to explain the ex vivo expansion of hematopoietic cells *(153)*. For example, Peng et al. developed a kinetic description of single-lineage hematopoietic cell expansion based on self-renewal responses to cytokine supplementation, the growth rates of different progenitor cell populations, and mature cell death *(154)*. The model is consistent with experimental observations of cytokine-supplemented hematopoietic cultures *(154)* and predicted a self-renewal probability of 0.62 to 0.73 under these conditions.

Mackey and colleagues used deterministic modeling approaches to develop multicompartment models to reveal control mechanisms that may be at the root of some types of hematopoietic disease *(155,156)*. Stem cell growth has also been mathematically described by defining cell growth in terms of the proliferation responses of subpopulations of cells *(157)*, in some cases taking into account symmetric and asymmetric division by defining differentiation as a state that is attained after a certain number of symmetric mitotic cycles have occurred *(158)*. Although stochastic and deterministic models utilize different approaches, both can be applied and fit to experimental data (and thus are somewhat limited in their ability to provide new insight into the mechanisms that regulate stem cell responses).

Using ES cells as a model stem cell system, we are developing novel models that incorporate the known variability in receptor expression between individual cells into a deterministic cell population-based model *(136)*. This generalized model illustrates how quantitative variations in ligand–receptor interactions, arising from interactions of the cell with its microenvironment, can result in alteration in cell fate choices. Our approach is distinct from stochastic models of stem cell differentiation control, which typically assume that cell fate processes are random and are best described by statistical probability distributions. This comprehensive approach, which attempts to incorporate molecular events in the description of macroscopic cellular behavior (i.e., ligand concentration and receptor expression to generate predictions of self-renewal) should be valuable to adult stem cell models *(55,88,136)*.

To be useful, mathematical models should not only be consistent with the observed data, but also be able to predict new experimental observations and to determine system-controlling parameters. Statistical models that aim to analyze stem cell gene expression and to correlate such information with stem cell hierarchy will also be useful in revealing common mechanisms between stem cell types. We have started to explore such an approach on clones of osteoblast progenitors to determine which genes are expressed as they develop into mature bone. Our analysis indicated that the adult progenitor cells can use several developmental routes to get to the same end stage *(142)*. This unexpected plasticity in the genetic paths used to generate the same stable differentiated state may likewise be a property of multiple stem cell systems. The modeling of highly complex molecular interactions and gene regulatory networks has already been successfully applied to predict system behavior in intracellular signaling networks comprised of hundreds of components *(159–162)*. The application of these and other approaches to stem cell systems should prove fruitful.

4. DEVELOPING SCALABLE STEM CELL BIOREACTORS

In addition to developing strategies to control and manipulate the cellular microenvironment, bioengineers must devise bioprocesses to implement this microenvironmental control at a clinically relevant scale. For some stem cell-based applications, current bioreactors must be scaled up to industrial size units (>10 L), and others (such as purified HSC) may require much smaller volumes (i.e., <100 mL); each poses significant process control challenges. Bioreactors can be designed with two goals: generation of large quantities of differentiated cells and expansion of transplantable stem cells. The former may find their implementation in the treatment of acute disease and injury (such as acute liver failure or burns), and the latter may be useful for the treatment of chronic disorders (such as diabetes or gene therapy for sickle cell anemia). A number of culture systems have been developed for the production of stem cell-based therapeutics. These include stirred or attachment-based culture techniques *(163)*.

Stirred cultures have a number of advantages, such as scalability, culture homogeneity, and simplicity *(163)*. Therefore, parameters such as oxygen tension, cytokine concentration, and serum components may be easily regulated in these cultures *(164–166)*. Stirred bioreactors have been successfully used to culture hematopoietic *(164)*, neural *(73)*, and bone marrow populations capable of reading out as fibroblast and bone progenitors (MSCs) adult stem cells *(165)*. Suspension culture systems may also be useful for controlling the ratios between differentiated and undifferentiated cells during in vitro culture. In addition, inhibitory signals, generated by the differentiated progeny (reviewed in ref. *60*) may regulate the yield of stem cells in such cultures. Mathematical models *(167)* have predicted and experiments have confirmed *(168)* that the control of this differentiated subpopulation dynamic can be used to influence culture output *(168)*.

Despite the simplicity of stirred cultures, these cultures may not be suitable for all types of adult stem cells. For example, epithelial progenitor cells may require three-dimensional (3-D) signals for expansion or directed differentiation. In such cases, the use of culture conditions that enhance adhesion-based interactions may be important. Some investigators have combined these requirements with suspension culture systems, for example, using simulated microgravity conditions that result in the maintenance of in vivo-like gene expression *(169)* and cellular organization *(170–173)*.

Adhesion-based cultures are typically used to create bioprocesses that have characteristics of the in vivo microenvironment. These may involve the use of scaffolds or beads as the templates onto which progenitor cells grow *(174)* and often utilize feeder or stomal cells as delivery vehicles for stimulatory signals (thereby overcoming the difficulties associated with insufficient knowledge of factors that influence stem cell self-renewal and differentiation). Bioengineering approaches for positioning anchorage-dependent cells on surfaces with control over size and spatial arrangements (cellular "micropatterning"; *see* refs. *175–178*) can create a high level of complexity in the cocultures and may be useful for the analysis of stem cell behavior under defined conditions and geometries.

These and other microfabrication techniques (reviewed in ref. *179*) may become important tools in creating bioreactors that mimic in vivo conditions. Soft lithography and photolithography techniques have become widely available tools for biological applications *(179)*. The particular advantage of these techniques is evident in biological applications that require length scales of 10 µm or greater *(180)*.

Microfluidics may be used in combination with these techniques to control the delivery of cytokines and growth factors to cells *(179)*. Microfluidic systems take advantage of the laminar flow of fluids within narrow channels (<100 µm) to allow for the formation of concentration gradients of soluble factors and therefore allow for direct control of cell responses at length scales that are developmentally relevant *(179,181)*.

Microfabricated bioreactors may also be used to study and expand stem cells under perfused conditions *(182,183)*. Such cultures maintain differentiated phenotypes of hepatocytes in vitro *(183)*; however, their feasibility in expanding stem cells has yet to be determined.

A critical property of ideal stem cell bioreactors is the ease of periodic medium replacement. Replenishing medium not only eliminates nutrient and cytokine depletion and end product inhibition, but also allows the transient presentation of signals specific to the differentiation stage of the cells. The use of such signaling techniques has been particularly important in inducing the differentiation of ES cells into hepatocytes by exposing the cells to transient conditions that mimic the embryonic development *(184)*. Such techniques may provide a valuable tool in adult stem cell therapies. Furthermore, metabolic properties such as cell-specific glucose consumption and lactate production increase and inhibitory factors such as medium acidification decrease in fresh medium *(185)*.

5. CONCLUSIONS

To utilize adult stem cells fully in cell therapy applications, understanding the molecular cues that regulate their behavior is crucial. The lack of suitable in vitro models has hindered stem cell research and limited much experimentation to in vivo models. The challenge is to design controlled systems that will deliver proper microenvironmental cues at optimal doses. Bioengineering approaches, including the modeling, analysis, and manipulation of microenvironmental cues, as well as the design of novel bioreactors, should facilitate the generation of therapeutically significant amounts of stem cells. Differentiated human tissues may provide a basis for the detailed understanding of the molecular mechanisms that control stem cell responses.

ACKNOWLEDGMENTS

We would like to acknowledge Dave Custer and Siddhartha Jain for their valuable comments regarding the chapter. Funding for A. K. is provided by the Natural Sciences and Engineering Research Council (NSERC) of Canada and Division of Biological Engineering at Massachusetts Institute of Technology, Cambridge. Results summarized in this work were supported by grants from NSERC, the Whitaker Foundation, the Stem Cell Network, Network of Centres of Excellence of Canada, and the Biotechnology Process Engineering Centre at the Massachusetts Institute of Technology. P. W. Z. is the Canada Research Chair in Stem Cell Bioengineering.

REFERENCES

1. Fuchs, E., and Segre, J. A. (2000). Stem cells: a new lease on life. Cell 100, 143–155.
2. Weissman, I. L. (2000). Stem cells: units of development, units of regeneration, and units in evolution. Cell 100, 157–168.
3. Watt, F. M., and Hogan, B. L. (2000). Out of Eden: stem cells and their niches. Science 287, 1427–1430.

4. Barrett, A. J., and Treleaven, J. (1998). The Clinical Practice of Stem-Cell Transplantation. Oxford, UK: Isis Medical Media, Mosby-Year Book.

5. Green, H. (1989). Regeneration of the skin after grafting of epidermal cultures. Lab Invest 60, 583–584.

6. Nordon, R., and Schindhelm, K. (1999). Ex Vivo Cell Therapy. San Diego, CA: Academic Press.

7. Michalopoulos, G. K., and DeFrances, M. C. (1997). Liver regeneration. Science 276, 60–66.

8. Sell, S. (1978). Distribution of alpha-fetoprotein- and albumin-containing cells in the livers of Fischer rats fed four cycles of N-2-fluorenylacetamide. Cancer Res 38, 3107–3113.

9. Thorgeirsson, S. S. (1996). Hepatic stem cells in liver regeneration. FASEB J 10, 1249–1256.

10. Potten, C. S., and Loeffler, M. (1990). Stem cells: attributes, cycles, spirals, pitfalls and uncertainties. Lessons for and from the crypt. Development 110, 1001–1020.

11. Tropepe, V., Coles, B. L., Chiasson, B. J., et al. (2000). Retinal stem cells in the adult mammalian eye. Science 287, 2032–2036.

12. Watt, F. M. (1998). Epidermal stem cells: markers, patterning and the control of stem cell fate. Philos Trans R Soc Lond B Biol Sci 353, 831–837.

13. Seale, P., and Rudnicki, M. A. (2000). A new look at the origin, function, and "stem-cell" status of muscle satellite cells. Dev Biol 218, 115–124.

14. Gage, F. H. (2000). Mammalian neural stem cells. Science 287 (5457), 1433–1438.

15. Ormerod, E. J., and Rudland, P. S. (1986). Regeneration of mammary glands in vivo from isolated mammary ducts. J Embryol Exp Morphol 96, 229–243.

16. Potten, C. S., Martin, K., and Kirkwood, T. B. (2001). Ageing of murine small intestinal stem cells. Novartis Found Symp 235, 66–79; discussion 79–84, 101–104.

17. De Haan, G., and Van Zant, G. (1999). Genetic analysis of hemopoietic cell cycling in mice suggests its involvement in organismal life span. FASEB J 13, 707–713.

18. Zandstra, P. W., Conneally, E., Piret, J. M., and Eaves, C. J. (1998). Ontogeny-associated changes in the cytokine responses of primitive human haemopoietic cells. Br J Haematol 101, 770–778.

19. Rufer, N., Brummendorf, T. H., Kolvraa, S., et al. (1999). Telomere fluorescence measurements in granulocytes and T lymphocyte subsets point to a high turnover of hematopoietic stem cells and memory T cells in early childhood. J Exp Med 190, 157–167.

20. Peters, A. H., O'Carroll, D., Scherthan, H., et al. (2001). Loss of the Suv39h histone methyltransferases impairs mammalian heterochromatin and genome stability. Cell 107, 323–337.

21. Phillips, R. L., Ernst, R. E., Brunk, B., et al. (2000). The genetic program of hematopoietic stem cells. Science 288, 1635–1640.

22. Danet, G. H., Lee, H. W., Luongo, J. L., Simon, M. C., and Bonnet, D. A. (2001). Dissociation between stem cell phenotype and NOD/SCID repopulat-

ing activity in human peripheral blood CD34(+) cells after ex vivo expansion. Exp Hematol 29, 1465–1473.

23. Szilvassy, S. J., Humphries, R. K., Lansdorp, P. M., Eaves, A. C., and Eaves, C. J. (1990). Quantitative assay for totipotent reconstituting hematopoietic stem cells by a competitive repopulation strategy. Proc Natl Acad Sci U S A 87, 8736–8740.

24. Dick, J. E., Magli, M. C., Huszar, D., Phillips, R. A., and Bernstein, A. (1985). Introduction of a selectable gene into primitive stem cells capable of long-term reconstitution of the hemopoietic system of W/Wv mice. Cell 42, 71–79.

25. Keller, G., and Snodgrass, R. (1990). Life span of multipotential hematopoietic stem cells in vivo. J Exp Med 171, 1407–1418.

26. Lemischka, I. R., Raulet, D. H., and Mulligan, R. C. (1986). Developmental potential and dynamic behavior of hematopoietic stem cells. Cell 45, 917–927.

27. Iscove, N. N., and Nawa, K. (1997). Hematopoietic stem cells expand during serial transplantation in vivo without apparent exhaustion. Curr Biol 7, 805–808.

28. Wakitani, S., Saito, T., and Caplan, A. I. (1995). Myogenic cells derived from rat bone marrow mesenchymal stem cells exposed to 5-azacytidine. Muscle Nerve 18, 1417–1426.

29. Makino, S., Fukuda, K., Miyoshi, S., et al. (1999). Cardiomyocytes can be generated from marrow stromal cells in vitro. J Clin Invest 103, 697–705.

30. Ferrari, G., Cusella-De Angelis, G., Coletta, M., et al. (1998). Muscle regenera-tion by bone marrow-derived myogenic progenitors. Science 279, 1528–1530.

31. Bittner, R. E., Schofer, C., Weipoltshammer, K., et al. (1999). Recruitment of bone-marrow-derived cells by skeletal and cardiac muscle in adult dystrophic mdx mice. Anat Embryol (Berl) 199, 391–396.

32. Gussoni, E., Soneoka, Y., Strickland, C. D., et al. (1999). Dystrophin expression in the mdx mouse restored by stem cell transplantation. Nature 401, 390–394.

33. Petersen, B. E., Bowen, W. C., Patrene, K. D., et al. (1999). Bone marrow as a potential source of hepatic oval cells. Science 284, 1168–1170.

34. Theise, N. D., Badve, S., Saxena, R., et al. (2000). Derivation of hepatocytes from bone marrow cells in mice after radiation-induced myeloablation. Hepatology 31, 235–240.

35. Alison, M. R., Poulsom, R., Jeffery, R., et al. (2000). Hepatocytes from non-hepatic adult stem cells. Nature 406, 257.

36. Theise, N. D., Nimmakayalu, M., Gardner, R., et al. (2000). Liver from bone marrow in humans. Hepatology 32, 11–16.

37. Lagasse, E., Connors, H., Al-Dhalimy, M., et al. (2000). Purified hematopoietic stem cells can differentiate into hepatocytes in vivo. Nat Med 6, 1229–1234.

38. Pereira, R. F., Halford, K. W., O'Hara, M. D., et al. (1995). Cultured adherent cells from marrow can serve as long-lasting precursor cells for bone, cartilage, and lung in irradiated mice. Proc Natl Acad Sci U S A 92, 4857–4861.

39. Brazelton, T. R., Rossi, F. M., Keshet, G. I., and Blau, H. M. (2000). From marrow to brain: expression of neuronal phenotypes in adult mice. Science 290, 1775–1779.

40. Mezey, E., Chandross, K. J., Harta, G., Maki, R. A., and McKercher, S. R. (2000). Turning blood into brain: cells bearing neuronal antigens generated in vivo from bone marrow. Science 290, 1779–1782.

41. Sanchez-Ramos, J., Song, S., Cardozo-Pelaez, F., et al. (2000). Adult bone marrow stromal cells differentiate into neural cells in vitro. Exp Neurol 164, 247–256.

42. Woodbury, D., Schwarz, E. J., Prockop, D. J., and Black, I. B. (2000). Adult rat and human bone marrow stromal cells differentiate into neurons. J Neurosci Res 61, 364–370.

43. Eglitis, M. A., and Mezey, E. (1997). Hematopoietic cells differentiate into both microglia and macroglia in the brains of adult mice. Proc Natl Acad Sci U S A 94, 4080–4085.

44. Kopen, G. C., Prockop, D. J., and Phinney, D. G. (1999). Marrow stromal cells migrate throughout forebrain and cerebellum, and they differentiate into astrocytes after injection into neonatal mouse brains. Proc Natl Acad Sci U S A 96, 10,711–10,716.

45. Krause, D. S., Theise, N. D., Collector, M. I., et al. (2001). Multi-organ, multi-lineage engraftment by a single bone marrow-derived stem cell. Cell 105, 369–377.

46. Galli, R., Borello, U., Gritti, A., et al. (2000). Skeletal myogenic potential of human and mouse neural stem cells. Nat Neurosci 3, 986–991.

47. Jackson, K. A., Mi, T., and Goodell, M. A. (1999). Hematopoietic potential of stem cells isolated from murine skeletal muscle. Proc Natl Acad Sci U S A 96, 14,482–14,486.

48. Ferrari, G., Stornaiuolo, A., and Mavilio, F. (2001). Bone-marrow transplantation failure to correct murine muscular dystrophy. Nature 411, 1014–1015.

49. Morshead, C. M., Benveniste, P., Iscove, N. N., and van der Kooy, D. (2002). Hematopoietic competence is a rare property of neural stem cells that may depend on genetic and epigenetic alterations. Nat Med 8, 268–273.

50. Terada, N., Hamazaki, T., Oka, M., et al. (2002). Bone marrow cell adopt the phenotype of other cells by spontaneous cell fusion [advance on-line publication]. Nature.

51. Ying, Q., Nichols, J., Evans, E. P., and Smith, A. G. (2002). Changing potency spontaneous fusion [advance on-line publication]. Nature.

52. Koller, M. R., Manchel, I., and Palsson, B. O. (1997). Importance of parenchymal:stromal cell ratio for the ex vivo reconstitution of human hematopoiesis. Stem Cells 15, 305–313.

53. Koller, M. R., Bender, J. G., Miller, W. M., and Papoutsakis, E. T. (1993). Expansion of primitive human hematopoietic progenitors in a perfusion bioreactor system with IL-3, IL-6, and stem cell factor. Biotechnology (N Y) 11, 358–363.

54. Zandstra, P. W., Jervis, E., Haynes, C. A., Kilburn, D. G., Eaves, C. J., and Piret, J. M. (1999). Concentration-dependent internalization of a cytokine/

cytokine receptor complex in human hematopoietic cells. Biotechnol Bioeng 63, 493–501.

55. Zandstra, P. W., Lauffenburger, D. A., and Eaves, C. J. (2000). A ligand-receptor signaling threshold model of stem cell differentiation control: a biologically conserved mechanism applicable to hematopoiesis. Blood 96(4), 1215–1222.

56. Zandstra, P. W., Conneally, E., Petzer, A. L., Piret, J. M., and Eaves, C. J. (1997). Cytokine manipulation of primitive human hematopoietic cell self-renewal. Proc Natl Acad Sci U S A 94, 4698–4703.

57. Koller, M. R., Oxender, M., Brott, D. A., and Palsson, B. O. (1996). flt-3 ligand is more potent than c-kit ligand for the synergistic stimulation of ex vivo hema-topoietic cell expansion. J Hematother 5, 449–459.

58. Audet, J., Miller, C. L., Rose-John, S., Piret, J. M., and Eaves, C. J. (2001). Distinct role of gp130 activation in promoting self-renewal divisions by mitogenically stimulated murine hematopoietic stem cells. Proc Natl Acad Sci U S A 98, 1757–1762.

59. Petzer, A. L., Hogge, D. E., Landsdorp, P. M., Reid, D. S., and Eaves, C. J. (1996). Self-renewal of primitive human hematopoietic cells (long-term-culture-initiating cells) in vitro and their expansion in defined medium. Proc Natl Acad Sci U S A 93, 1470–1474.

60. Madlambayan, G. J., Rogers, I., Casper, R. F., and Zandstra, P. W. (2001). Controlling culture dynamics for the expansion of hematopoietic stem cells. J Hematother Stem Cell Res 10(4), 481–492.

61. Audet, J., Zandstra, P. W., Eaves, C. J., and Piret, J. M. (1998). Advances in hematopoietic stem cell culture. Curr Opin Biotechnol 9, 146–151.

62. Yagi, M., Ritchie, K. A., Sitnicka, E., Storey, C., Roth, G. J., and Bartelmez, S. (1999). Sustained ex vivo expansion of hematopoietic stem cells mediated by thrombopoietin. Proc Natl Acad Sci U S A 96, 8126–8131.

63. Bodine, D. M., Seidel, N. E., Zsebo, K. M., and Orlic, D. (1993). In vivo admin-istration of stem cell factor to mice increases the absolute number of pluripotent hematopoietic stem cells. Blood 82, 445–455.

64. Li, C. L., and Johnson, G. R. (1994). Stem cell factor enhances the survival but not the self-renewal of murine hematopoietic long-term repopulating cells. Blood 84, 408–414.

65. Petzer, A. L., Zandstra, P. W., Piret, J. M., and Eaves, C. J. (1996). Differential cytokine effects on primitive (CD34$^+$CD38$^-$) human hematopoietic cells: novel responses to Flt3-ligand and thrombopoietin. J Exp Med 183(6), 2551–2558.

66. Audet, J., Miller, C. L., Rose-John, S., Piret, J. M., and Eaves, C. J. (2001). Distinct role of gp130 activation in promoting self-renewal divisions by mitogenically stimulated murine hematopoietic stem cells. Proc Natl Acad Sci U S A 98, 1757–1762.

67. Gritti, A., Parati, E. A., Cova, L., et al. (1996). Multipotential stem cells from the adult mouse brain proliferate and self-renew in response to basic fibro-blast growth factor. J Neurosci 16, 1091–1100.

68. Reynolds, B. A., and Weiss, S. (1992). Generation of neurons and astrocytes

from isolated cells of the adult mammalian central nervous system. Science 255, 1707–1710.

69. Reynolds, B. A., Tetzlaff, W., and Weiss, S. (1992). A multipotent EGF-responsive striatal embryonic progenitor cell produces neurons and astrocytes. J Neurosci 12, 4565–4574.

70. Gensburger, C., Labourdette, G., and Sensenbrenner, M. (1987). Brain basic fibroblast growth factor stimulates the proliferation of rat neuronal precursor cells in vitro. FEBS Lett 217, 1–5.

71. Richards, L. J., Kilpatrick, T. J., and Bartlett, P. F. (1992). De novo generation of neuronal cells from the adult mouse brain. Proc Natl Acad Sci U S A 89, 8591–8595.

72. Ray, J., Peterson, D. A., Schinstine, M., and Gage, F. H. (1993). Proliferation, differentiation, and long-term culture of primary hippocampal neurons. Proc Natl Acad Sci U S A 90, 3602–3606.

73. Kallos, M. S., and Behie, L. A. (1999). Inoculation and growth conditions for high-cell-density expansion of mammalian neural stem cells in suspension bioreactors. Biotechnol Bioeng 63, 473–483.

74. Toma, J. G., Akhavan, M., Fernandes, K. J., et al. (2001). Isolation of multi-potent adult stem cells from the dermis of mammalian skin. Nat Cell Biol 3, 778–784.

75. Carpenter, M. K., Cui, X., Hu, Z. Y., et al. (1999). In vitro expansion of a multipotent population of human neural progenitor cells. Exp Neurol 158, 265–278.

76. Rathjen, P. D., Toth, S., Willis, A., Heath, J. K., and Smith, A. G. (1990). Differentiation inhibiting activity is produced in matrix-associated and diffusible forms that are generated by alternate promoter usage. Cell 62, 1105–1114.

77. Neildez-Nguyen, T. M., Wajcman, H., Marden, M. C., et al. (2002). Human erythroid cells produced ex vivo at large scale differentiate into red blood cells in vivo. Nat Biotechnol 20, 467–472.

78. Fambrough, D., McClure, K., Kazlauskas, A., and Lander, E. S. (1999). Diverse signaling pathways activated by growth factor receptors induce broadly overlapping, rather than independent, sets of genes. Cell 97, 727–741.

79. Billia, F., Barbara, M., McEwen, J., Trevisan, M., and Iscove, N. N. (2001). Resolution of pluripotential intermediates in murine hematopoietic differentiation by global complementary DNA amplification from single cells: confirmation of assignments by expression profiling of cytokine receptor transcripts. Blood 97, 2257–2268.

80. Brail, L. H., Jang, A., Billia, F., Iscove, N. N., Klamut, H. J., and Hill, R. P. (1999). Gene expression in individual cells: analysis using global single cell reverse transcription polymerase chain reaction (GSC RT-PCR). Mutat Res 406, 45–54.

81. Lauffenburger, D. A., and Linderman, J. J. (1996). Receptors: Models for Binding, Trafficking, and Signaling. New York: Oxford University Press.

82. Starr, R., Willson, T. A., Viney, E. M., et al. (1997). A family of cytokine-inducible inhibitors of signalling. Nature 387, 917–921.

83. Lumelsky, N., Blondel, O., Laeng, P., Velasco, I., Ravin, R., and McKay, R.

(2001). Differentiation of embryonic stem cells to insulin-secreting structures similar to pancreatic islets. Science 292, 1389–1394.

84. Pluen, A., Boucher, Y., Ramanujan, S., et al. (2001). Role of tumor–host interactions in interstitial diffusion of macromolecules: cranial vs subcutaneous tumors. Proc Natl Acad Sci U S A 98, 4628–4633.

85. Wittrup, K. D. (2001). Protein engineering by cell-surface display. Curr Opin Biotechnol 12, 395–399.

86. Saga, T., Neumann, R. D., Heya, T., et al. (1995). Targeting cancer micrometastases with monoclonal antibodies: a binding-site barrier. Proc Natl Acad Sci U S A 92, 8999–9003.

87. Fallon, E. M., Liparoto, S. F., Lee, K. J., Ciardelli, T. L., and Lauffenburger, D. A. (2000). Increased endosomal sorting of ligand to recycling enhances potency of an interleukin-2 analog. J Biol Chem 275, 6790–6797.

88. Zandstra, P. W., Le, H. V., Daley, G. Q., Griffith, L. G., and Lauffenburger, D. A. (2000). Leukemia inhibitory factor (LIF) concentration modulates embryonic stem cell self-renewal and differentiation independently of proliferation. Biotechnol Bioeng 69, 607–617.

89. Lauffenburger, D. A., Fallon, E. M., and Haugh, J. M. (1998). Scratching the (cell) surface: cytokine engineering for improved ligand/receptor trafficking dynamics. Chem Biol 5, R257–R263.

90. Fallon, E. M., and Lauffenburger, D. A. (2000). Computational model for effects of ligand/receptor binding properties on interleukin-2 trafficking dynamics and T cell proliferation response. Biotechnol Prog 16, 905–916.

91. Maynard, J., and Georgiou, G. (2000). Antibody engineering. Annu Rev Biomed Eng 2, 339–376.

92. Chirumamilla, R. R., Muralidhar, R., Marchant, R., and Nigam, P. (2001). Improving the quality of industrially important enzymes by directed evolution. Mol Cell Biochem 224, 159–168.

93. French, A. R., Tadaki, D. K., Niyogi, S. K., and Lauffenburger, D. A. (1995). Intracellular trafficking of epidermal growth factor family ligands is directly influenced by the pH sensitivity of the receptor/ligand interaction. J Biol Chem 270, 4334–4340.

94. MacBeath, G., and Schreiber, S. L. (2000). Printing proteins as microarrays for high-throughput function determination. Science 289, 1760–1763.

95. Kuhl, P. R., and Griffith-Cima, L. G. (1996). Tethered epidermal growth factor as a paradigm for growth factor-induced stimulation from the solid phase. Nat Med 2, 1022–1027.

96. Wehrle-Haller, B., and Weston, J. A. (1995). Soluble and cell-bound forms of steel factor activity play distinct roles in melanocyte precursor dispersal and survival on the lateral neural crest migration pathway. Development 121, 731–742.

97. Kurosawa, K., Miyazawa, K., Gotoh, A., et al. (1996). Immobilized anti-KIT monoclonal antibody induces ligand-independent dimerization and activation of Steel factor receptor: biologic similarity with membrane-bound form of Steel factor rather than its soluble form. Blood 87, 2235–2243.

98. Miyazawa, K., Williams, D. A., Gotoh, A., Nishimaki, J., Broxmeyer, H. E., and Toyama, K. (1995). Membrane-bound Steel factor induces more persistent tyrosine kinase activation and longer life span of c-kit gene-encoded protein than its soluble form. Blood 85, 641–649.
99. Doheny, J. G., Jervis, E. J., Guarna, M. M., Humphries, R. K., Warren, R. A., and Kilburn, D. G. (1999). Cellulose as an inert matrix for presenting cytokines to target cells: production and properties of a stem cell factor-cellulose-binding domain fusion protein. Biochem J 339(Pt. 2), 429–434.
100. Nakashima, T., Kobayashi, Y., Yamasaki, S., et al. (2000). Protein expression and functional difference of membrane-bound and soluble receptor activator of NF-kappaB ligand: modulation of the expression by osteotropic factors and cytokines. Biochem Biophys Res Commun 275, 768–775.
101. Barille, S., Collette, M., Thabard, W., Bleunven, C., Bataille, R., and Amiot, M. (2000). Soluble IL-6R alpha upregulated IL-6, MMP-1 and MMP-2 secretion in bone marrow stromal cells. Cytokine 12, 1426–1429.
102. Slater, M. (1996). Dynamic interactions of the extracellular matrix. Histol Histopathol 11, 175–180.
103. Flaumenhaft, R., and Rifkin, D. B. (1991). Extracellular matrix regulation of growth factor and protease activity. Curr Opin Cell Biol 3, 817–823.
104. Rafii, S., Mohle, R., Shapiro, F., Frey, B. M., and Moore, M. A. (1997). Regulation of hematopoiesis by microvascular endothelium. Leuk Lymphoma 27, 375–386.
105. Friedenstein, A. J., Gorskaja, J. F., and Kulagina, N. N. (1976). Fibroblast precursors in normal and irradiated mouse hematopoietic organs. Exp Hematol 4, 267–274.
106. Prosper, F., and Verfaillie, C. M. (2001). Regulation of hematopoiesis through adhesion receptors. J Leukoc Biol 69, 307–316.
107. Ploemacher, R. E., Mayen, A. E., De Koning, A. E., Krenacs, T., and Rosendaal, M. (2000). Hematopoiesis: gap junction intercellular communication is likely to be involved in regulation of stroma-dependent proliferation of hemopoietic stem cells. Hematology 5, 133–147.
108. Potten, C. S., and Loeffler, M. (1987). A comprehensive model of the crypts of the small intestine of the mouse provides insight into the mechanisms of cell migration and the proliferation hierarchy. J Theor Biol 127, 381–391.
109. Potten, C. S. (1998). Stem cells in gastrointestinal epithelium: numbers, characteristics and death. Philos Trans R Soc Lond B Biol Sci 353, 821–830.
110. Evans, G. S., and Potten, C. S. (1991). Stem cells and the elixir of life. Bioessays 13, 135–138.
111. Potten, C. S. (1991). The role of stem cells in the regeneration of intestinal crypts after cytotoxic exposure. Prog Clin Biol Res 369, 155–171.
112. Paulus, U., Potten, C. S., and Loeffler, M. (1992). A model of the control of cellular regeneration in the intestinal crypt after perturbation based solely on local stem cell regulation. Cell Prolif 25, 559–578.
113. Li, Y. Q., Roberts, S. A., Paulus, U., Loeffler, M., and Potten, C. S. (1994). The crypt cycle in mouse small intestinal epithelium. J Cell Sci 107 (Pt 12), 3271–3279.

114. Potten, C. S. (1986). Cell cycles in cell hierarchies. Int J Radiat Biol Relat Stud Phys Chem Med 49, 257–278.
115. Loeffler, M., Birke, A., Winton, D., and Potten, C. (1993). Somatic mutation, monoclonality and stochastic models of stem cell organization in the intestinal crypt. J Theor Biol 160, 471–491.
116. Marshman, E., Booth, C., and Potten, C. S. (2002). The intestinal epithelial stem cell. Bioessays 24, 91–98.
117. Simmons, P. J., Levesque, J. P., and Zannettino, A. C. (1997). Adhesion molecules in haemopoiesis. Baillieres Clin Haematol 10, 485–505.
118. Gassmann, M., Fandrey, J., Bichet, S., et al. (1996). Oxygen supply and oxygen-dependent gene expression in differentiating embryonic stem cells. Proc Natl Acad Sci U S A 93, 2867–2872.
119. Sauer, H., Rahimi, G., Hescheler, J., and Wartenberg, M. (2000). Role of reactive oxygen species and phosphatidylinositol 3-kinase in cardiomyocyte differentiation of embryonic stem cells. FEBS Lett 476, 218–223.
120. McAdams, T. A., Miller, W. M., and Papoutsakis, E. T. (1997). Variations in culture pH affect the cloning efficiency and differentiation of progenitor cells in ex vivo haemopoiesis. Br J Haematol 97, 889–895.
121. Wang, G. L., Jiang, B. H., Rue, E. A., and Semenza, G. L. (1995). Hypoxia-inducible factor 1 is a basic-helix-loop-helix-PAS heterodimer regulated by cellular O2 tension. Proc Natl Acad Sci U S A 92, 5510–5514.
122. Broxmeyer, H. E., Cooper, S., Lu, L., Miller, M. E., Langefeld, C. D., and Ralph, P. (1990). Enhanced stimulation of human bone marrow macrophage colony formation in vitro by recombinant human macrophage colony-stimulating factor in agarose medium and at low oxygen tension. Blood 76, 323–329.
123. Bradley, T. R., Hodgson, G. S., and Rosendaal, M. (1978). The effect of oxygen tension on haemopoietic and fibroblast cell proliferation in vitro. J Cell Physiol 97, 517–522.
124. Cipolleschi, M. G., Dello Sbarba, P., and Olivotto, M. (1993). The role of hypoxia in the maintenance of hematopoietic stem cells. Blood 82, 2031–2037.
125. Ivanovic, Z., Belloc, F., Faucher, J. L., Cipolleschi, M. G., Praloran, V., and Dello Sbarba, P. (2002). Hypoxia maintains and interleukin-3 reduces the pre-colony-forming cell potential of dividing CD34(+) murine bone marrow cells. Exp Hematol 30, 67–73.
126. Koller, M. R., Emerson, S. G., and Palsson, B. O. (1993). Large-scale expansion of human stem and progenitor cells from bone marrow mononuclear cells in continuous perfusion cultures. Blood 82, 378–384.
127. Palsson, B. O., Paek, S. H., Schwartz, R. M., et al. (1993). Expansion of human bone marrow progenitor cells in a high cell density continuous perfusion system. Biotechnology (N Y) 11, 368–372.
128. Csete, M., Walikonis, J., Slawny, N., et al. (2001). Oxygen-mediated regulation of skeletal muscle satellite cell proliferation and adipogenesis in culture. J Cell Physiol 189, 189–196.
129. Lennon, D. P., Edmison, J. M., and Caplan, A. I. (2001). Cultivation of rat marrow-derived mesenchymal stem cells in reduced oxygen tension: effects on in vitro and in vivo osteochondrogenesis. J Cell Physiol 187, 345–355.

130. Li, C., and Xu, Q. (2000). Mechanical stress-initiated signal transductions in vascular smooth muscle cells. Cell Signal 12, 435–445.
131. Burger, E. H., and Klein-Nulen, J. (1999). Responses of bone cells to biomechanical forces in vitro. Adv Dent Res 13, 93–98.
132. Fisher, A. B., Chien, S., Barakat, A. I., and Nerem, R. M. (2001). Endothelial cellular response to altered shear stress. Am J Physiol Lung Cell Mol Physiol 281, L529–L533.
133. Ingber, D. E. (1997). Integrins, tensegrity, and mechanotransduction. Gravit Space Biol Bull 10, 49–55.
134. Pittenger, M. F., Mackay, A. M., Beck, S. C., et al. (1999). Multilineage potential of adult human mesenchymal stem cells. Science 284, 143–147.
135. Till, J. E., McCulloch, E. A., and Siminovitch, L. (1964). A stochastic model of stem cell proliferation, based on the growth of spleen colony-forming cells. Proc Natl Acad Sci U S A 51, 29.
136. Viswanathan, S., Benatar, T., Rose-John, S., Lauffenburger, D. A., and Zandstra, P. W. (2002). Ligand/receptor signaling threshold (LIST) model accounts for gp130-mediated embryonic stem cell self-renewal responses to LIF and HIL-6. Stem Cells 20, 119–138.
137. Mayani, H., Dragowska, W., and Lansdorp, P. M. (1993). Cytokine-induced selective expansion and maturation of erythroid vs myeloid progenitors from purified cord blood precursor cells. Blood 81, 3252–3258.
138. Fairbairn, L. J., Cowling, G. J., Reipert, B. M., and Dexter, T. M. (1993). Suppression of apoptosis allows differentiation and development of a multipotent hemopoietic cell line in the absence of added growth factors. Cell 74, 823–832.
139. Hu, M., Krause, D., Greaves, M., et al. (1997). Multilineage gene expression precedes commitment in the hemopoietic system. Genes Dev 11, 774–785.
140. Liu, L., and Roberts, R. M. (1996). Silencing of the gene for the beta subunit of human chorionic gonadotropin by the embryonic transcription factor Oct-3/4. J Biol Chem 271, 16,683–16,689.
141. Enver, T., and Greaves, M. (1998). Loops, lineage, and leukemia. Cell 94, 9–12.
142. Madras, N., Gibbs, A. L., Zhou, H., Zandstra, P. W., and Aubin, J. E. (2002). Modeling stem cell development by retrospective analysis of gene expression profiles in single progenitor-derived colonies. Stem Cells, 20(3), 230–240.
143. Mayani, H., Dragowska, W., and Lansdorp, P. M. (1993). Lineage commitment in human hemopoiesis involves asymmetric cell division of multipotent progenitors and does not appear to be influenced by cytokines. J Cell Physiol 157, 579–586.
144. Ogawa, M. (1993). Differentiation and proliferation of hematopoietic stem cells. Blood 81, 2844–2853.
145. Humphries, R. K., Eaves, A. C., and Eaves, C. J. (1981). Self-renewal of hemopoietic stem cells during mixed colony formation in vitro. Proc Natl Acad Sci U S A 78, 3629–3633.
146. Kimmel, M., and Axelrod, D. E. (1991). Unequal cell division, growth regulation and colony size of mammalian cells: a mathematical model and analysis of experimental data. J Theor Biol 153, 157–180.

147. Lemischka, I. (2001). Stem cell dogmas in the genomics era. Rev Clin Exp Hematol 5, 15–25.

148. Francis, G. E., and Leaning, M. S. (1985). Stochastic model of human granulocyte-macrophage progenitor cell proliferation and differentiation. I. Setting up the model. Exp Hematol 13, 92–98.

149. Nakahata, T., Gross, A. J., and Ogawa, M. (1982). A stochastic model of self-renewal and commitment to differentiation of the primitive hemopoietic stem cells in culture. J Cell Physiol 113, 455–458.

150. Le Douarin, N. M., and Dupin, E. (1993). Cell lineage analysis in neural crest ontogeny. J Neurobiol 24, 146–161.

151. Yakovlev, A. Y., Mayer-Proschel, M., and Noble, M. (1998). A stochastic model of brain cell differentiation in tissue culture. J Math Biol 37, 49–60.

152. Boucher, K., Yakovlev, A. Y., Mayer-Proschel, M., and Noble, M. (1999). A stochastic model of temporally regulated generation of oligodendrocytes in cell culture. Math Biosci 159, 47–78.

153. Fokas, A. S., Keller, J. B., and Clarkson, B. D. (1991). Mathematical model of granulocytopoiesis and chronic myelogenous leukemia. Cancer Res 51, 2084–2091.

154. Peng, C. A., Koller, M. R., and Palsson, B. O. (1996). Unilineage model of hematopoiesis predicts self-renewal of stem and progenitor cells based on ex vivo growth data. Biotechnol Bioeng 52, 24–33.

155. Milton, J. G., and Mackey, M. C. (1989). Periodic haematological diseases: mystical entities or dynamical disorders? J R Coll Physicians Lond 23, 236–241.

156. Haurie, C., Dale, D. C., and Mackey, M. C. (1998). Cyclical neutropenia and other periodic hematological disorders: a review of mechanisms and mathematical models. Blood 92, 2629–2640.

157. Sherley, J. L., Stadler, P. B., and Stadler, J. S. (1995). A quantitative method for the analysis of mammalian cell proliferation in culture in terms of dividing and non-dividing cells. Cell Prolif 28, 137–144.

158. Yakovlev, A. Y., Boucher, K., Mayer-Proschel, M., and Noble, M. (1998). Quantitative insight into proliferation and differentiation of oligodendrocyte type 2 astrocyte progenitor cells in vitro. Proc Natl Acad Sci U S A 95, 14,164–14,167.

159. Haugh, J. M., Schooler, K., Wells, A., Wiley, H. S., and Lauffenburger, D. A. (1999). Effect of epidermal growth factor receptor internalization on regulation of the phospholipase C-gamma1 signaling pathway. J Biol Chem 274, 8958–8965.

160. Bhalla, U. S., and Iyengar, R. (1999). Emergent properties of networks of biological signaling pathways. Science 283, 381–387.

161. Weng, G., Bhalla, U. S., and Iyengar, R. (1999). Complexity in biological signaling systems. Science 284, 92–96.

162. Schoeberl, B., Eichler-Jonsson, C., Gilles, E. D., and Muller, G. (2002). Computational modeling of the dynamics of the MAP kinase cascade activated by surface and internalized EGF receptors. Nat Biotechnol 20, 370–375.

163. Zandstra, P. W., and Nagy, A. (2001). Stem cell bioengineering. Annu Rev Biomed Eng 3, 275–305.
164. Zandstra, P. W., Eaves, C. J., and Piret, J. M. (1994). Expansion of hematopoietic progenitor cell populations in stirred suspension bioreactors of normal human bone marrow cells. Biotechnology (NY) 12(9), 909–914.
165. Baksh, D., Davies, J. E., and Zandstra, P. W. (2003). Adult human bone marrow-derived progenitor cells are capable of adhesion-independent survival and expansion. Exp Hematol 31(8) 723–732.
166. Cashman, J. D., Clark-Lewis, I., Eaves, A. C., and Eaves, C. J. (1999). Differentiation stage-specific regulation of primitive human hematopoietic progenitor cycling by exogenous and endogenous inhibitors in an in vivo model. Blood 94, 3722–3729.
167. Furusawa, C., and Kaneko, K. (2000). Origin of complexity in multicellular organisms. Phys Rev Lett 84, 6130–6133.
168. Sandstrom, C. E., Bender, J. G., Papoutsakis, E. T., and Miller, W. M. (1995). Effects of CD34+ cell selection and perfusion on ex vivo expansion of peripheral blood mononuclear cells. Blood 86, 958–970.
169. Hammond, T. G., Lewis, F. C., Goodwin, T. J., et al. (1999). Gene expression in space. Nat Med 5, 359.
170. Molnar, G., Schroedl, N. A., Gonda, S. R., and Hartzell, C. R. (1997). Skeletal muscle satellite cells cultured in simulated microgravity. In Vitro Cell Dev Biol Anim 33, 386–391.
171. Freed, L. E., Langer, R., Martin, I., Pellis, N. R., and Vunjak-Novakovic, G. (1997). Tissue engineering of cartilage in space. Proc Natl Acad Sci U S A 94, 13,885–13,890.
172. Ingram, M., Techy, G. B., Saroufeem, R., et al. (1997). Three-dimensional growth patterns of various human tumor cell lines in simulated microgravity of a NASA bioreactor. In Vitro Cell Dev Biol Anim 33, 459–466.
173. Margolis, L., Hatfill, S., Chuaqui, R., et al. (1999). Long term organ culture of human prostate tissue in a NASA-designed rotating wall bioreactor. J Urol 161, 290–297.
174. Mitteregger, R., Vogt, G., Rossmanith, E., and Falkenhagen, D. (1999). Rotary cell culture system (RCCS): a new method for cultivating hepatocytes on microcarriers. Int J Artif Organs 22, 816–822.
175. Bhatia, S. N., Toner, M., Tompkins, R. G., and Yarmush, M. L. (1994). Selective adhesion of hepatocytes on patterned surfaces. Ann N Y Acad Sci 745, 187–209.
176. Bhatia, S. N., Yarmush, M. L., and Toner, M. (1997). Controlling cell interactions by micropatterning in co-cultures: hepatocytes and 3T3 fibroblasts. J Biomed Mater Res 34, 189–199.
177. Bhatia, S. N., Balis, U. J., Yarmush, M. L., and Toner, M. (1998). Microfabrication of hepatocyte/fibroblast co-cultures: role of homotypic cell interactions. Biotechnol Prog 14, 378–387.
178. Bhatia, S. N., Balis, U. J., Yarmush, M. L., and Toner, M. (1999). Effect of cell-cell interactions in preservation of cellular phenotype: cocultivation of hepatocytes and nonparenchymal cells. FASEB J 13, 1883–1900.

179. Whitesides, G. M., Ostuni, E., Takayama, S., Jiang, X., and Ingber, D. E. (2001). Soft lithography in biology and biochemistry. Annu Rev Biomed Eng 3, 335–373.
180. Folch, A., and Toner, M. (2000). Microengineering of cellular interactions. Annu Rev Biomed Eng 2, 227–256.
181. Takayama, S., Ostuni, E., LeDuc, P., Naruse, K., Ingber, D. E., and Whitesides, G. M. (2001). Subcellular positioning of small molecules. Nature 411, 1016.
182. Powers, M. J., and Griffith, L. G. (1998). Adhesion-guided in vitro morphogenesis in pure and mixed cell cultures. Microsc Res Tech 43, 379–384.
183. Griffith, L. G., Wu, B., Cima, M. J., Powers, M. J., Chaignaud, B., and Vacanti, J. P. (1997). In vitro organogenesis of liver tissue. Ann N Y Acad Sci 831, 382–397.
184. Hamazaki, T., Iiboshi, Y., Oka, M., et al. (2001). Hepatic maturation in differentiating embryonic stem cells in vitro. FEBS Lett 497, 15–19.
185. Caldwell, J., Locey, B., Clarke, M. F., Emerson, S. G., and Palsson, B. O. (1991). Influence of medium exchange schedules on metabolic, growth, and GM-CSF secretion rates of genetically engineered NIH-3T3 cells. Biotechnol Prog 7, 1–8.

16

Stem Cells As Common Ancestors

Somatic Cell Phylogenies From
Somatic Sequence Alterations

Darryl Shibata

1. INTRODUCTION

Stem cell definitions vary and often reflect how their properties are measured *(1–4)*. Here, it is illustrated that stem cells can also be defined phylogenetically as common ancestors or "mothers" of cell groups. This approach is essentially fate mapping, but uses sequences instead of histologically visible markers to trace ancestry. An advantage of using sequences to trace somatic cell fates is the lack of a requirement for prior experimental intervention. Unlike definitions based on potential stem cell behaviors in experimental settings, a phylogenetic approach reconstructs how stem cells behave in unmanipulated intact organisms.

Sequence comparisons are commonly employed to reconstruct phylogeny among species, populations, and individuals (Fig. 1). Sequences obtained from present-day organisms are analyzed to infer ancestry. Phylogenetic trees consist of branches and nodes, with a node representing a last common ancestor. These trees may be complex, with multiple branches and nodes. Moving backward in time, branch numbers become smaller as they coalesce to more remote common ancestors. Sequence comparisons between species eventually infer an ultimate common ancestor that existed billions of years ago.

Somatic cells in individual metazoan organisms can be organized into similar phylogenetic trees, with the zygote representing the ultimate somatic cell common ancestor (Fig. 1). Early in life, these trees reflect development, and later in life, these trees reflect aging. Developmental fates can be visibly dissected because undifferentiated cells become increasingly numerous and differentiated. However, microscopic fates of adult cells are more uncertain because it is difficult to distinguish between morphologically similar cells. Random sequence errors likely occur independently in different cells during

From: *Adult Stem Cells*
Edited by: K. Turksen © Humana Press Inc., Totowa, NJ

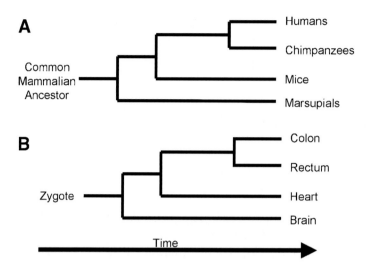

Fig. 1. Phylogenetic trees. (**A**) A simple tree for mammals. (**B**) A simple tree for somatic human cells. Phylogenetic relationships between individual cells can be inferred by comparing sequences between somatic cells.

aging, and these sequence difference may potentially distinguish between the fates of otherwise morphologically identical cells.

2. DEFINITIONS

A phylogenetic approach reconstructs lineages through time and defines "stem" cells in terms of branches and nodes that produce related or clonal cell groups. Different types of stem cell renewal will yield different shaped trees. Nodes or common ancestors are defined as *stem cells*, and branches connecting nodes are *stem cell lineages* (Fig. 2). Nodes that represent stem cells are usually connected to present-day cells by long branches, reflecting the persistence of stem cell lineages over time. Branch lengths are proportional to number of sequence differences between cells, which should increase with the number of divisions that separate cells from a common ancestor or stem cell.

A phylogenetic definition of stem cells may differ from functional definitions of stem cells. Instead of a physical description or isolation of a stem cell, a phylogenetic analysis is inferential and often identifies stem cells (such as the zygote) that no longer physically exist. Many of the stem cells identified by sequences may not function as stem cells as defined by certain experimental criteria. Instead of a prospective test of stem cell potential, a phylogenetic analysis reconstructs ancestors to current cells. These remote common ancestors fit a retrospective definition of stem cells because they produced all the cell types of a clonal group. Of note, quiescent cells with stem cell potentials will not be easily detected by a phylogenetic analysis because they produce few progeny.

3. EXPERIMENTAL CONSIDERATIONS

3.1. Advantages of a Phylogenetic Approach

The primary advantage of a phylogenetic approach is that systematic studies of human stem cells are possible because no prior experimental manipulations are required. Tissue histories are reconstructed from sequence differences that inherently accumulate during life. General shapes of somatic human cell trees are already known. For example, the last common ancestor between two epithelial cells on different hands likely existed before birth. Such known relationships can be used to calibrate the molecular clocks essential for a phylogenetic analysis.

3.2. Disadvantages of a Phylogenetic Approach

The use of sequences to reconstruct the past is controversial and complicated. It requires sophisticated quantitative methods that are well developed, but poorly understood by the mathematically challenged. The studies of my group represent one of the first attempts to apply these controversial methods to reconstruct somatic cell phylogenies *(5)*. To appeal to a general audience, the mathematical aspects of this approach (*see* ref. 5 for details) are largely ignored in this review.

Aside from general controversies over phylogenetic techniques, the primary disadvantage for somatic cell studies is the short lifetimes of many organisms relative to the high genetic fidelity of normal cells. Somatic alterations are needed to reconstruct trees, and few alterations will accumulate in short-lived organisms. Therefore, a phylogenetic approach is most practical for long-lived organisms like humans because, over decades, alterations may accumulate even with low error rates.

Another general problem for phylogenetic studies is that multiple trees are often possible with the same data. This same problem can be expected for somatic cell trees. However, instead of relying solely on sequence data to reconstruct somatic cell trees, sequence data may be used to distinguish between stem cell renewal mechanisms deduced from other experimental systems. For example, one type of tree is expected with stem cell niches, and another very different tree is expected with immortal stem cells (Fig. 2).

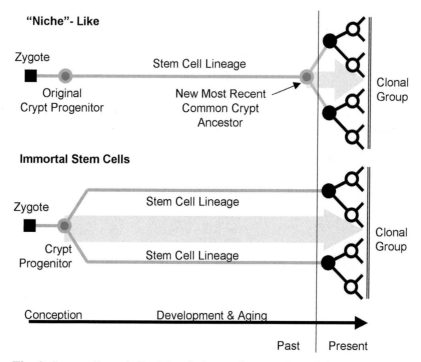

Fig. 2. Stem cells as defined by phylogenetic trees. Branch lengths are proportional to times between last common ancestors. All somatic cells trace their origins to the zygote (*black square*). Small clonal cell groups such as single colon crypts are maintained by multiple stem cells (*black circles*). The last common ancestor between immortal stem cells is a common progenitor (*gray circle*) that existed around birth. Therefore, immortal stem cells produce multiple long-lived stem cell lineages. In contrast, the last common ancestor for niche stem cells will be more recent because random loss with replacement will homogenize a niche. Therefore, niche stem cells produce a single long-lived lineage. Times to a last most recent common ancestor (*gray arrows*) are more recent for niches compared with immortal stem cells. Note that only present-day cells (those cells on the right connected by *black lines*) are visible. Ancestors (represented by *gray lines*) can be inferred by comparing sequences between current cells, with greater sequence differences expected for clonal groups maintained by immortal stem cells compared to niches.

3.3. Requirements

There are three general basic experimental requirements to reconstruct somatic stem cell lineages. The first is the ability to identify clonal cell groups related to single common ancestors (Fig. 2). This requirement facilitates subsequent quantitative analysis because all sequences obtained from

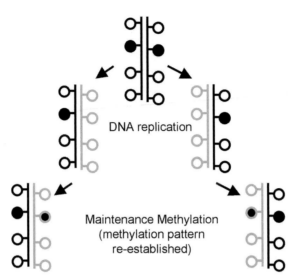

Fig. 3. Methylation (*black lollipops*) is inherited. CpG sites on newly synthesized DNA strands (*gray*) are unmethylated (*open lollipops*) after replication. Hemimethylated CpG sites are methylated after DNA replication to maintain the original epigenetic pattern. Somatic cell histories may be recorded by random epigenetic errors that lead to *de novo* methylation or demethylation.

a clonal cell group should coalesce to a single ancestor. Niches are common in mammals *(2)*, and cells derived from each niche (such as a single colon crypt) are clonal cell groups.

The second requirement is measurement of a locus that accumulates alterations. Such a locus is often referred to as a *molecular clock*. Mutation rates in normal cells are low and are estimated at about 10^{-9} per base per division *(6)*. Assuming one division per day, approx 1900 yr would be required for a 50% probability of a mutation in a 1000-basepair sequence. Therefore, this approach is currently impractical for sequences because age-related somatic mutations will be extremely infrequent.

For this reason, the initial approach of my group *(5)* examined methylation changes in CpG-rich islands because epigenetic fidelity is likely to be less than genetic fidelity. Methylation in mammals occurs primarily on cytosine next to guanine, and CpG dinucleotides tend to cluster in CpG islands *(7,8)*. Methylation, like base sequences, exhibits somatic inheritance; therefore, phylogenetic approaches should also apply to methylation patterns (Fig. 3). The exact mechanisms responsible for *de novo* methylation (adding a methyl group to an unmethylated CpG), maintenance methylation (adding a methyl group to a hemimethylated C after deoxyribonucleic acid [DNA] replication), and demethylation (loss of a methyl group from a previously methylated CpG) are currently unknown *(9)*, but in general, the same methylation pattern is faithfully copied after DNA replication *(8,10)*.

Methylation of certain CpG islands occurs rapidly during development (such as with imprinting), but most CpG islands remain unmethylated *(7,8,11)*. However, some CpG islands become methylated in adult colons *(12,13)*, suggesting some methylation changes occur during normal human aging. These epigenetic changes should record somatic cell fates. Although methylation of CpG islands may be correlated with gene expression *(8)*, methylation in the studies of my group *(5)* is unlikely to confer changes in selection or phenotype because these genes are not normally expressed in colon.

Unlike histologic markers, which have only two possible states (present or absent), multiple patterns or tags are possible with methylation markers. Typically, 5 to 9 CpG sites are present on the molecular clock or locus of interest (a stretch of about 150 bases within a CpG-rich sequence), and a methylation pattern or tag is defined by the 5'-to-3' order of methylation. For a locus with 9 CpG sites, 512 tags (2^N where N = number of CpG sites) are possible. These tags can be represented as a binary code with a methylated CpG = 1, and an unmethylated CpG = 0. For example, 000000000 would be a fully unmethylated tag, and 111111111 would be a fully methylated tag.

Also, unlike histologic markers, methylation tags are dynamic, changing and recording cell fates continuously during life. Changes are postulated in my group's studies to occur stepwise (one at a time) and independently between sites (Fig. 4). Distances or the amount of drift between tags can be quantified from the absolute number of site differences between tags. For example, the differences between tag pairs 0010 and 0100 and between 1101 and 0111 are the same and equal to 2. The rate of change inferred *(5)* was 2 $\times$ 10^{-5} changes per CpG site per division, with both methylation and demethylation possible. Epigenetic fidelity at these CpG sites is about 10,000-fold less than genetic fidelity, allowing them to function as practical somatic cell molecular clocks.

The final requirement is the study of adults of long-lived organisms. The resolution of phylogenetic techniques is poor for recent events, and the study of adults increases the differences expected for different stem cell renewal patterns.

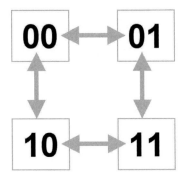

Fig. 4. Methylation tags can be represented by a binary code (0 = unmethylated, and 1 = methylated) based on the 5'-to-3' order of CpG sites. A tag with two CpG sites is illustrated. Changes are postulated to be independent between CpG sites and usually occur stepwise (one change per tag per division) because error rates are low (estimated at 2×10^{-5} per division per site). Unlike histologic markers, methylation tags have multiple states, change continuously, and require no prior experimental manipulations. More complex changes (which store more information) are possible with tags containing more of CpG sites.

4. APPLICATION TO HUMAN COLON CRYPTS

4.1. Colon Crypt Stem Cell Niches

Murine studies *(1,14)* demonstrated that intestinal crypts are maintained by small numbers of stem cells located near or at crypt bases (Fig. 5). Differentiated cells migrate upward and are lost at the surface. Turnover of differentiated epithelial cells is rapid, and essentially all cells but stem cells are lost within a week. Human colon crypts contain about 2000 cells, and individual crypts can be isolated from fresh colons *(15,16)*. These individual crypts qualify as clonal units. Methylation patterns within these crypts (Fig. 5) can be determined after bisulfite sequencing *(17)*.

Sample data are illustrated in Fig. 5. To simplify the analysis, a locus on the X chromosome called BGN was examined in a male individual. With only one X chromosome per cell, each tag represents a single cell. In this case, 24 methylation tags were randomly sampled from each crypt. There are multiple different tags in each crypt, and tags are similar within each crypt and different between crypts. The process of aging can be examined by collecting tags from multiple crypts from multiple individuals of different ages.

The short lifetimes of differentiated crypt cells (about a week) relative to methylation fidelity (2×10^{-5} per division) imply that most methylation differences represent changes acquired in the long-lived stem cells. Methyla-

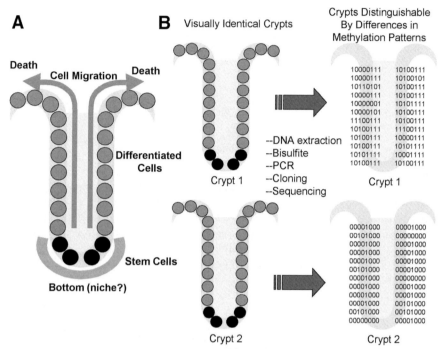

Fig. 5. Colon crypts are maintained by multiple stem cells present at or near their bases. Individual crypts can be isolated, and methylation tags can be sampled after bisulfite treatment, polymerase chain reaction (PCR), cloning of PCR products, and sequencing of individual PCR clones. In this way, methylation tags are digital representations of crypt structures and histories. Comparisons of tags between and within crypts can recreate phylogenetic trees.

tion changes do not appear to arise frequently during epithelial cell maturation because tags are similar when sampled from either the upper or the lower portions of crypts *(5)*. Therefore, and similar to histologic markers, although most crypt genomes come from differentiated cells, they essentially represent the status of stem cells. Intracrypt methylation tag differences therefore are proportional to numbers of divisions because current stem cells shared a common ancestor.

Possible models for crypt stem cells are immortal stem cells, which always divide asymmetrically, and a niche with a population-type mechanism *(1–4,18)*. These two models yield very different trees, with multiple long branches with immortal stem cells and a single long branch with a niche (Fig. 2). Immortal stem cells should persist for the lifetime of an individual; therefore, tags within a crypt should become increasingly different with age. In contrast, stem cell loss with replacement occurs in niches. Similar to stud-

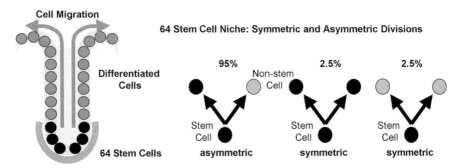

Fig. 6. A 64-stem-cell human colon crypt niche scenario consistent with a methylation tag analysis. Asymmetric division occurs 95% of the time, and symmetric division occurs 5% of the time. Symmetric division eventually homogenizes a niche such that all crypt cells are related to a new, more recent common ancestor. A new common crypt ancestor arises in this scenario about every 8.2 yr. Niches with between 4 and 512 stem cells and different percentages of symmetric division are also consistent with observed methylation tags.

ies using histologically visible markers *(1,4)*, eventually tissues maintained by niches become homogeneous for a single methylation tag because all other stem cells are lost. This replacement creates a new common crypt ancestor, which is represented phylogenetically by propagation along a single branch and a new node. Unlike histologic markers, methylation tags can record multiple crypt niche replacement cycles because replacement to homogeneity is counteracted by further tag drift.

Tag patterns collected from multiple individuals and crypts are complex, reflecting stochastic changes in methylation and, for crypt niches, stochastic stem cell loss with replacement. A quantitative approach, however, can interpret tag patterns to exclude certain crypt scenarios. Tag patterns expected with either immortal stem cells or crypt niches can be determined by computer simulation. Tag patterns such as in Fig. 5 are more consistent with crypt niches maintained by multiple stem cells *(5)*. The exact numbers of stem cells in human crypt niches are uncertain, with between 4 and 512 stem cells possible with the data of my group. One possible scenario consistent with murine studies are niches maintained by 64 stem cells, with asymmetric divisions 95% of the time and symmetric divisions 5% of the time (Fig. 6). Under this scenario, crypt niche replacement to a single, new common ancestor recurs on average every 8.2 yr (95% confidence intervals of 2.7 to 19 yr).

Differences between the 100% asymmetric division of immortal stem cells and the 95% asymmetric divisions of crypt stem cell niches are relatively small and would be difficult to detect with short observation periods. However, differences become readily apparent over the decades recorded by methylation changes in adult colons. For example, niche succession to a single common ancestor has recurred on average 10 times for each crypt in an 82-yr-old colon.

4.2. Stability of Crypt Niches

The turnover of stem cells in niches implies the physical stability of a niche depends not on the stem cells, but rather on the cells that define such niches. Crypts may divide by fission *(19)* and, possibly like their stem cells, crypt niches may also be continuously created and destroyed. Relationships between normal crypts are difficult to define based on physical criteria because they are morphologically alike. One approach to determine whether crypts are continuously created is to measure whether methylation patterns are alike in physically adjacent crypts. Similar to crypt stem cell trees (Fig. 2), if crypt niches are immortal, then adjacent crypts should have very different methylation patterns. In contrast, more closely related methylation patterns are expected if crypt niches continuously divide because adjacent crypts should be recently related.

Preliminary data suggest crypt niches are long-lived because physically adjacent crypts are no more related to each other than crypts taken more than 15 cm apart in the same colon (ref. 5 and unpublished data). In other words, the most recent common ancestor between two crypts is older than the most recent common stem cell ancestor within a crypt. The lack of methylation tag similarity between physically adjacent adult crypts suggests crypt fission has not occurred recently, or that crypts migrate after fission. Normal human colon appears to become mosaic with respect to methylation, and each adult crypt may eventually attain methylation patterns completely distinct from millions of other crypts.

4.3. Crypt Evolution (Muller's Ratchet)

Continuous crypt niche succession cycles have interesting implications for stem cell evolution. Stem cells are likely targets for neoplastic transformation, and tumor progression is usually associated with a selective advantage and visible clonal succession *(20,21)*. However, a stem cell niche is also synonymous with clonal succession except that it occurs invisibly without a change in phenotype and is not driven by mutation. This natural niche rhythm provides a mechanism other than selection to accumulate alterations passively through sequential clonal successions because the genotype of the dominant stem cell becomes the genotype of the crypt. Instead of driving clonal succession, early alterations, including those critical to tumorigenesis that do not yet confer growth advantages in normal-appearing cells, may passively hitchhike to clonal dominance along with these niche succession cycles.

For example, cells heterozygous for an APC mutation (APC+/–) are phenotypically normal. By niche succession, a single APC+/– stem cell could attain dominance and convert a niche into a tumor-prone familial polyposis-type crypt. In this case, the APC mutation hitchhiked to dominance rather conferring a growth advantage.

Crypts could become increasing "fit" if less-fit stem cells are lost during niche succession cycles. However, how alterations accumulate in small populations has been long examined by population biologists, and a somewhat intuitively opposite conclusion is that, despite selection, fitness usually decreases over time without sex. Most mutations are deleterious, and one benefit of sex may be to reduce mutation burden. A niche with its small and finite asexual population size would be prone to Muller's ratchet *(22)*.

Muller proposed that an asexual population inevitably accumulates deleterious mutations because of a ratchetlike irreversible loss of individuals with fewer mutations *(22–25)*. Muller's ratchet is exacerbated in small populations *(25)* because selection effects (positive or negative) on fixation are decreased relative to *drift* (defined as random sampling in a finite population). Niches decrease the onerous requirement that each new mutation associated with tumor progression confers a selective growth advantage and potentially fix even deleterious alterations that only later contribute to a tumor phenotype. The physical partition of stem cells into small niches may overall decrease the risk of cancer *(20)* because of the fitness decline predicted by Muller's ratchet, although asexual populations sometimes become more fit *(26)*.

5. IMPLICATIONS AND LIMITATIONS

The finding that crypt niches maintain human colon crypts is not surprising considering murine studies *(1–4,14,18)* and limited human observations *(27)*. However, because prior experimental manipulations are not required for methylation studies, the phylogenetic studies of my group suggest crypt niches are not artifacts of experimental manipulations, and crypt niche replacement recurs continuously throughout life. Unlike the turnover of niche stem cells, most crypt niches appear to persist the lifetime of an individual. Although some stem cells may undergo a limited number of divisions to preserve their replicative potential and reduce the accumulation of mutations *(1–3)*, our analysis suggested crypt niche stem cells normally replicate daily. A lower stem cell mitotic rate would imply a higher methylation error rate, and further studies to characterize epigenetic replication fidelity in adult somatic cells could yield better estimates of stem cell division rates.

A limitation of my group's phylogenetic approach is that much is unknown about how methylation tags change through time *(28)*. Only a small subset of CpG islands exhibits the age-related methylation *(12,13)* that allows them to function as molecular clocks. The methylation tags do not fit any single model exactly, and it is possible that much more complicated stem cell behaviors are present. Exact crypt parameters, such as the number of stem cells per niche, are also uncertain, and more precise characterization may be difficult. Potentially, stem cell numbers may be variable between crypts or within a single crypt over time. Nevertheless, phylogenetic approaches represent practical experimental alternatives for the systematic examinations of adult human tissues. Although quantitative approaches based on passive observations cannot verify if a given stem cell scenario is correct, some scenarios will be rendered very unlikely. Organizing somatic adult cells into phylogenetic trees represents another approach to classify and characterize human aging and disease.

ACKNOWLEDGMENT

This work was supported by grant DK61140 from the National Institute of Diabetes and Digestive and Kidney Disease.

REFERENCES

1. Potten, C. S., and Loeffler, M. (1990). Stem cells: attributes, cycles, spirals, pitfalls, and uncertainties. Lessons for and from the crypt. Development 110, 1001–1020.
2. Watt, F. M., and Hogan, B. L. (2000). Out of Eden: stem cells and their niches. Science 287, 1427–1430.
3. Slack, J. M. (2000). Stem cells in epithelial tissues. Science 287, 1431–1433.
4. Spradling, A., Drummond-Barbosa, D., and Kai, T. (2001). Stem cells find their niche. Nature 414, 98–104.
5. Yatabe, Y., Tavare, S., and Shibata, D. (2001). Investigating stem cells in human colon by using methylation patterns. Proc Natl Acad Sci U S A 98, 10,839–10,844.
6. Loeb, L. A. (1991). Mutator phenotype may be required for multistage carcinogenesis. Cancer Res 51, 3075–3079.
7. Cross, S. H., and Bird, A. P. (1995). CpG islands and genes. Curr Opin Genet Dev 5, 309–314.

8. Bird, A. (2002). DNA methylation patterns and epigenetic memory. Genes Dev 6, 6–21.

9. Hsieh, C. L. (2000). Dynamics of DNA methylation pattern. Curr Opin Genet Dev 10, 224–228.

10. Pfeifer, G. P., Steigerwald, S. D., Hansen, R. S., Gartler, S. M., and Riggs, A. D. (1990). Polymerase chain reaction-aided genomic sequencing of an X chromosome-linked CpG island: methylation patterns suggest clonal inheritance, CpG site autonomy, and an explanation of activity state stability. Proc Natl Acad Sci U S A 87, 8252–8256.

11. Reik, W., Dean, W., and Walter, J. (2001). Epigenetic reprogramming in mammalian development. Science 293, 1089–1093.

12. Ahuja, N., Li, Q., Mohan, A. L., Baylin, S. B., and Issa, J. P. (1998). Aging and DNA methylation in colorectal mucosa and cancer. Cancer Res 58, 5489–5494.

13. Issa, J. P. (2000). CpG-island methylation in aging and cancer. Curr Top Microbiol Immunol 249, 101–118.

14. Booth, C., and Potten, C. S. (2000). Gut instincts: thoughts on intestinal epithelial stem cells. J Clin Invest 105, 1493–1499.

15. Potten, C. S., Kellett, M., Roberts, S. A., Rew, D. A., and Wilson, G. D. (1992). Measurement of in vivo proliferation in human colorectal mucosa using bromodeoxyuridine. Gut 33, 71–78.

16. Cheng, H., Bjerknes, M., and Amar, J. (1984). Methods for the determination of epithelial cell kinetic parameters of human colonic epithelium isolated from surgical and biopsy specimens. Gastroenterology 86, 78–85.

17. Clark, S. J., Harrison, J., Paul, C. L., and Frommer, M. (1994). High sensitivity mapping of methylated cytosines. Nucleic Acids Res 22, 2990–2997.

18. Williams, E. D., Lowes, A. P., Williams, D., and Williams, G. T. (1992). A stem cell niche theory of intestinal crypt maintenance based on a study of somatic mutation in colonic mucosa. Am J Pathol 141, 773–776.

19. Park, H. S., Goodlad, R. A., and Wright, N. A. (1995). Crypt fission in the small intestine and colon. A mechanism for the emergence of G6PD locus-mutated crypts after treatment with mutagens. Am J Pathol 147, 1416–1427.

20. Cairns, J. (1975). Mutation selection and the natural history of cancer. Nature 255, 197–200.

21. Nowell, P. C. (1976). The clonal evolution of tumor cell populations. Science 194, 23–28.

22. Muller, H. J. (1964). The relation of recombination to mutational advance. Mutat Res 1, 2–9.

23. Felsenstein, J. (1974). The evolutionary advantage of recombination. Genetics 78, 737–756.

24. Smith, J. M., and Nee, S. (1990). Clicking into decline? Nature 348, 391–392.

25. Chao, L. (1997). Evolution of sex and the molecular clock in RNA viruses. Gene 205, 301–308.

26. Chao, L. (1990). Fitness of RNA virus decreased by Muller's ratchet. Nature 348, 454–455.

27. Campbell, F., Williams, G. T., Appleton, M. A., Dixon, M. F., Harris, M., and Williams, E. D. (1996). Post-irradiation somatic mutation and clonal stabilisation time in the human colon. Gut 39, 569–573.
28. Ro, S., and Rannala, B. (2001). Methylation patterns and mathematical models reveal dynamics of stem cell turnover in the human colon. Proc Natl Acad Sci U S A 98, 10,519–10,521.

Index